Tourism and Environmental Change

Global environmental change is one of the major issues confronting humankind. *Tourism and Global Environmental Change* is the first comprehensive analysis of the economic, social and political relationships between tourism and global environmental change. In this book, tourism is seen to be both a significant contributor to global environmental change as well as one of the economic sectors that will be the most impacted by such changes.

Tourism and Global Environmental Change is divided into three sections. The first section examines the tourism and global environmental change relationship in specific environments, including polar regions, mountains, rivers, forests, coastal regions, reefs, deserts and the urban environment. The second section looks at specific global issues related to environmental change and includes the spread of disease and its potential effects on tourism, biodiversity, water resources and extreme weather events. The final section discusses some of the different perceptions held by tourists and the tourist industry on global environmental change. It concludes by investigating some of the potential responses to global environmental change by the tourism industry and government.

This indispensable collection of essays from leading scholars in the field, *Tourism and Global Environmental Change*, concludes that there is a major crisis facing tourism. It argues that impacts are real and are potentially extremely serious both for tourism and for the communities that depend upon the tourism industry.

Stefan Gössling is Associate Professor, Department of Service Management, Lund University, Sweden.

C. Michael Hall is Professor, Department of Tourism, University of Otago, Dunedin, New Zealand, and Docent Department of Geography, University of Oulu, Finland.

Contemporary Geographies of Leisure, Tourism and Mobility
Series Editor: C. Michael Hall
Professor at the Department of Tourism, University of Otago, New Zealand.

The aim of this series is to explore and communicate the intersections and relationships between leisure, tourism and human mobility within the social sciences.

It will incorporate both traditional and new perspectives on leisure and tourism from contemporary geography, e.g. notions of identity, representation and culture, while also providing for perspectives from cognate areas such as anthropology, cultural studies, gastronomy and food studies, marketing, policy studies and political economy, regional and urban planning, and sociology, within the development of an integrated field of leisure and tourism studies.

Also, increasingly, tourism and leisure are regarded as steps in a continuum of human mobility. Inclusion of mobility in the series offers the prospect to examine the relationship between tourism and migration, the sojourner, educational travel, and the second home and retirement travel phenomena.

The series comprises two strands:

Contemporary Geographies of Leisure, Tourism and Mobility aims to address the needs of students and academics, and the titles will be published in hardback and paperback. Titles include:

The Moralisation of Tourism
Sun, sand ... and saving the world?
Jim Butcher

The Ethics of Tourism Development
Mick Smith and Rosaleen Duffy

Tourism in the Caribbean
Trends, development, prospects
Edited by David Timothy Duval

Qualitative Research in Tourism
Ontologies, epistemologies and methodologies
Edited by Jenny Phillimore and Lisa Goodson

The Media and the Tourist Imagination
Converging cultures
Edited by David Crouch, Rhona Jackson and Felix Thompson

Tourism and Global Environmental Change
Ecological, social, economic and political interrelationships
Edited by Stefan Gössling and C. Michael Hall

Routledge Studies in Contemporary Geographies of Leisure, Tourism and Mobility is a forum for innovative new research intended for research students and academics, and the titles will be available in hardback only. Titles include:

1. **Living with Tourism**
   Negotiating Identities in a Turkish Village
   Hazel Tucker

2. **Tourism, Diaspora and Space**
   Tim Coles and Dallen J. Timothy

3. **Tourism and Postcolonialism**
   Contested discourses, identities and representations
   C. Michael Hall and Hazel Tucker

Tourism and Global Environmental Change

Ecological, social, economic and political interrelationships

Edited by Stefan Gössling and C. Michael Hall

Routledge
Taylor & Francis Group

LONDON AND NEW YORK

First published 2006
by Routledge
2 Park Square, Milton Park, Abingdon, Oxon OX14 4RN

Simultaneously published in the USA and Canada
by Routledge
270 Madison Ave, New York, NY 10016

Reprinted 2007

Routledge is an imprint of the Taylor & Francis Group

Typeset in Times New Roman by
Bookcraft Ltd, Stroud, Gloucestershire
Printed and bound in Great Britain by
MPG Books Ltd, Bodmin

British Library Cataloguing in Publication Data
A catalogue record for this book is available from the British Library

Library of Congress Cataloging in Publication Data
Gössling, Stefan.
 Tourism and global environmental change : ecological, social,
 economic, and political interrelationships / Stefan Gössling and
 C. Michael Hall.
  p. cm. — (Contemporary geographies of leisure, tourism, and
  mobility) Includes bibliographical references and index.
  1. Tourism—Environmental aspects. 2. Global environmental change.
  I. Hall, Colin Michael, 1961– II. Title. III. Series.
 G155.A1G67 2005   338.4791–dc22
 2005011328

ISBN10: 0-415-36131-1(hbk)
ISBN10: 0-415-36132-X(pbk)

ISBN13: 9-78-0-415-36131-6 (hbk)
ISBN13: 9-78-0-415-36132-3 (pbk)

Contents

Tables

Figures

Contributors

Erika Andersson Cederholm, Department of Service Management, Lund University, Box 882, 25108 Helsingborg, Sweden.

Shirley Brooks, School of Life and Environmental Sciences, Memorial Tower Building, University of KwaZulu-Natal, Durban 4041, South Africa.

Rolf Bürki, University of Higher Education, Notkerstr. 27, 9000 St Gallen, Switzerland.

Stephen J. Craig-Smith, School of Tourism and Leisure Management, The University of Queensland, Ipswich Campus, 11 Salisbury Road, Ipswich, Queensland 4305, Australia.

Chris R. de Freitas, School of Geography and Environmental Science, City Campus, University of Auckland, Private Bag 92019, Auckland, New Zealand.

William Ellery, School of Life and Environmental Sciences, Memorial Tower Building, University of KwaZulu-Natal, Durban 4041, South Africa.

Hans Elsasser, Department of Geography, University of Zurich, Winterthurerstrasse 190, 8057 Zurich, Switzerland.

Xavier Font, Tourism Hospitality and Events School, Leeds Metropolitan University, Leeds LS1 3HE, United Kingdom.

Stefan Gössling, Department of Service Management, Lund University, Box 882, 25108 Helsingborg, Sweden.

Szilvia Gyimóthy, Department of Service Management, Lund University, Box 882, 25108 Helsingborg, Sweden.

C. Michael Hall, Department of Tourism, School of Business, University of Otago, PO Box 56, Dunedin, New Zealand.

Jacqueline M. Hamilton, Center for Marine and Climate Research, Sustainability and Global Change, University of Hamburg, Bundestrasse 55, 20146 Hamburg, Germany.

Thomas Hickler, Department of Physical Geography and Ecosystems Analysis, Lund University, Sölvegatan 12, 223 62 Lund, Sweden.

Johan Hultman, Department of Service Management, Lund University, Box 882, 25108 Helsingborg, Sweden.

Margaret E. Johnston, School of Outdoor Recreation, Parks and Tourism, Lakehead University, Thunder Bay, Ontario, Canada.

Brenda E. Jones, Department of Geography, University of Waterloo, 200 University Avenue West, Waterloo, Ontario, Canada N2L 3G1.

Maren A. Lau, Center for Marine and Climate Research, Sustainability and Global Change, University of Hamburg, Bundestrasse 55, 20146 Hamburg, Germany.

Christian J. Nöthiger, Ruetschistrasse 27, 8037 Zurich, Switzerland.

Robert Preston-Whyte, School of Life and Environmental Sciences, Memorial Tower Building, University of KwaZulu-Natal, Durban 4041, South Africa.

Daniel Scott, Department of Geography, University of Waterloo, 200 University Avenue West, Waterloo, Ontario, Canada N2L 3G1.

Richard Tapper, Environment Business and Development Group, 16 Glenville Road, Kingston upon Thames, KT2 6DD, United Kingdom.

Preface

Global environmental change undoubtedly represents one of the major challenges to humanity and the planet, but also to the academy as well.

Concern over the environmental impacts of human actions at a global scale is not new. Arguably, such concerns can be traced back at least to the work of George Perkins Marsh in 1864 and have been an ongoing thread in debates on conservation and resource management ever since. However, for all the scientific and academic writing on the need for sound resource management and the expressions of interest in sustainable development, the scale and severity of human impact on natural bio-physical processes has continued to grow, as has the loss of biodiversity.

Over the past 25 years the tourism industry has often sought to portray itself as a relatively benign contributor to the conservation of the environment, while simultaneously providing positive benefits in terms of employment, economic development and wealth generation. Much of this contribution has been described within the rubric of sustainable tourism and is often highlighted in the portrayal of tourism as a relatively environmentally friendly industry. It is for these reasons that one would, therefore, assume tourism to be at the forefront of efforts to promote more positive approaches towards managing global environmental change. Unfortunately, that is not the case.

As this volume demonstrates, tourism is both a gross contributor to and increasingly affected by global environmental change. Although tourism can be a contributor to environmental conservation at the local scale, particularly through instances of ecotourism or nature-based tourism – where specific charismatic species or ecosystems are conserved – on a global scale tourism is a significant element in environmental change.

The editors intend that this book highlight some of the complexities and issues in the relationship between tourism and global environmental change. We also believe that *Tourism and Global Environmental Change* reinforces the need for a sense of urgency, from students of tourism, as well as the industry and government, to take positive actions to curb the more undesirable elements of global environmental change, even if that means fundamental changes in the consumption and practices of tourism itself. Finally, we believe that the book highlights the need for the tourism research community to look at tourism impacts on a far wider scale than just what occurs at the level of the local destination.

Acknowledgements

The editors would like to thank Robert Bockermann, Peter Burns, Dick Butler, Jean-Paul Ceron, Tim Coles, Arthur Conacher, Dave Crag, Ross Dowling, David Duval, Monica Gilmour, Mathias Gößling, Szilvia Gyimóthy, Tuija Härkönen, Cecilia Hegarty, Nadine Heck, James Higham, Johan Hultman, Magnus Jirström, Bruno Jansson, Donna Keen, Timo Kunkel, Alan Lew, Madelaine Mattson, Geoff McBoyle, Dieter Müller, Stephen Page, Paul Peeters, Julie Pitcher, Robert Preston-Whyte, Greg Richards, Meike and Linnea Rinsche, Jarkko Saarinen, Anna-Dora Saetorsdottir, Chrissy Schriber, Geoff Wall, Brian Wheeler, Allan Williams, Sandra Wall and Andrea Valentin (even if she did kill Michael's mulberry tree!), who have all contributed in various ways to this book, as have various graduate and undergraduate classes in Dunedin, Helsingborg, Oulu, and Umeå.

Gavin Bryars, Jeff Buckley, Nick Cave, Bruce Cockburn, Elvis Costello, *Friends*, Stephen Cummings, Hoodoo Gurus, The Sundays, Ed Kuepper, Jackson Code, Sarah McLachlan, Monty Python, Morphine, Vinnie Reilly, David Sylvian, Jennifer Warnes, Chris Wilson and BBC 6 Music were also essential to the writing process.

The constantly rainy and cold weather in Sweden gave Stefan a great opportunity to work continuously. Monica Gilmour was a huge help, as usual, in getting this book to the publishers, while acknowledgement must also be given to the convivial atmosphere of the Department of Social and Economic Geography, Umeå University, Sweden, where research for the book was undertaken by Michael during sabbatical. On a personal level Michael would like to give very great thanks to Jody, as well as to his own global collection of significant others, some of whom are mentioned above, for their continued support at a personal and professional level. Finally, the authors would like to express their appreciation to Andrew Mould and Zoe Kruze at Routledge for their continued interest in the project.

1 An introduction to tourism and global environmental change

Stefan Gössling and C. Michael Hall

Introduction

Tourism is largely dependent on natural resources. For example, the provision of fresh water for drinking, taking showers, swimming pools or the irrigation of hotel gardens seem self-evident preconditions for tourism all around the world. Beaches and coastlines, mountains, forests, lakes, oceans and the scenery provided by landscapes containing these elements are central to the attraction potential of most destinations. Similarly, biodiversity is a tourist magnet in many regions, including a wide variety of bird and fish species, as well as charismatic mammals such as moose or deer, whales, dolphins or the 'big five' (leopard, lion, rhino, elephant, hippopotamus) in national parks in eastern and southern Africa. In mountainous areas, snow cover is a *conditio sine qua non* for winter sports, including skiing, snowboarding, snowmobiling and dog sledding, and many areas would lose their tourist appeal without snow – for instance, what would impressive mountain ranges like the Alps or tropical Mount Kilimanjaro be without their white-covered tops? Clearly, most tourism is based on stable and, for tourism, favourable environmental conditions.

Global environmental change (GEC) threatens these very foundations of tourism through climate change, modifications of global biogeochemical cycles, land alteration, the loss of non-renewable resources, unsustainable use of renewable resources and gross reductions in biodiversity. Elements of the global environment are always changing although change is never uniform across time and space. Nevertheless, 'all changes are ultimately connected with one another through physical and social processes alike' (Meyer and Turner 1995: 304). The scale and rate of change has increased dramatically because of human actions within which tourism is deeply embedded.

Human impacts on the environment can have a global character in two ways. First, 'global refers to the spatial scale or functioning of a system' (Turner *et al.* 1990: 15). Here, the climate and the oceans have the characteristic of a global system and both influence and are influenced by tourism production and consumption. A second kind of GEC occurs if a change 'occurs on a worldwide scale, or represents a significant fraction of the total environmental phenomenon or global resource' (Turner *et al.* 1990: 15–16). Tourism is significant for both types of change.

This volume takes a systematic approach to understanding the environmental, social, economic and political interrelationships of tourism and GEC. In the first section, environmental change in each of the environments of importance for tourism is analysed, including the polar regions, mountains, lakes and streams, forests, coastal zones, islands and reefs, deserts and savannah regions, as well as urban environments. The volume's second section focuses on four aspects of GEC that might become particularly important for tourism: availability of fresh water, existence of diseases, biodiversity, and frequency and intensity of extreme weather events. In the final section, the book discusses issues of adaptation to GEC. Although this is a relatively unexplored research field of great uncertainty, the contributors present options for tourism adaptation in those environments that are likely to face the most rapid environmental changes – mountains – and presents a number of new avenues to the discussion of the consequences of GEC for tourism from sociology and business perspectives.

Tourism development

Any adequate conceptualisation of tourism demands a comprehensive approach that involves the relationships between tourism, leisure and other social practices and behaviours related to human movement (Coles *et al.* 2004; 2005). Such an assessment is necessary in the analysis of contemporary human mobility given the extent to which time-space convergence has made it easier for those with sufficient time and economic budgets to move over time and space. Travel which once took two or three days to accomplish may now be completed as a daytrip. Convergence through physical travel is also complemented by convergence in communications (Janelle and Hodge 2000). Clearly, such shifts in mobility have implications for a wide range of human activities both within and outside of tourism. They can be seen alongside ideas of accessibility, extensibility, distance and proximity, significant elements of global socio-cultural change (Johnston *et al.* 1995), as well as underlying social contributions to tourism related consumption of the environment (Hall 2005a).

The relative lack of interplay and cross-fertilisation between the fields that study human mobility is remarkable (Williams and Hall 2002). This is especially evidenced by the difficulties to be encountered in finding overlap between national and international surveys of tourism and migration, and studies of short- and long-term travel undertaken in transport studies, a factor that also has considerable impact on the understanding of the contribution of tourism to GEC. Nevertheless, as Coles *et al.* (2004) argued, the conceptualisation and development of theoretical approaches to tourism should consider the relationships of tourism to other forms of mobility, including the creation of extended transnational networks that also promote human movement. Figure 1.1 presents a model for describing different forms of temporary mobility in terms of three dimensions of space, time and number of trips. The fact that the number of movements declines the further one travels in time and space away from the point of origin is well recognised in the study of spatial interaction. However, it has not been used as a means to illustrate the totality of trips that are undertaken by individuals.

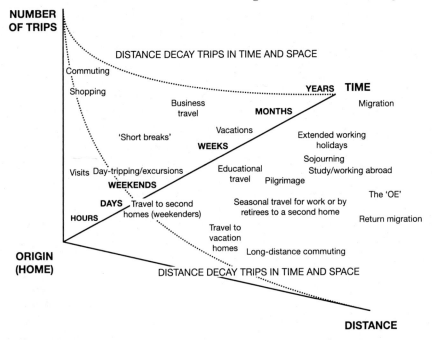

Figure 1.1 Extent of mobility in time and space
Source: after Hall 2003

The relationship represented in Figure 1.1 holds whether one is describing the totality of movements of an individual over their life span from a central point (home), or whether one is describing the total characteristics of a population (Hall 2005a; 2005b). The figure illustrates the relationship between tourism and other forms of temporary mobility including various forms of what is often regarded as migration or temporary migration (Bell and Ward 2000). Such activities, which have increasingly come to be discussed in the tourism literature, include travel for work and international or 'overseas' experiences (e.g. Rosenkopf and Almeida 2003), education (e.g. Field 1999), health (e.g. Goodrich 1994), as well as travel to second homes (e.g. Hall and Müller 2004), return migration (e.g. Duval 2003) and diaspora (e.g. Coles and Timothy 2004). Arguably, some of these categories could be described as 'partial tourists' (Cohen 1974), or even as 'partial migrants', although the amenity or leisure dimension remains important as a motivating factor in their voluntary mobility (Frändberg 1998; Coles *et al*. 2004; Hall 2005a).

Focusing on the range of mobilities also provides an important dimension with respect to the examination of tourism's impacts. Mobilities need to be examined over the duration of the lifecourse, so that the linkages and relationships between different forms of 'temporary' and 'permanent' mobilities, particularly tourism and amenity migration, are better understood. Such a lifecourse approach may also have extremely practical applications in terms of studying how people switch modes of consumption between locations and activities. For example, while people may attempt to be green consumers in one aspect of their life, their consumption may actually increase in others, thereby leading to no net improvement in their overall rate of consumption.

Most research on tourism impacts has also tended to occur at the destination (Lew *et al.* 2004). This has meant that research has often examined local factors rather than the totality of the impact of tourism in time and space by also considering effects at the tourism-generating region, and travel to and from destinations. As significant as destination impacts might be, the study of tourism impacts therefore needs to be undertaken over the totality of the tourism consumption and production system, rather than just at the destination (Bach and Gößling 1996; Frändberg 1998; Høyer 2000; Gössling 2002; Frändberg and Vilhelmson 2003; Hall and Higham 2005).

Consideration of the total movement of humans in time and space is important because the extent of space-time compression that has occurred for many people in the developed world has led to fundamental changes in individual mobility and in assumptions about personal mobility in recent years. Travel time budgets have not changed substantially but the ability to travel further at a lower per unit cost within a given time budget has led to a new series of social interactions and patterns of production and consumption (Schafer 2000). For many people leisure mobility is now routine. Advances in transport and communication technology that have been adopted by a substantial, relatively affluent, proportion of the population enable such people to travel long distances to satisfy demands for amenities – what one would usually describe as tourism. Indeed, Hall (2005a) has argued that one interpretation of this perspective is that the study of tourism is intrinsically the study of the mobile consumption patterns of the wealthier members of society.

The means of transport used in international tourism changed fundamentally with the rise of civil aviation in the 1960s. From being an option only for wealthy tourists in the 1960s and 1970s, aviation soon became one of the most popular means of transport in international travel, with growth rates in the order of 5–6 per cent per year from 1970 to 1993, and 7.1–7.8 per cent per year from 1994 to 1996. Some 40 per cent of all international tourist arrivals might now be by air (Gössling 2000). At the end of 2002, airlines were actively operating some 10,789 passenger jets with 100 seats or more, representing a total of 1.9 million installed seats.

Boeing (2003) and Airbus (2003) predict that air travel will continue to grow rapidly, with average annual growth rates of 5.0–5.2 per cent to 2022/23. By 2022, the number of aircraft is predicted to increase by 90 per cent to about 20,500, while the number of installed seats will more than double to reach 4.5 million (Airbus 2003). Boeing (2003) also predicts that competition for markets will be strong, leading to more airline entrants, lower fares and improved networks. Simultaneously, it is anticipated that governments will continue to deregulate air travel markets. Consequently, air travel is one of the key factors in international tourism development, outpacing growth in surface-bound means of transport.

Two other aspects of air travel deserve mention in the context of tourism and GEC. First, flight distances are predicted to increase, with the average distance flown growing from 1,437km in 2002 to 1,516km in 2022 (Airbus 2003). Second, a large proportion of the current fleet of aircraft (60 per cent) will still be in operation 20 years from now (Boeing 2003), indicating long operation times for aircraft and concomitant high fuel use for older models still in use. The estimated scale of international tourism growth, as well as the enthusiasm from the tourism industry that seemingly accompanies such growth, is well illustrated in the World Tourism Organization's (WTO) 2020 vision:

> By the year 2020, tourists will have conquered every part of the globe as well as engaging in low orbit space tours, and maybe moon tours. The Tourism 2020 Vision study forecasts that the number of international arrivals worldwide will increase to almost 1.6 billion in 2020. This is 2.5 times the volume recorded in the late 1990s ... Although the pace of growth will slow down to a forecast average 4 per cent a year – which signifies a doubling in 18 years, there are no signs at all of an end to the rapid expansion of tourism ... Despite the great volumes of tourism forecast for 2020, it is important to recognise that international tourism still has much potential to exploit ... the proportion of the world's population engaged in international tourism is calculated at just 3.5 per cent.
>
> (WTO 2001: 9, 10)

Whether such rates of growth are possible is debatable given the potential impacts of future increased fuel prices or 'wildcard' events such as an economic or political crisis, such as occurred in Asia in 1997/98 (Hall 2005a). However, the broader and probably more important debate, as to whether such rates of growth are sustainable or even desirable given the potential environmental effects of such growth, is not really being fully entered into in the academic field and certainly not by leading tourism bodies such as the WTO and the World Travel and Tourism Council (WTTC). For example, the *Blueprint for New Tourism* published by the WTTC in 2003 does not even acknowledge the potential relationships between tourism and global climate and environmental change. The WTO has recently described the interaction of tourism and climate change as a 'two-way relationship' (WTO 2003), but there is no official document, as yet, dealing with this problematic interaction. Quite the contrary, the WTO's STEP programme (WTO 2005) fully ignores the environmental consequences of leisure transport.

Global environmental change

There is comprehensive evidence for GEC caused by human activities: land-use changes, altered biogeochemical cycles, climate change, biotic exchange, as well as disturbance regimes (e.g. tropical storms) have been described in a wide range of publications (Klein Goldewijk 2001, Sala *et al.* 2000, IPCC 2001, Loh and Wackernagel 2004). GEC will have complex consequences for ecosystems, as well as for social and economic systems. Biodiversity will mostly be affected by land-use changes, followed by climate change (increasing temperatures), nitrogen deposition, biotic exchange and atmospheric CO_2 concentration increases (Sala *et al.* 2000). Note, however, that different biomes will be affected to varying degrees by these changes – for example, Arctic environments will mostly suffer from increasing temperatures (ACIA 2004). Global environmental change will also affect human social and economic systems though increasing temperatures, sea-level rise, changing precipitation patterns and weather extremes, which in turn will cause other environmental changes, such as new disease frontiers, coastal erosion and new patterns of urbanization and mobility. We shall now discuss the observed and predicted patterns of some key parameters of GEC.

Temperature increase

Global average surface temperatures, the average of near-surface air temperature over land and sea surface temperature, are changing. As documented by the Inter-governmental Panel of Climate Change (IPCC) – a scientific body created in 1988 by the World Meteorological Organization and the United Nations Environmental Programme – global average surface temperatures have increased by $0.6 \pm 0.2°C$ over the twentieth century (IPCC 2001). Globally, the IPCC anticipates that the 1990s were the warmest decade and 1998 the warmest year since 1861, when temperature measurements were introduced.

There is even more evidence for ongoing climate change. For example, since the late 1950s, when weather balloons were introduced, global temperature increases in the lowest 8km of the atmosphere and in surface temperature have been about 0.2°C per decade. Since 1979, information is completed by satellite records, which show that the global average temperature of the lowest 8km of the atmosphere has changed by $0.05 \pm 0.01°C$ per decade; however, global average surface temperature has increased even more rapidly, by $0.15 \pm 0.05°C$ per decade. Satellite data also shows that the extent of snow cover might have decreased by about 10 per cent since the late 1960s, while ground-based observations show that the annual dura-tion of lake and river ice cover in the mid- and high latitudes of the northern hemi-sphere might have reduced by two weeks over the twentieth century (IPCC 2001). Other observable changes include the retreat of glaciers all over the world, and there is now evidence that the Western Arctic Ice Shield is melting (Payne *et al.* 2004). For example, between 1980 and 2004 Chinese scientists measured a 5.5 per cent shrinkage by volume in China's 46,298 glaciers, a loss equivalent to more than $3,000km^2$ (1,158 square miles) of ice; with there being a noticeable accelera-tion in recent years. The effects are so dramatic that it is predicted that two-thirds of China's glaciers – which are estimated to account for 15 per cent of the Earth's ice – would disappear by the end of the 2050s, and almost all would have melted by 2100 (Watts 2004).

Overall, there is thus compelling evidence that climate change is real and, furthermore, that it is primarily caused by human emissions of greenhouse gases, mainly as a result of the burning of fossil fuels (IPCC 2001). The warming effect of greenhouse gases is measured in terms of radiative forcing, which in turn can be translated into an increase in global mean temperature. From 1750 to 2000, all anthropogenic emissions of greenhouse gases have caused an additional radiative forcing of $2.43 \ Wm^{-2}$, of which $1.46 \ Wm^{-2}$ is from CO_2 – the most important anthropogenic greenhouse gas – 0.48 Wm^{-2} from CH_4, 0.34 Wm^{-2} from the halocarbons and 0.15 Wm^{-2} from N_2O (IPCC 2001). The observed depletion of the stratospheric ozone layer from 1979 to 2000 is estimated to have caused a negative radiative forcing of –0.15 Wm^{-2}. On the other hand, ozone in the troposphere is estimated to have increased, leading to a positive radiative forcing of 0.35 Wm^{-2}. The human dimension of climate change needs to be highlighted because policy makers and the media – particularly in countries such as Australia and the USA which have not signed the Kyoto accord – frequently assert that climate change is highly uncertain. These assertions have been used as an argument against adopting strong measures to reduce greenhouse gas emissions (O'Riordan and Jäger 1996; Sarewitz and Pielke Jr. 2000; Brown and Oliver 2004; Gow 2004a, b).

Generally, climate change is analysed through climate models. Such models cannot be used to make exact statements on the character of future climate change, because there is a range of uncertainties concerning the parameters used, their interaction and feedback processes, as well as future emission levels of greenhouse gases. Clearly, modelling the world's climate is an extremely complicated exercise. Hence, models still need to be improved (von Storch *et al.* 2004) and might for years to come imply a degree of uncertainty. Nevertheless, they are already now deemed to be sufficiently reliable to address the question of how the climate will respond to increasing greenhouse gases.

For example, while there is still controversy over the cause and extent of natural climate variability in the past (Moberg *et al.* 2005, see also Figure 1.2), warming as observed since the mid-1980s is unprecedented and unquestioned in terms of its human cause. Natural fluctuations, if underestimated in current models, are of importance because they might 'either amplify or attenuate anthropogenic climate change significantly' (Moberg *et al.* 2005: 617). In other words: uncertainty increases the risk factor implicit in current models (Challenor *et al.* 2005). In the past, climate models have pointed at temperature increases in the range of 1.4 – 5.8°C by 2100 (IPCC 2001), with a likely scenario of a 3°C warming by 2100 (Kerr 2004: 932; see also Figure 1.2). Recent research indicates, however, that the range might very well be larger, with up to 11.5°C warming by 2100 (Challenor *et al.* 2005; Stainforth *et al.* 2005). Any warming beyond 3–4°C is assumed to have adverse impacts for coastal resources, biodiversity, marine and terrestrial ecosystem productivity and agriculture (Hitz and Smith 2004; Parry 2005).

It is also worth noting that temperature increases are not likely to be linear. A recent example is the summer 2003, which was probably the hottest in Europe since AD1500 (Stott *et al.* 2004), and which is a good example of increasing temperature variability in European summer heatwaves as a result of climate change (Schär *et al.* 2004). The social, environmental, economic and political consequences of the heatwave were complex. Schär and Jendritzky (2004) report that, according to reinsurance estimates, drought conditions caused crop losses worth US$12.3 billion in Europe, plus an additional US$1.6 billion in forest fire damage in Portugal. European electricity markets reacted with increasing energy prices, because power plants had to curtail production owing to the lack of cooling water, with electricity spot prices rising beyond €100 (US$130) per

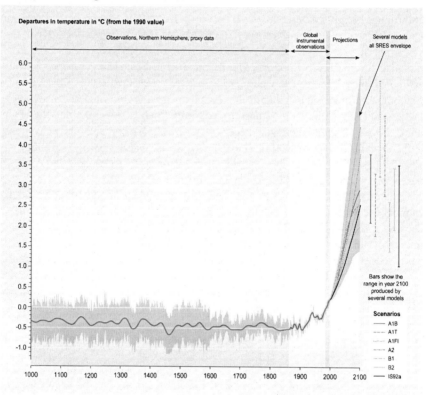

Figure 1.2 Variations in the Earth's surface temperature, 1000–2100

Source: reproduced by permission of IPCC (2005)

MWh. Schär and Jendritzky (2004: 559) continue their list of fatal consequences of the heatwave:

> In the Alps, many glaciers underwent unprecedented melting, and the thawing of permafrost led to a series of severe rock falls. But it was the unusual number of deaths during 1–15 August that caught the headlines. Estimates based on the statistical excess over mean mortality rates amount to between 22,000 and 35,000 heat-related deaths across Europe as a whole. In France the mortality rate increased by 54 per cent during those two weeks, and the increase was statistically significant in all 22 French regions and for all age groups above 45 years.

The consequences of the heatwave even had political dimensions. In France, for example, the government's mishandling of the heatwave – old people were not taken care off because of the country's prolonged summer vacations, causing the death of approximately 15,000 people – was suggested as being part of the reason for the election of the moderate-right government over the socialist party in later elections (Caldwell 2003). Stott *et al.* (2004) analysed whether the heatwave was already a result of human interference with the climate system; this is whether it was caused by human greenhouse gas emissions. They conclude that it is impossible to say whether any such weather event might have occurred by chance in an unmodified climate, but that it is nevertheless possible to estimate 'by how much human activities may have increased the risk of the occurrence of such a heatwave' (Stott *et al.* 2004: 610). At a confidence level of greater than 90 per cent, more than half of the risk of the 2003 heatwave is attributable to anthropogenic changes to the climate system, and summers like the one of 2003 might become the norm by the middle of the twenty-first century.

Sea-level rise

One of the most important consequences of temperature increases will be sea-level rise, a result of the thermal expansion of the oceans as well as the melting glaciers and ice caps. According to the IPCC (2001), the average global sea level has risen by between 0.1 and 0.2m during the twentieth century, and the IPCC's scenarios predict a sea-level rise of 0.09–0.88m by 2100. The current state of the West Arctic Ice Sheet is of great concern because if it melted completely it could raise global sea levels by approximately seven metres (Oppenheimer and Alley 2004). While such a scenario is unlikely (however, see Rapley 2005), there still is potential for a marked increase in the rate of sea-level rise due to accelerated ice loss (Payne *et al.* 2004).

Sea-level rise will have substantial socio-economic consequences, because it will lead to changes in flooding by storm surges, lead to the loss of coastal wetlands and biodiversity (Nicholls 2004), cause land inundation and coastal erosion (Nicholls 2004, Zhang *et al.* 2004). Zhang *et al.* estimate that, given a sea-level rise towards the higher end of the IPCC's 2001 scenarios (i.e. a 90cm sea-level rise), at least 100 million people living within one metre of mean sea level will be affected.They also point out that some of the most heavily developed and economically valuable real estate will be threatened by coastal erosion, which includes many areas of high tourism and recreational significance (Zhang *et al.* 2004). As coastlines are lost, constructions adjacent to such areas will be damaged or destroyed, with the rate of long-term sandy beach erosion being two orders of magnitude greater than the rate of sea-level rise. Human vulnerability will, however, largely depend on economic and political opportunities to invest, for instance, in coastline protection (Nicholls 2004). Nevertheless, in several countries, such as the Netherlands and the United Kingdom, there is now substantial policy pressure to remove coastal protection measures from some areas because the cost of maintaining them is extremely high, while wetlands may be able to be restored in areas that have previously been drained. Nicholls (2004) also demonstrates that the loss of coastal wetlands due to sea-level rise might be in the order 5–20 per cent by the 2080s. Obviously, sea-level rise will have the most severe consequences for low-lying countries, particularly if coupled with other GECs, such as weather extremes. Small Island Developing States have thus been identified as the areas most significantly impacted by sea-level rise: ' … vulnerabilities already exist [in SIDS] that will only be exacerbated by accelerated global warming' (London 2004: 500). For example, 85 per cent of the population, 80 per cent of the infrastructure and 90 per cent of the economic activity of St Vincent are located on a coastal strip less than 5km from the high-water mark and 5m above sea level (SVG 2000, cited in London 2004). Similar is true for a wide range of other islands.

Land-use change

Land use is changing on a global scale (Richards 1990; Klein Goldewijk 2001). A growing world population and changes to diet associated with demands for better living standards mean that increasing amounts of land are being converted from a relatively natural state to agricultural land use. The land area needed to feed people depends on factors such as diet, soil productivity, water availability, and chemical and energy inputs into the agricultural system.

The average area of arable land needed to support the diet of a person from the developed countries is estimated at 0.5ha per person (ha/p) (Giampetro and Pimentel 1994). The average arable land area for a person on a mainly vegetarian diet has been estimated at 0.2ha/p (Engelman et al. 2000), although the borderline of arable land scarcity has been estimated at 0.07ha/p (Smil 1993). Nielsen (2005) estimates that the global average available in 2003 was 0.24ha/p and that, based on low-level projections of world population growth, the average available in 2025 will be 0.16ha/p, and 0.11ha/p in 2050. The biologically productive surface area required to support the average consumption of Western Europe has been estimated at 6.0ha/p and for very poor countries at 1.0ha/p. In 2003, the global average available was estimated at 1.8ha/p with a forecast of 0.7ha/p for 2050 (Nielsen 2005; also see Wackernagel et al. 2002). Urban areas only occupy between 1 and 2 per cent of all land resources. However, expanding urban areas often directly occupy highly productive agricultural land, while the ecological footprint of urban centres means that they impact an area far wider than the immediate city.

The available land area per person is decreasing not only because of population growth but also because of land degradation and the impacts of climate change (Bugmann 1997; IPCC 2001; Hannah et al. 2002; Williams et al. 2003; Cox et al. 2004; Parry et al. 2004; Parry 2005). Causes of land degradation include deforestation, poor agricultural practices, urbanisation and industrialisation, pollution, salination, acidification and waterlogging. It is estimated that, between 1945 and 1990, 1,965 million hectares was lost through soil degradation, a rate of 44 million hectares per year (Nielsen 2005). The rate of degradation of arable land has been estimated at approximately 10 million hectares per year (Nielson 2005). In the case of Western Australia, for example, one of the world's biodiversity 'hotspots', 18 million of the 25 million hectares originally covered by native vegetation has been cleared. Of this, 1.8 million hectares is salt affected to some degree, with 80 per cent of the waterways in the south-west also significantly affected by salt (Water and Rivers Commission 2000). Indeed, on a global scale Parry (2005) noted that some thresholds – with respect to such issues as water stress, global cereal production and flooding – have already been exceeded at a regional level and warned that, at a global level, while there may be some positive impacts for agriculture, terrestrial ecosystem productivity and forestry at low levels of climate change, there will be negative and increasing damages at higher levels of climate change. However, negative biodiversity and dispersion impacts are expected to occur alongside of positive net ecosystem productivity even at low rates of change (Leemans and van Vliet 2005; Parry 2005).

The interrelationship of land-use change with population increase means that ecological capacities are increasingly being stretched, and are exceeded in a number of the developed countries.

> If the combined global footprint is smaller than global ecological capacity, we are still within the limits of sustainability. However, if our combined global footprint is larger, we are living beyond our means – that is, with a global ecological deficit.
>
> (Nielsen 2005: 36–7)

Tourism is a part of the processes of land-use change because of its contribution to consumption and its own ecological footprint (Gössling *et al.* 2002). However, tourism is also affected by land-use change because of the loss of resources, such as biodiversity and amenity landscapes, that serve to attract visitors.

Precipitation patterns

Global precipitation patterns have changed substantially in recent decades, both with respect to amount and intensity. According to the IPCC (2001), precipitation might have increased by 0.5–1 per cent per decade in the twentieth century over most mid- and high latitudes of the northern hemisphere continents, and rainfall by 0.2–0.3 per cent per decade over the tropical land areas (10°N to 10°S). However, increases in rainfall in the tropics are not evident over the past few decades and, in the northern hemisphere sub-tropical land areas (10°N to 30°N), rainfall may have decreased during the twentieth century by about 0.3 per cent per decade. No changes have been observed over the southern hemisphere, and data is insufficient to establish trends over the oceans.

So far as the intensity of precipitation patterns are concerned, there seems to be a 2–4 per cent increase in the frequency of heavy precipitation events in the mid- and high latitudes of the northern hemisphere over the latter half of the twentieth century. Heavy precipitation events can be a result of changes in atmospheric moisture, thunderstorm activity and large-scale storm activity. Precipitation patterns can also be affected by El Niño-Southern Oscillation phenomena (see below, *Extreme climate and weather events*). Globally, runoff is expected to increase by 4 per cent given a 1°C global temperature rise, a result of more intense evaporation above oceans coupled with continental precipitation (Labat *et al.*

2004). Note, however, that there are increasing and decreasing runoff trends on intercontinental and regional scales as indicated above.

In Europe, models predict substantial changes in precipitation patterns (Xu 2000). More intense precipitation, most of which is projected to occur in winter, will contribute to increased lake inflows, lake levels and runoff (see Palmer and Räisänen 2002). During the summer, drier conditions, exacerbated by greater evaporation, will reduce lake inflows and lake levels. Higher temperatures and decreasing water levels in summer may also affect thermal stratification, evaporation and species composition of lakes (Hulme *et al.* 2003). Sweden is one country where changes in precipitation patterns have been investigated quite thoroughly (SWECLIM 2002). Models for Sweden predict increases in precipitation in the order of 30–60 per cent by 2100, which will be unevenly distributed over the year. In winter, precipitation will increase and might fall as rain rather than snow. This, in turn, will cause faster snowmelt (see ACIA 2004). During the summer, rainfall will decrease and lake waterlevels will fall. Over the whole year, rainfall is predicted to be more intense (SWECLIM 2002). For streams, this implies that winter runoff will increase and spring and summer runoff decrease (Xu 2000). Increasing river discharge to the ocean has already been observed over much of the Arctic, with spring peak river flows now occurring earlier (ACIA 2004). In the northern latitudes, melting glaciers might also contribute to greater runoff.

Extreme climate and weather events

Extreme climate and weather events have cost millions of human lives in the past decades, and there is evidence that some extremes have become more frequent. For example, warm episodes of the El Niño-Southern Oscillation phenomenon have been more frequent, persistent and intense since the mid-1970s, compared with the previous 100 years (IPCC 2001). Furthermore, in some regions such as parts of Asia and Africa, the frequency and intensity of droughts have been observed to increase in recent decades. While these patterns seem to be connected to GEC, it should be noted that alterations in tropical and extra-tropical storm intensity and frequency are dominated by inter-decadal to multi-decadal variations, and no significant trends are evident over the twentieth century. Similarly, no systematic changes in the frequency of tornadoes, thunder days, or hail events were observed (IPCC 2001).

Table 1.1 shows that changes in climate and weather phenomena have, with high probability, already occurred in the latter half of the twentieth century and will, with even higher probability, continue in the twenty-first century. For example, higher maximum and minimum temperatures over land areas have been observed in the latter half of the twentieth century and will with high probability continue to be seen in the future. The diurnal temperature range will be reduced, while the heat index will increase. More intense precipitation events are likely over many northern hemisphere mid- and high latitude land areas, and very likely over many areas in the future. Increased summer continental drying and risk of drought is likely to have occurred in some areas in the latter half of the twentieth century, and is likely to occur over most mid-latitude continental interiors in the future. An increase in tropical cyclone peak wind intensities has not been observed in the analyses available as yet; however, such increases are likely over some areas in the future. Finally, there is insufficient data for the assessment of the increase in tropical cyclone mean and peak precipitation intensities, but these are likely over some areas in the future (IPCC 2001).

Weather extremes are costly, as exemplified by well-documented events in the USA. In the period 1980–2004, the USA experienced 62 weather extremes: 53 of these disasters occurred since 1988, and seven events were in 1998 alone. These weather extremes, mostly hurricanes, caused hundreds of deaths, and the total normalised losses from the 62 events are over US$390 billion – including wide-scale, long-lasting events such as drought (NOAA 2005). Another potentially serious impact of extreme weather events is their potential impacts on species and ecosystems (Leemans and Eickhout 2004). Although climate change has an impact on species through such factors as timing of life cycle events and impacts on the structure and dynamics of geographic ranges, it has been recognised that since 1990 there has been greater ecological response than could be expected from the observed average 0.7°C warming trend alone. Observed responses of ecosystems and species correlate better with changes in extreme weather events, such as heat waves, heavier precipitation, fewer cold extremes and high-magnitude storm events, than with 'normal' climate characteristics (Leemans and van Vliet 2005). As a result of such findings, the projected impacts of climate change (e.g. IPCC 2001) on species and ecosystems are likely to be substantially underestimated.

Tourism and environment

The relationship between tourism and the environment has been the object of scientific research for more than 30 years. Figure 1.3 illustrates how 'the environment' made its entry into the field of tourism studies. In the 1950s, when tourism started to grow rapidly in the post-World War II era, it was mainly seen as an economic sector with great potential for national economies, opening up opportunities for recreation and leisure for large parts of the population in the industrialised countries. It was not until the rise of the green movement in the 1960s and 1970s that environmental impacts of tourism were realised. However, these were generally 'local' in character – focusing on erosion problems or beach crowding – and it was not until the publication of Swiss scientist and environmentalist Jost Krippendorf's *Die Landschaftsfresser* (*The landscape devourers*) (1975) that the environmental impacts of tourism were perceived by a broader public in Europe. Although the work of Mathieson and Wall (1982) charts the development of early awareness of the impacts of tourism in specific environments, there has not been a publication – similar to that of Krippendorf – in the English language literature that has encouraged a wider public debate on tourism's environmental impacts. Instead, tourism often remains something that impacts negatively 'somewhere else'.

During the 1960s and 1970s, the evaluation of environmental 'impacts' was usually based on aesthetic judgements and/or questionable scientific methods, and

Table 1.1 Changes in climate and weather phenomena

Changes in phenomenon	Confidence in observed changes (latter half of the 20th century)[1]	Confidence in projected changes (during the 21st century)[1]
Higher maximum temperatures and more hot days over nearly all land areas	Likely	Very likely
Higher minimum temperatures, fewer cold days and frost days over nearly all land areas	Very likely	Very likely
Reduced diurnal temperature range over most land areas	Very likely	Very likely
Increase of heat index[a] over land areas	Likely, over many areas	Very likely, over most areas
More intense precipitation events	Likely, over many northern hemisphere mid- to high latitude land areas	Very likely, over many areas
Increased summer continental drying and associated risk of drought	Likely, in a few areas	Likely, over most mid-latitude continental interiors (lack of consistent projections in other areas)
Increase in tropical cyclone peak wind intensities	Not observed in the few analyses available	Likely, over some areas
Increase in tropical cyclone mean and peak precipitation intensities	Insufficient data for assessment	Likely, over some areas

Source: IPCC 2001

a Heat index: A combination of temperature and humidity that measures effects on human comfort.

1 The IPCC uses the following expressions to indicate confidence levels: *virtually certain* (greater than 99 per cent chance that a result is true); *very likely* (90–99 per cent chance); *likely* (66–90 per cent chance); *medium likelihood* (33–66 per cent chance); *unlikely* (10–33 per cent chance); *very unlikely* (1–10 per cent chance); *exceptionally unlikely* (less than 1 per cent chance)

often forwarded in rather popular assessments with the aim of being thought-provoking and reaching large parts of the population. Criticism was maintained during the 1980s, its perspective now becoming global (see Mathieson and Wall 1982). Environmental problems were assessed by newly developed methods and solutions presented. For example, Butler (1980) presented his famous destination lifecycle model, suggesting that any destination would inevitably face decline if social, economic or environmental conditions became less favourable (a work that was based on earlier publications that highlighted destination decline, see Wolfe

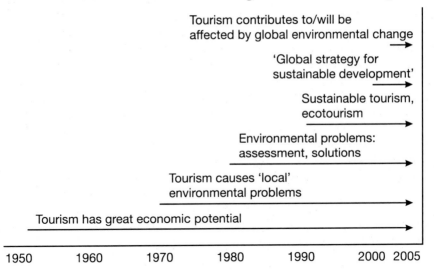

Figure 1.3 Themes in the context of tourism and environmental change

1952; Christaller 1963; Stansfield 1978; Hall 2005b). O'Reilly (1986) emphasised the concept of interrelated environmental, social, economic and perceptional carrying capacities. Since then, a wide range of tools have been developed to assess and cope with environmental change, including the Level of Acceptable Change concept (LAC) and the Environmental Impact Assessment (EIA). These concepts are used frequently, for example, an EIA is a prerequisite for tourist infrastructure development in many countries. Likewise, the carrying capacity concept has been the focal point of recent research (Cocossis and Mexa 2004). However, all these concepts and models remained local in character.

Following the publication of the Brundtland Report in 1987 (WCED 1987), the term sustainable development became widely used in relation to tourism (see Butler 1993). However, the term 'sustainable development' had been used in relation to tourism and the World Conservation Strategy prior to the WCED, and 'sustainable tourism' became a new paradigm in the quest for minimised environmental, social and economic impacts of travel and recreation. The term 'ecotourism' also emerged in the mid-1980s, but was not as yet understood in its current meaning of 'responsible travel to natural areas that conserves the environment and improves the well-being of local people' (International Ecotourism Society 2004; note that a wide range of similar definitions exist). The emergence of sustainable tourism and ecotourism resulted in a shift towards a more positive view of tourism's environmental and economic contributions.

This positive view is expressed on several levels, including an openness to acknowledge the environmental consequences of tourism that were earlier denied or disregarded by the tourist industry and its organisations, but also a more self-confident position that tourism is both sustainable and more beneficial than other industries. For example, the German company TUI, the world's largest tour operator, suggested in 1997 that tourism should become 'a global strategy for sustainable development'. Similar views are mediated by the industry's own organisations such as Green Globe 21, a worldwide certification scheme for 'sustainable' tourism developed by the WTTC.

Scientifically, this is challenged by the insight that the global environmental consequences of tourism have been neglected in past assessments, and that tourism is an important contributor to land-use changes, loss of biodiversity, emission of greenhouse gases, resource depletion, and so on (e.g. Høyer, 2000; Gössling, 2002; Gössling et al., 2002; Ceron, 2003; Peeters, 2003; Hall 2005a; Hall and Higham 2005). This insight has recently been acknowledged by the WTO, which notes that tourism and climate change need to be seen as a 'two-way relationship' (WTO 2003), although as noted above the potential implications of climate change or GEC have not been incorporated into forecasts for future tourism growth. For example, the WTTC (2003: 5) blueprint for 'new tourism' states:

> New Tourism looks beyond short-term considerations. It focuses on benefits not only for people who travel, but also for people in the communities they visit, and for their respective natural, social and cultural environments.

However, neither climate change nor GEC is mentioned in the document. The major concern of an increasing number of observers, however, is that tourism is affecting and will be affected by GEC.

Tourism and global environmental change

> Imagine Davos as a long street, which, 150 years ago, was just a country road leading to a village with no more than 30 houses, surrounded by alpine hills, behind which, as a supreme promise to skiers, genuine Swiss rock formations loomed. Here, in this classic Thomas Mann Magic Mountain landscape, where one could cure his tuberculosis or pass away in all decadence, a quite unalpine shoe-box architecture has spread in the past five decades, box next to box – hotels that, when planned, could not as yet be interpreted as memorials of tourism against the undeniable change of the global climate. In other words: the past five years were the hottest since weather statistics have been kept in Europe. Glaciers are melting, snow cannons work in January, the snow cover is thin, and at some point in time all of these hotels will be standing vacant and be admitted to social purposes.
>
> Naumann (2005) [authors' translation]

The importance of environmental assets for tourism is well understood, because a wide range of recreational activities are dependent on climatic and natural resources. Climatic resources include the thermal characteristics of destinations, with warmer climates generally being more attractive for the broad majority of holidaymakers. Coastal zones are the loci of tourism activities throughout the world, followed by mountain environments. The Mediterranean alone sees more than 100 million international tourist arrivals per year, and the European demand for long-haul sun and sea tourism was estimated to be in the order of 10–12 million in 1998 (WTO 2001). On the other hand, low temperatures guarantee lasting snow cover in winter sport resorts. The WTO estimates that there are some 25 million skiers worldwide, plus another 10 million snowboarders, cross-country skiers, etc. (WTO 2001). Other climatic parameters such as rainfall, hours of sunshine and wind speed, as well as their temporal pattern, greatly influence the appeal of particular destinations (for a recent collection of conference papers see Scott and Matzarakis 2004). Natural assets of importance for tourism include essential resources such as fresh water (see Chapter 10, this volume) and food availability, as well as the physical environments attracting tourists, often high-value amenity landscapes (see Chapters 2 through to 7, this volume). Likewise, biodiversity is also of great importance as a tourist attraction at both the ecosystem and species level (see Chapter 12, this volume).

Climatic assets

Tourist preferences vary throughout the world, and it seems difficult to make generalising statements on climatic requirements for tourism. It is widely recognised, however, that tourism is subject to weather and climate, with travel decisions being based to a large extent on images of sun, sand and sea, or the availability of snow, and thus on climate variables such as temperature, rain and humidity (e.g. Smith, 1993; de Freitas, 2001). Because of this, it is expected that climate change will affect travel behaviour, both as a result of altering conditions for holidaymaking at the destination level and climate variables perceived as less or more comfortable by the tourists. In particular, the effects of increasing temperatures and related parameters (such as rain) on the choice of a destination and time of departure have been the focus of recent research. For example, in an attempt to identify 'optimal' temperatures, Maddison (2001) analysed travel patterns of British tourists and found that the maximum daytime temperature perceived as comfortable was 30.7°C, with even small increases above this level leading to decreasing numbers of visits. Maddison also found that greater rainfall would deter tourists. In another study, Lise and Tol (2002) analysed a cross-section of destinations of Organisation of Economic Co-operation and Development (OECD) tourists. Using factor and regression analysis, they found that OECD tourists preferred an average temperature of 21°C at the hottest month of the year at their destinations. Both studies conclude that tourists may shift to other destinations or travel during other periods of the year under a scenario of climate change.

Statistical models express the behaviour of tourists as a function of weather, climate and other factors – such as travel costs – and thus need to be seen as top-down approaches to understanding the interaction of travel choice and climate. Few studies have chosen to analyse tourist perceptions of weather conditions from a bottom-up perspective. Using a combined strategy of climate variable measurements and stated weather perceptions based on structured interviews, Mansfeld *et al*. (2003) assessed the biometeorological comfort of beach tourists in Eilat, Israel. The results show that differences in wind velocity and cloudiness had a significant influence on the tourists' comfort perception which, in this case study with rather moderate temperatures (20–24°C), was negative. Temperature differences also had an influence on the tourists' comfort perception, but the importance of this variable was generally much lower. Mansfeld *et al*. thus suggest that perceptions might be very different under summer conditions, when both wind velocity and cloudiness might be perceived as rather positive. Furthermore, the study revealed that domestic tourists were more sensitive to weather conditions than tourists from overseas, hinting at the importance of other aspects, such as whether the tourists usually live in warm, temperate or cold climates. The study concludes that weather conditions shape the tourists' comfort perception, even though the importance of single variables depends on the background conditions at the destination level – relatively extreme (high or low) weather variables – and the conditions experienced before going on holiday.

Other recent research indicates that the role of weather parameters is not easily understood. For example, Gössling *et al*. (2005a) found that the majority of tourists in an *in situ* case study in Zanzibar, Tanzania, expressed no aversion to a theoretical increase in temperatures, despite high average day temperatures around 30–31°C. The case study confirms that climate is an important aspect of travel decisions, even though many tourists seem to rather implicitly consider 'temperature' in their travel choices. In their *in situ* perception, rain and weather extremes have a far greater effect on travel decisions. Climate variables such as more rain, storms and higher humidity are more likely to negatively influence travel decisions under a scenario of climate change than higher temperatures, which are not necessarily perceived as negative by all tourists. It is acknowledged, however, that the role of weather and climate information in the choice of destinations remains insufficiently investigated, and that the results of this particular case study cannot be extrapolated to the global tourist population (also see chapters in Hall and Higham 2005). Hence, qualitative research indicates a degree of complexity in weather perceptions that is as yet not fully considered in statistical analyses of changing travel patterns under conditions of climate change.

Natural assets

A wide range of publications has sought to assess the consequences of climate change for the natural assets of tourist industries of destinations (e.g. König 1999; Richardson and Loomis 2003; Scott 2003). Most of these publications have warned that tourist destinations might lose part of their attractiveness through climate change, for example, as a result of loss of snow in ski resorts. In some areas, however, there might also be 'gains' in terms of less rain or extended summer seasons. Overall, the impact of GEC on tourism development has still been little explored because the complexity of changing environmental conditions and tourist perceptions is not well understood. In the following section, some of the expected changes are outlined for the environments that are most central to tourism – coastal zones and mountains, as well as biodiversity. More specific and detailed discussions can be found in the following chapters of the book dealing with all environments of importance for tourism.

Coastal zones are the most attractive areas for tourism all around the world. As noted above, the Mediterranean alone receives some 116 million international tourist arrivals per year plus an unknown number of domestic tourist arrivals (WTO 2003). There are also a great number of both cold and warm water islands receiving a growing number of tourists (e.g. Iceland, New Zealand, Caribbean islands, Indian Ocean islands, Pacific islands), who usually account for a substantial share of foreign exchange earnings (Gössling 2003). It is clear that GEC will have severe consequences for coastal environments because it will lead to sea-level rise, coastal erosion, land inundation, flooding by storm surges, and loss of coastal wetlands and biodiversity (Nicholls 2004; Zhang *et al.* 2004). In particular, weather extremes such as El Niño-Southern Oscillation (ENSO) phenomena or tropical storms can result in severe damage. For example, increasing water temperatures, as observed during ENSO phenomena, have been more frequent and intense in recent decades (IPCC 2001). The 1997–98 ENSO, for instance, had a severe impact on the climate of the Indian Ocean. In March and April 1998, seawater temperatures increased by 1.5°C above average values measured during the same period in 1997. Following the event, coral mortality ranged from 50–90 per cent over extensive areas of shallow reefs in the Seychelles (Lindén and Sporrong 1999).

However, while sea-level rise and coastal erosion will have a direct impact on the physical preconditions for leisure, jeopardising leisure opportunities in many areas, it is not clear how tourists will perceive the loss of other resources, such as marine biodiversity. One recent study in Mauritius, for instance, found that snorkelling and diving were important tourist activities with, for example, 62 per cent of the tourists in the sample participating in snorkelling activities. While both snorkellers and divers stated that the abundance of species, unsullied reefs (no broken corals) and good visibility were important criteria for their underwater experience, the results also indicate that only few of the tourists have the knowledge to judge whether environmental conditions in Mauritius are good and reefs healthy. Indeed, there is evidence that environmental conditions are deteriorating (see Gössling *et al.* 2005b). Thus, it remains unclear when environmental conditions reach a state that no longer appeals to tourists. It also raises the question of whether or not tourists are informed about these environmental conditions before going on holiday – that is, whether they are part of the tourists' decision-making process or not.

After coastal zones, mountains are the most important regions for tourism, attracting millions of tourists worldwide. It is clear that mountain ecosystems are among the most vulnerable to GEC, including glacial retreat, melting of permafrost, loss of biodiversity and changing treelines (see Chapter 3, this volume). In particular, increasing temperatures have already had severe consequences for ski tourism in many areas, and will be of even greater importance in the future.

Consequently, the concept of 'snow-reliability' has been introduced, which defines ski-resorts with a 100-day ski season and a minimum 30–50cm snow depth in at least seven out of ten years (Elsasser and Bürki 2002). In some countries such as Norway, high-altitude resorts have already started to advertise snow security. Opportunities to use technical devices such as snow cannons will, in many areas, define whether or not resorts remain economically viable. Scott (Chapter 3, this volume) also reminds us that:

> It is increasingly recognised that the value of the mountain landscape for tourism depends not just on the presence and quality of tourism infrastructure, but also on the quality of the mountain landscape. Consequently, if climate change adversely affects the natural setting (for example, loss of glaciers, reduced biodiversity, fire or disease impacted forest landscape, reduced snow cover) at a destination, the quality of the tourism product could be diminished with implications for visitation and local economies.

Biodiversity can be of great importance for tourism. Bird watchers, for example, might travel around the world to see a particular bird species (Blondel 2002). Salmon is an important species for fishing, along with a wide range of other salt and fresh water species (Svenson *et al.* 2001). Whale watching is now an activity offered in 87 countries in the world (Hoyt and Hvenegaard 2002). Wildlife is generally an attraction, particularly to tourists from industrialised countries where people have fewer experiences with nature. Tourists may often be attracted by environments that differ from those experienced at home, and the exoticism of biodiversity is thus of great importance to tourism, particularly for those destinations that focus on ecotourism and other forms of nature-based tourism that are often connected to national parks and public and private reserves.

Biodiversity can also have an important symbolic function, as exemplified by moose in Sweden, koalas in Australia or the lion in Africa. Without these species, these countries and regions would lose much of their mythical power. However, many species are threatened by GEC. Within a wide range of taxa, such as plants, birds, reptiles and butterflies, it has been estimated that climate change alone will cause 15–37 per cent of species to become committed to extinction by 2050 (Thomas *et al.* 2004). Estimates vary in relation to the climate change scenario and assumptions concerning species' ability to disperse to new habitats if the climate in their old habitat is not suitable anymore (Thomas *et al.* 2004). However, as the chapter on biodiversity illustrates (see Chapter 12), pressures on biodiversity include not just climate change, but also land clearance, changing land use, pollution and tourism urbanisation.

Although much of the focus on biodiversity is on endemic biodiversity, usually in relatively natural ecosystems, it must be noted that agricultural landscapes also provide significant tourism opportunities. Wine and food tourism is a rapidly growing form of special interest tourism that emphasises local foodways. It is regarded as having some potential to maintain heirloom varieties of plants and animals and provide a basis for more sustainable forms of agriculture.

Nevertheless, agricultural landscapes are also subject to the impacts of climate change as well as other processes such as urbanisation. For example, climate change is resulting in stress on some wine regions while improving prospects for others, such as the United Kingdom. Some areas that are cool climate wine regions may trend towards growing warmer region grape varieties by 2100, while some wine styles, such as ice wine, made in Canada, Germany, Austria and Switzerland, will also be threatened by warmer weather conditions. In addition, climate change will mean new pests and diseases to manage, as well as issues with respect to water supply for irrigation (Margolis and Pape 2004). The loss of certain agricultural landscapes because of climate change will undoubtedly have impacts on tourism flows because of the loss of amenity values.

In summary, tourism and recreation are likely to be severely affected by GEC, because both climatic and natural resources will be altered or lost. In particular, it should be noted that tourism usually takes place in areas that are at high risk of being affected by GEC, such as coastal or mountainous zones. It seems clear that low-lying mountain resorts in the temperate zones are facing the most immediate risk of losing their winter tourism potential. Such losses already occur in tourist resorts throughout the Alps and in North America. Potential gains, on the other hand, might occur in summer, even though this remains highly speculative. Similarly, low-lying tropical islands and coastal zones will be at great risk. Even here, this will primarily depend on the extent of extremes. It is also clear that environmental change and vulnerability are interdependent. The tsunami hitting Asia in late 2004, for instance, seems to have hit particularly hard in those regions where environmental change has been most substantial. Even though it was a natural disaster and not a result of GEC, the tsunami seems to have had more severe consequences in areas where coastal zones were more densely populated and where natural ecosystems, such as mangroves and coral reefs, had been damaged, converted or destroyed. Such areas might often be tourist zones, where job opportunities induce coastal zone-directed migration (see Gössling and Schulz 2005) and where natural resources are exploited and/or ecosystems degraded (e.g. Gössling 2001).

Future tourist flows

An increasing number of publications have sought to analyse tourist travel flows in relation to climatic and other parameters such as per capita income. The ultimate goal is to develop scenarios for future travel flows, possibly including 'most at-risk destinations'. Such scenarios are meant to help the tourist industry in planning future operations, and they are of importance in developing plans for adaptation. While such scenarios are greatly needed, it needs to be acknowledged that the predictability of tourist flows under a scenario of GEC seems difficult (see Table 1.2).

Statistical databases used to predict travel flows are generally insufficient. For example, data provided by the WTO do not distinguish between business and leisure tourists. Instead, statistics refer to 'international arrivals of tourists by country of residence', 'arrivals by nationality', 'arrivals in all establishments' and

Table 1.2 Weaknesses of current models in predicting travel flows

---

- Statistical database insufficient
- Temperature assumed to be the most important weather parameter
- Importance of other weather parameters largely unknown (rain, storms, humidity, hours of sunshine)
- Role of weather extremes unknown
- Role of information in decision making unclear
- Role of non-climatic parameters unclear (for example, social unrest, political instability, risk perceptions)
- Existence of fuzzy variables problematic (terrorism, war, epidemics)
- Assumed linearity of change in behaviour unrealistic

'arrivals of tourists in hotels'. None of these databases is consistent for all countries in the world. Predicting travel flows on such generalised databases is thus likely to have substantial influence on the reliability of results, because business travellers constitute a large share of international tourists and usually do not travel for reasons related to climate. Business travellers might also influence other correlations, such as the one between a country's poverty line and the number of tourist arrivals, because it seems plausible that there are more business co-operations between wealthier countries, and hence more travel between these. In addition, statistical information is usually at a scale which is insufficient to detail tourist flows between and within regions that would be differently affected by climate change or other dimensions of GEC.

Temperature is often assumed to be the most important weather parameter in the analysis of tourism flows. Outside a certain temperature range, weather perceptions become unfavourable and problems of discomfort arise (McGregor *et al.* 2002). For example, Maddison (2001), in analysing travel patterns of British tourists, found that the maximum daytime temperature perceived as comfortable was 30.7°C, with even small increases above this level leading to decreasing numbers of visits. Based on such analyses, statistical models have sought to predict how travel flows will change in the future with increasing temperatures.

While there is no question that temperature is an extremely important weather parameter with major influence on travel decisions, it should nevertheless be considered that the issue of perceptions is little explored. For example, the expectation of warm destinations might follow a general logic of 'warm is good – warmer is better'. An advertisement campaign by the tour operator Resfeber in Sweden in January/February 2005 portrays well-known cities in the tropics in association with temperatures. The list starts with: 'Bangkok – 32°C', suggesting that temperatures beyond 30°C might not be understood as 'too hot' after all. Perceptions of 'too hot' would also imply that tourists are i) usually informed on the climatic conditions at their holiday destinations and ii) that they are able to interpret this information. While a decent level of information on climatic conditions can only be confirmed for a share of tourists (for the first case study on this issue see Hamilton and Lau, Chapter 13, this volume), it seems questionable whether a tourist is capable of interpreting a 1°C temperature increase in terms of comfort – notably in the absence of information on other parameters such as humidity or wind-speed. Rather, climate in the sense of warm weather conditions at the destination level might often be implicitly considered in travel decisions; there might, for example, be the notion that the tropics are warm. This notion might also be 'broad', in the sense that tourists do not distinguish between 'warm and dry' climates such as, for example, in Tunisia or 'warm and humid' climates such as, for example, in the Indian Ocean islands. This is of importance, because it is generally expected that there is a linear relationship between increasing temperatures and changing travel flows (Lise and Tol 2001, Maddison 2001). Note as well that warming will be more pronounced in the northern latitudes and less pronounced in the tropics, where tourists might expect to find 'hot' climates anyway (see Gössling *et al.* 2005a).

Following this line of reasoning, one might expect *perceptions* to play the most central role in decision making. For example, should the perception of 'warm' countries change toward one of being 'too warm', this might cause rather sudden changes in travel flows on broad regional scales. It is also clear that temperature increase is only linear when measured over long periods, and that there might be great differences from year to year. The fuzzy reaction of tourists to such changes could, for example, be felt in central Europe in summer 2003 (prolonged period of temperatures reaching 42°C peaks) and 2004 (cold and rainy, temperatures generally not exceeding 25°C), when travel decisions were rather random in Germany. Obviously, many people had expected warm summers even in 2004, and made a decision to stay at home. When the weather remained cold and rainy, there was a last-minute rush on charter trips to virtually any 'warm' destination in late July 2004. Newspaper headlines read:

> Hamburg escapes weather-frustration. Last minute record bookings. Stress on airports. Almost all trips booked out.
>
> (*Hamburger Morgenpost*, 14 July 2004)

Weather extremes thus remain the fuzzy variable in the modelling of travel flows. We are warned that:

> ... due to the complexity of the Earth-system, it is possible that climate change will evolve differently from the gradually changing scenarios. ... For example, storm intensities and tracks could change in unforeseen ways or temperatures could rise or fall abruptly due to unexpected disturbances of global weather systems.
>
> (ACIA 2004: 125)

With more incidences such as experienced during the summers of 2003 and 2004, tourists might become more aware of weather extremes – and either adapt to these or adjust their travel behaviour. Overall, little is known as yet about the complexity of these issues, and it seems unjustified to assume simple relationships.

Outline of chapters

The book is structured in three sections. The first section focuses on the environments of relevance for tourism, followed by a section on factors of particular relevance: water availability, diseases, biodiversity and weather extremes. The final section on adaptation and response opens the floor for discussions by questioning some of the assumed relationships between tourism and GEC. It also contains an overview and an example of adaptation in what can be assumed to be the most vulnerable environments to GEC in the short-term future, mountain environments.

Environments

In Chapter 2, Margaret Johnston outlines threats to polar environments. Northern and southern high latitudes appear to be those that will be most intensively affected by GEC, with average winter temperatures in some parts of the Arctic having increased by 3°C over the past 60 years and Arctic sea ice having diminished by 10 per cent over the past 30 years. Such rapid changes affect landscapes, vegetation patterns and animal behaviour, and thus tourism, which is dependent on scenic landscapes of snow and ice, as well as access to wildlife populations. In particular, climate change might affect options to participate in traditional hunting and fishing, expedition-style and destination cruising, wildlife viewing, northern lights tourism, skiing and snowmobiling, dogsledding, as well as cultural and aboriginal tourism. Hence, Johnston wonders how tourism in the polar regions will change: 'What will it mean for tourists when the inaccessible becomes accessible, and the "inhospitable" climate appears more hospitable?'

In Chapter 3, Daniel Scott examines tourism in one of the most vulnerable environments in terms of GEC. As Scott recognises, this is an environment in which, arguably, some of the most detailed reseach on tourism–GEC relationships have been conducted because of the potential impacts of climate change on alpine and winter tourism.

The importance of lakes, streams, reservoirs, canals and wetlands for a wide variety of leisure activities, including bathing and swimming, recreational boating, sport fishing, ice-skating and bird watching is sketched by Jones, Scott and Gössling in Chapter 4. Global environmental change is likely to affect a number of properties of water bodies, such as water levels, water properties, biodiversity and water supply. Water abstraction, land-use changes, the introduction of alien species and weather extremes may be additional stressors for freshwater systems. Increasing temperatures will also reduce the average number of days that lakes and streams can be used for winter sports activities. Overall, in the case of both mountain environments and water bodies, GEC is likely to have negative effects for recreation and tourism, and many communities depending on tourism might experience substantial economic losses when the resources that attract tourists are scarce, deteriorating and contested.

Boreal, temperate and tropical forest ecosystems are the focus of Chapter 5 by Stefan Gössling and Thomas Hickler. Forest ecosystems are of major importance for recreation and tourism, particularly in Europe, Japan, and other industrialised countries, and attract millions of visitors daily. Forest-based activities include, for example, walking, hiking, mountain biking, cross-country skiing, dog-sledding, fishing, hunting, bird watching, or mushroom and berry collection. Global environmental change will affect forest ecosystems through land-use changes, changing temperatures and precipitation patterns, increasing CO_2 concentrations, and nitrogen deposition. Increasing temperatures, in particular, will have consequences for biodiversity, which is under severe stress through global climate change. Even though these changes will only be felt in the medium-term future, consequences for tourism and recreation might be fundamental.

In Chapter 6, Stephen Craig-Smith, Richard Tapper and Xavier Font focus on the most important natural assets for tourism: coastal zones, islands and reefs. These are the foundations for much of the traditional sun, sand and sea tourism, and these environments also serve as the playgrounds for what the authors call 'post-modern tourists', tourists who are engaged in more active and/or educational activities such as 'surfing, diving, sailing, walking, bird spotting, nature viewing or general exploring'. Clearly, coastlines and low-lying islands in particular will suffer from sea-level rise, rain and cloudiness, storminess and coastal erosion, habitat degeneration and other changes related to GEC, including the potential of increased high magnitude weather events. However, some coasts could also become more attractive to tourism, because they might profit from increasing temperatures.

Robert Preston-Whyte, Shirley Brooks and William Ellery present a GEC scenario for tourism in deserts and savannah regions in Chapter 7. Arid landscapes are fragile and their 'extreme' nature is one of the factors attracting tourists. However, climate change will affect desert environments as there is a trend towards warmer conditions, which will increase evapotranspiration rates in arid areas and lead to a decrease in potential water availability. Together with other pressures, this might substantially change desert and savannah environments and affect in particular those elements that are currently of importance for tourism. For example, eutrophic or arid savannah in South Africa may spread at the expense of dystrophic or moist savannah. It is the latter, however, that is of importance for tourism, because these environments favour a high diversity of spectacular plants and large mammals.

Chapter 8 by C. Michael Hall focuses on tourism urbanisation processes and their relationship to GEC. Hall notes that although urbanisation is a major process in land-use change at the global level, tourism is only responsible for a very small amount of total urbanisation. Nevertheless, the spatial concentration of tourism urbanisation in specific environments, especially coastal areas, means that it still exerts substantial influence on biodiversity in those areas.

Global issues

As outlined in Chapter 4, water bodies are of great amenity value for tourism. Chapter 10 by Gössling focuses on the importance of fresh water for tourism, which is of essential importance for the maintenance of swimming pools, irrigated gardens and bathrooms. Visitors in areas with limited water resources are substantial contributors to water scarcity. Usually, tourism is concentrated in regions such as islands and coastal zones with few fossil water resources, low aquifer renewal rates and few surface water sources. Tourism also causes a shift in global water consumption from regions of relative water abundance to those that are water scarce. Related to these aspects, water quality may often decrease through tourism, as a result of the discharge of untreated sewage, nutrient loads and toxic substances into adjacent water bodies. In many areas, water for tourism has already become scarce, and might even become scarcer through changing precipitation patterns under a scenario of GEC. This may, in many areas, demand technological solutions to cope with the problem – these are usually energy intense and thus accelerate climate change.

In Chapter 9 Hall charts the role of mobility in the spread of disease and argues that the present period of contemporary globalisation presents substantial new challenges with respect to emergent disease control that tourism mobility is exacerbating.

Extreme weather events may be one of the most central aspects of tourism in the future. In Chapter 11, Chris de Freitas investigates whether extreme weather conditions will become more frequent and intense in the future, including hurricanes and other storms, heatwaves, cold waves and high temperatures. Obviously, intense storms are the most severe form of extreme weather conditions and pose a threat to tourists and tourist infrastructure. This is of essential importance in the tropics, where tourism is often concentrated in coastal areas that are at times hit by hurricanes. However, de Freitas argues that there is no clear evidence from climate modelling whether the frequency and intensity of storms will increase in the future.

Chapter 12 by Hall examines tourism's role in conserving biodiversity. Hall notes that tourism is often portrayed as an environmentally friendly industry because of nature-based tourism activities. Yet Hall argues that such an image does not sit easily with the realities of tourism's net contribution to biodiversity maintenance and he argues that species ranges and ecosystem dynamics have been only poorly integrated into wider thinking about ecotourism and its relationship to biological conservation through the establishment of national parks and conservation reserves.

Stakeholder adaptation and perceptions

In Chapter 13 Jacqueline Hamilton and Maren Lau investigate the role of climate information in destination choice decision making, noting that there are a wide range of information sources that people use, as well as a range of values that people use to assess destinations. Shifts in perceptions of climate are also significant for tourism destinations and this issue is taken up in Chapter 14 by Szilvia Gyimóthy who discusses the development of new marketing perspectives in high latitudes.

As Scott points out in the introduction to Chapter 15, even if international agreements to limit emissions of greenhouses gases are successfully established, some climate change is inevitable. Realising this, options and opportunities for adaptation become significant in tourism planning and management. Focusing on the ski industry in North America, Scott presents three general categories of adaptation options: i) hard technological developments (for example, snowmaking); ii) soft business practices (for example, market diversification); and iii) government and industry policy (for example, environmental regulatory frameworks). However, not all ski areas will be able to adapt strategies to cope with GEC because climate conditions will simply become too unfavourable to continue operations.

Issues of adaption in alpine locations are also taken up by Christian Nöthiger, Rolf Bürki and Hans Elsasser (Chapter 16) who discuss the importance of integrating costs of natural disasters into tourism planning based on the example of the avalanches winter 1999 and the storm Lothar in the Swiss Alps. These incidences had substantial costs for the tourist industry and are an inherent danger for any resort operation.

Finally in this section, Erika Andersson Cederholm and Johan Hultman (Chapter 17) present a sociological perspective on tourism and GEC that provides a useful counterpoint to many of the other chapters in the book. They argue that tourism has changed in that it is increasingly built on the marketing of experiences. Experiences, however, are placeless because destinations and their authenticity are social constructions created through TV, websites, catalogues and brochures. Hence, the impact of GEC might primarily be an issue of the geographical re-organisation of experiences, and thus ultimately no more than a spatial redistribution of tourists. Andersson Cederholm and Hultman conclude: 'as long as nature is constructed as an exclusive experiential product, the consumption of nature will have the same communicative function as before even if anthropogenic climate change rearranges nature on a global scale.'

The conclusion by Gössling and Hall argues that GEC presents a major challenge to the tourism industry, not only in terms of adapting to change, but also in finding ways and means by which the impacts of tourism can be mediated. Furthermore, they argue that the tourism–GEC relationship provides significant challenges to the study of tourism that need to be addressed if the global embedding of tourism in systems of production and consumption are to be adequately understood.

For too long the focus of tourism research has been on the destination. However, there is little doubt that anthropogenic environmental change presents a significant challenge to tourism and tourism studies on a global scale. Whether or not the field of tourism studies as well as the tourism industry is able to effectively respond to the practical, intellectual and policy implications of GEC is highly debatable. However, we hope that this book will at least contribute to such a debate, if not to a more profound awareness of the need to respond to the fundamental necessity of conserving our natural capital.

References

ACIA (Arctic Climate Impact Assessment) (2004) *Impacts of a Warming Arctic*. Cambridge: Cambridge University Press. www.acia.uaf.edu

Airbus (2003) *Global Market Forecast 2003–2022*. Blagnac, France: Airbus S.A.S.

Bach, W. and Gößling, S. (1996) 'Klimaökologische Auswirkungen des Flugverkehrs', *Geographische Rundschau*, 48: 54–9.

Bell, M. and Ward, G. (2000) 'Comparing temporary mobility with permanent migration', *Tourism Geographies*, 2: 87–107.

Blondel, J. (2002) 'Birding in the sky: Only fun, a chance for ecodevelopment, or both?' In F. di Castri and V. Balaji (eds) *Tourism, Biodiversity and Information*. Leiden: Backhuys Publishers, pp.307–17.

Brown, P. and Oliver, M. (2004) 'Top scientist attacks US over global warming', *The Guardian*, January 9.

Bugmann, H. (1997) 'Sensitivity of forests in the European Alps to future climatic change', *Climate Research*, 8: 35–44.

Butler, R.W. (1980) 'The concept of a tourist area cycle of evolution: implications for management of resources', *Canadian Geographer*, 24(1): 5–12.

Butler, R. (1993) 'Tourism – an evolutionary perspective', in J.G. Nelson, R. Butler and G. Wall (eds), *Tourism and Sustainable Development: Monitoring, Planning, Managing*, Publication Series no. 37, Waterloo: University of Waterloo, Department of Geography.

Caldwell, C. (2003) 'The decline of France', *Weekly Standard*, 12–08–2003, Volume 009, Issue 13. Available at: www.weeklystandard.com/Content/Public/Articles/000/000/003/429zmcyt.asp?pg=1 (accessed 20 February 2005).

Ceron, J.P. (2000) 'Tourisme et changement climatique', in *Impacts potentiels du changement climatique en France au XXIème siècle*. Premier ministre. Ministère de l'aménagement du territoire et de l'environnement 1998, pp.104–11.

Challenor, P., Hankin, R. and Marsh, B. (2005) 'The probability of rapid climate change', paper presented at International Symposium on the Stabilisation of Greenhouse Gases, Hadley Centre, Met Office, Exeter, 1–3 February.

Christaller, W. (1963) 'Some considerations of tourism location in Europe: the peripheral regions – underdeveloped countries – recreation areas', *Regional Science Association Papers*, 12: 95–105.

Cocossis, H. and Mexa, A. (2004) *The Challenge of Tourism Carrying Capacity Assessment: Theory and Practice*. Aldershot: Ashgate Publishing.

Cohen, E. (1974) 'Who is a tourist: a conceptual clarification', *Sociological Review*, 22: 527–55.

Coles, T. and Timothy, D. (eds) (2004) *Tourism, Diasporas and Space*. London: Routledge.

Coles, T., Duval, D. and Hall, C.M. (2004) 'Tourism, mobility and global communities: New approaches to theorising tourism and tourist spaces', in W. Theobold (ed.) *Global Tourism: The Next Decade*, 3rd ed. Oxford: Butterworth Heinemann, pp.463–81.

Coles, T., Hall, C.M. and Duval, D. (2005) 'Mobilising tourism: A post-disciplinary critique', *Tourism Recreation Research*, 30 (in press).

Cox, P.M., Betts, R.A., Collins, M., Harris, P.P., Huntingford, C. and Jones, C.D. (2004) 'Amazonian forest dieback under climate-carbon cycle projections for the 21st century', *Theoretical and Applied Climatology*, 78: 137–56.

de Freitas, C. (2001) 'Theory, Concepts and Methods in Tourism Climate Research', Proceedings of the First International Workshop on Climate, Tourism and Recreation, Porto Carras, Neos Marmaras, Halkidiki, Greece, 5 –10 October, pp. 3–20.

Duval, D.T. (2003) 'When hosts become guests: return visits and diasporic identities in a Commonwealth eastern Caribbean community', *Current Issues in Tourism*, 6: 267–308.

Elsasser, H. and Bürki, R. (2002) 'Climate change as a threat to tourism in the Alps', *Climate Research*, 20: 253–7.

Engelman, R., Cincotta, R.P., Dye, B., Gardner-Outlaw, T. and Wisnewski, J. (2000) *People in the Balance: Population and the Natural Resources at the Turn of the Millennium*. Washington, DC: Population Action International.

Field, A.M. (1999) 'The college student market segment: A comparative study of travel behaviors of international and domestic students at a Southeastern University', *Journal of Travel Research*, 37: 375–81.

Frändberg, L. (1998) *Distance Matters: An inquiry into the relation between transport and environmental sustainability in tourism*. Humanekologiska skrifter no.15. Göteborg: Section of Human Ecology, Göteborg University.

Frändberg, L. and Vilhelmson, B. (2003) 'Personal mobility – a corporeal dimension of transnationalisation. The case of long-distance travel from Sweden', *Environment and Planning*, A 35: 1751–68.

Giampetro, M. and Pimentel, D. (1994) 'Energy utilization', in C.J. Arntzen and E.M. Ritter (eds) *Encyclopedia of Agricultural Science*. San Diego: Academic Press, pp.73–6.

Goodrich, J.N. (1994) 'Health tourism: A new positioning strategy for tourist destinations', *Journal of International Consumer Marketing*, 6: 227–37.

Gössling, S. (2000) 'Sustainable tourism development in developing countries: some aspects of energy-use', *Journal of Sustainable Tourism*, 8(5): 410–25.

Gössling, S. (2001) 'Tourism, environmental degradation and economic transition: Interacting processes in a Tanzanian coastal community', *Tourism Geographies*, 3(4): 230–54.

Gössling, S. (2002) 'Global environmental consequences of tourism', *Global Environmental Change*, 12: 283–302.

Gössling, S. (ed.) (2003) *Tourism and Development in Tropical Islands. Political Ecology Perspectives*. Cheltenham: Edward Elgar Publishing.

Gössling, S. and Schulz, U. (2005) 'Tourism-Related Migration in Zanzibar, Tanzania', *Tourism Geographies*, 7(1): 43–62.

Gössling, S., Borgström-Hansson, C., Hörstmeier, O. and Saggel, S. (2002) 'Ecological footprint analysis as a tool to assess tourism sustainability', *Ecological Economics*, 43(2–3): 199–211.

Gössling, S., Bredberg, M., Randow, A., Svensson, P. and Swedlin, E. (2005a) 'Tourist perceptions of climate change', *Current Issues in Tourism,* in press.

Gössling, S., Helmersson, J., Liljenberg, J. and Quarm, S. (2005b) 'Diving tourism and global environmental change. A perception case study in Mauritius', *Tourism Management*, submitted.

Gow, D. (2004a) 'CO_2 cuts will raise prices, says industry', *The Guardian*, London, January 17.

Gow, D. (2004b) 'CO_2 limits suicidal for competitiveness, says industry', *The Guardian*, London, January 20.

Hall, C.M. (2003) 'Tourism and temporary mobility: Circulation, diaspora, migration, nomadism, sojourning, travel, transport and home'. Paper presented at International Academy for the Study of Tourism Conference, 30 June–5 July, Savonlinna, Finland.

Hall, C.M. (2005a) *Tourism: Rethinking the Social Science of Mobility*. Harlow: Prentice Hall.

Hall, C.M. (2005b) 'Space-time accessibility and the tourist area cycle of evolution: The role of geographies of spatial interaction and mobility in contributing to an improved understanding of tourism', in R. Butler (ed.) *The Tourism Area Life-Cycle*. Clevedon: Channelview (in press).

Hall, C.M. and Higham, J. (2005) 'Introduction: Tourism, recreation and climate change', in C.M. Hall and J. Higham (eds) *Tourism, Recreation and Climate Change*. Clevedon: Channelview.

Hall, C.M. and Müller, D.K. (eds) (2004) *Tourism, Mobility and Second Homes: Between Elite Landscape and Common Ground*. Clevedon: Channelview Press.

Hannah, L., Midgley, G.F., Lovejoy, T., Bond, W.J., Bush, M., Lovett, J.C., Scott, D. and Woodward, F.I. (2002) 'Conservation of biodiversity in a changing climate', *Conservation Biology*, 16: 264–8.

Hitz, S. and Smith, J. (2004) 'Estimating global impacts from climate change', *Global Environmental Change*, 14: 201–18.

Høyer, K.G. (2000) 'Sustainable tourism or sustainable mobility? The Norwegian case', *Journal of Sustainable Tourism*, 8: 147–61.

Hoyt, E. and Hvenegaard, G.T. (2002). 'A review of whale watching and whaling with applications for the Caribbean', *Coastal Management*, 30: 381–99.

Hulme, M., Conway, D. and Lu, X. (2003) *Climate Change: An Overview and Its Impact on the Living Lakes*. Report prepared for the 8th Living Lakes Conference 'Climate change and governance: managing impacts on lakes'. Zuckerman Institute for Connective Environmental Research, University of East Anglia, Norwich, UK, 7–12 September.

International Ecotourism Society (2004) www.ecotourism.org (accessed 1 January 2005).

IPCC (Intergovernmental Panel of Climate Change) (2001) *Climate change 2001: the Scientific Basis*, contribution of the working group I to the third assessment report of the Intergovernmental Panel of Climate Change. Cambridge: Cambridge University Press.

IPCC (Intergovernmental Panel of Climate Change) (2005) www.ipcc.ch (accessed 10 January 2005).

Janelle, D.G. and Hodge, D. (eds) (2000) *Information, Place, and Cyberspace. Issues in Accessibility*. Berlin: Springer-Verlag.

Johnston, R.J., Taylor, P.J. and Watts, M.J. (eds) (1995) *Geographies of Global Change: Remapping the World in the Late Twentieth Century*. Oxford: Blackwell.

Kerr, R.A. (2004) 'Three degrees of consensus', *Science*, 305: 932–4.

Klein Goldewijk, K. (2001) 'Estimating global land use change over the past 300 years: the HYDE database', *Global Biogeochemical Cycles*, 15(2): 417–34.

König, U. (1999) 'Climate change and snow tourism in Australia', *Geographica Helvetica*, 54(3): 147–57.

Krippendorf, J. (1975) *Die Landschaftsfresser: Tourismus und Erholungslandschaft, Verderben oder Segen?* Schönbühl, Switzerland: Hallwag.

Labat, D., Goddéris, Y., Probst, J.L. and Guyot, J.L. (2004) 'Evidence for global runoff increase related to climate warming', *Advances in Water Resources*, 27: 631–42.

Leemans, R. and Eickhout, B. (2004) 'Another reason for concern: regional and global impacts on ecosystems for different levels of climate change', *Global Environmental Change*, 14: 219–28.

Leemans, R. and van Vliet, A. (2005) 'Responses of species to changes in climate determine climate protection targets', paper presented at International Symposium on the Stabilisation of Greenhouse Gases, Hadley Centre, Met Office, Exeter, 1–3 February.

Lew, A., Hall, C.M. and Williams, A.M. (eds) (2004) *Companion to Tourism*. Oxford: Blackwell.

Lindén, O. and Sporrong, N. (1999) *Coral Reef Degradation in the Indian Ocean, status reports and project presentations*, SAREC Marine Science Program, Department of Zoology, Stockholm University.

Lise, W. and Tol, R. (2002) 'Impact of climate on tourist demand', *Climatic Change*, 55(4): 429–49.

Loh, J. and Wackernagel, M. (2004) *Living Planet Report 2004*. www.panda.org/downloads/general/lpr2004.pdf (accessed 10 February 2005)

London, J.B. (2004) 'Implications of climate change on small island developing states: experience in the Caribbean region', *Journal of Environmental Planning and Management*, 47(4): 491–501.

Maddison, D. (2001) 'In search of warmer climates? The impact of climate change on flows of British tourists', *Climatic Change*, 49: 193–208.

Mansfeld, Y., Freundlich, A. and Kutiel, H. (2003) 'The Relationship between Weather Conditions and Tourists' Perception of Comfort: The case of the winter sun resort of Eilat', Paper presented during the NATO Advanced Research on Climate Change and Tourism: Assessment and Coping Strategies, 6–8 November, Warsaw, Poland.

Margolis, M. and Pape, E. (2004) 'Vins d'Angleterre', *Newsweek*, 44–6.

Mathieson, A. and Wall, G. (1982) *Tourism: Economic, Physical and Social Impacts*. Harlow: Longman.

McGregor, G., Markone, M., Bartzokas, A. and Katsoulis, B. (2002) 'An evaluation of the nature and timing of summer human thermal discomfort in Athens, Greece', *Climate Research*, 20(1): 83–94.

Meyer, W.B. and Turner II, B.L. (1995) 'The Earth transformed: trends, trajectories, and patterns', in R.J. Johnston, P.J., Taylor, and M.J. Watts (eds) *Geographies of Global Change: Remapping the World in the Late Twentieth Century*. Oxford: Blackwell, pp.302–17.

Moberg, A., Sonechkin, D.M., Holmgren, K. Datsenko, N.M. and Karlén, W. (2005) 'Highly variable Northern Hemisphere temperatures reconstructed from low- and high-resolution proxy data', *Nature*, 433: 613–7.

Naumann, M. (2005) Weltwirtschaftsforum 2005. Verantwortungsbewusst. www.zeit.de/2005/05/davos3 (accessed 15 February 2005).

Nicholls, R.J. (2004) 'Coastal flooding and wetland loss in the 21st century: changes under the SRES climate and socio-economic scenarios', *Global Environmental Change*, 14(1): 69–86.

Nielsen, R. (2005) *The Little Green Handbook: A Guide to Critical Global Trends*. Carlton North: Scribe Publications.

NOAA (2005) http://lwf.ncdc.noaa.gov/oa/reports/billionz/html#LIST (accessed 15 January 2005)

Oppenheimer, M. and Alley, R.B. (2004) 'The Antarctic Ice Sheet and long term climate change policy', *Climatic Change*, 64: 1–10.

O'Reilly, A.M. (1986) 'Tourism carrying capacity-concept and issues', *Tourism Management*, 7: 254–8.

O'Riordan, T. and Jäger, J. (1996) *Politics of Climate Change: a European Perspective*. London: Routledge.

Palmer, T.N. and Räisänen, J. (2002) 'Quantifying the risk of extreme seasonal precipitation events in a changing climate', *Nature*, 415: 512–4.

Parry, M. (2005) 'Avoiding dangerous climate change: Overview of impacts', paper presented at International Symposium on the Stabilisation of Greenhouse Gases, Hadley Centre, Met Office, Exeter, 1–3 February.

Parry, M.L., Rosenzweig, C., Iglesias, A., Livermore, M. and Fischer, G. (2004) 'Effects of climate change on global food production', *Environmental Change*, 14: 53–67.

Payne A.J., Vieli, A., Shepherd, A.P., Wingham, D.J. and Rignot, E. (2004) 'Recent dramatic thinning of largest West Antarctic ice stream triggered by oceans', *Geophysics Research Letters*, 31.

Peeters, P. (2003) 'The tourist, the trip and the earth'. In NHTV Marketing and Communication Departments (ed.) *Creating a fascinating world.* Breda: NHTV, pp.1–8.

Rapley, C. (2005) 'Antarctic Ice Sheet and sea level rise', paper presented at International Symposium on the Stabilisation of Greenhouse Gases, Hadley Centre, Met Office, Exeter, 1–3 February.

Richards, J.F. (1990) 'Land transformation', in B.L. Turner II, W.C. Clark, R.W. Kates, J.F. Richards, J.T. Mathews and W.B. Meyer (eds) *The Earth as Transformed by Human Action.* New York: Cambridge University Press, pp.163–78.

Richardson, R.B. and Loomis, J.B. (2003) 'The effects of climate change on mountain tourism: a contingent behavior methodology', Paper presented at the First International Conference on Climate Change and Tourism, Djerba, Tunisia, 9–11 April 2003.

Rosenkopf, L. and Almeida, P. (2003) 'Overcoming local search through alliances and mobility', *Management Science*, 49: 751–66.

Sala, O.E., Chapin III, F.S., Armesto, J.J., Berlow, E., Bloomfield, J., Dirzo, R., Huber-Sanwald, E., Huenneke, L.F., Jackson, R.B., Kinzig, A., Leemans, R., Lodge, D.M., Mooney, H.A., Oesterheld, M., Poff, N.L., Sykes, M.T., Walker, B.H., Walker, M. and Wall, D.H. (2000) 'Global biodiversity scenarios for the year 2100', *Science*, 287: 1770–4.

Sarewitz, D. and Pielke Jr., R. (2000) 'Breaking the global-warming gridlock', *The Atlantic Monthly*, 286(1)(July): 54–64.

Schafer, A. (2000) 'Regularities in travel demand: An international perspective', *Journal of Transportation and Statistics*, 3(3): 1–31.

Schär, C. and Jendritzky, G. (2004) 'Hot news from summer 2003', *Nature*, 432: 559–60.

Schär, C., Vidale, P.L., Lüthi, D., Frei, C., Häberli, C., Liniger, M.A. and Appenzeller, C. (2004) 'The role of increasing temperature variability in European summer heatwaves', *Nature*, 427: 332–6.

Scott, D. (2003) 'Climate Change and Tourism and the mountain regions of North America', in *Climate Change and Tourism.* Proceedings of the First International Conference on Climate Change and Tourism, Djerba, Tunisia, 9–11 April.

Scott, D. and Matzarakis, C.R. (2004) *Advances in Tourism Climatology.* Berichte des Meteorologischen Insitutes der Universität Freiburg. Freiburg: Meteorologisches Institut der Universität Freiburg.

Smil, V. (1993) *Global Ecology: Environmental Change and Social Flexibility.* London: Routledge.

Smith, K. (1993) 'The influence of weather and climate on recreation and tourism', *Weather*, 48(12), 398–403.

Stainforth, D.A., Aina, T., Christensen, C., Collins, M., Faull, N., Frame, D.J., Kettleborough, J.A., Knight, S., Martin, A., Murphy, J.M., Piani, C., Sexton, D., Smith, L.A., Spicer, R.A., Thorpe, A.J., and Allen, M.R. (2005) 'Uncertainty in predictions of the climate response to rising levels of greenhouse gases', *Nature*, 433: 403–6.

Stansfield, C.A. (1978) 'Atlantic City and the resort cycle: background to the legalization of gambling', *Annals of Tourism Research*, 5(2): 238–51.

Storch, H. von , Zorita, E., Jones, J.M., Dimitriev, Y., González-Rouco, F. and Tett, S.F.B. (2004) 'Reconstructing past climate from noisy data', *Science*, 306: 679–82.

Stott, P.A., Stone, D.A. and Allen, M.R. (2004) 'Human contribution to the European heatwave of 2003', *Nature*, 432: 610–13.

Svenson, S., Scott, D., Wall, G. and McBoyle, G. (2001) 'Potential impacts of climate variability and change on ice fishing in the Lakelands tourism region of Ontario', Paper
presented at Annual Meeting of the Canadian Association of Geographers. Université de
Montréal, Montréal, Québec.

SWECLIM (Swedish Regional Climate Modelling Programme) (2002) *Arsrapport 2002.*
Norrköping: SMHI.

Thomas, C.D., Cameron, A., Green, R.E., Bakkenes, M., Beaumont, L.J., Collingham,
Y.C., Erasmus, B.F.N., de Siqueira, M.F., Grainger, A., Hannah, L., Hughes, L.,
Huntley, B., Van Jaarsveld, A.S., Midgley, G.F., Miles, L., Ortega-Huerta, M.A.,
Townsend Peterson, A., Phillips, O.L. and Williams, S.E. (2004) 'Extinction risk from
climate change', *Nature*, 427: 145–8.

Turner, B.L., Clark, W.C., Kates, R.W., Richards, J.F., Mathews, J.Y. and Meyer, W.B.
(eds) (1990) *The Earth as Transformed by Human Action*, Cambridge: Cambridge
University Press.

Wackernagel, M., Monfreda, C. and Deumling, D. (2002) *Ecological Footprint of Nations:
November 2002 Update: How Much Nature Do They Use? How Much Nature Do They
Have?* San Francisco: Redefining Progress.

Water and Rivers Commission (2000) *Water Facts 15: Salinity*, Perth: Water and Rivers
Commission.

Watts, J. (2004) 'Highest icefields will not last 100 years, study finds', *The Guardian*,
September 24.

WCED (World Comission on Environment and Development) (1987) *Our Common Future.*
New York: Oxford University Press.

Williams, A.M. and Hall, C.M. (2002) 'Tourism, migration, circulation and mobility: The
contingencies of time and place', in C.M. Hall and A.M. Williams (eds) *Tourism and
Migration: New Relationships Between Consumption and Production.* Dortrecht: Kluwer,
pp.1–52.

Williams, S.E., Bolitho, E.E. and Fox, S. (2003) 'Climate change in Australian tropical rainforests: an impending environmental catastrophe', *Proceedings Royal Society of London*,
B 270(1527): 1887–92.

Wolfe, R.J. (1952) 'Wasaga Beach: the divorce from the geographic environment', *Canadian Geographer*, 1(2): 57–65.

WTO (World Tourism Organization) (2001) *Tourism 2020 Vision – Global Forecasts and
Profiles of Market Segments*. Madrid: World Tourism Organization.

WTO (2003) *Climate Change and Tourism*. Proceedings of the First International Conference on Climate Change and Tourism. 9–11 April, Djerba, Tunisia. Madrid, Spain:
WTO. (CD-ROM).

WTO (2005) Liberalization with a human face. Poverty alleviation – Sustainability – Fair
trade, www.world-tourism.org/step/menu.html (accessed 10 January 2005).

WTTC (World Travel and Tourism Council) (2003) *Blueprint for New Tourism*. London:
World Travel and Tourism Council.

Xu, C.-Y. (2000) 'Modelling the effects of climate change on water resources in central
Sweden', *Water Resources Management*, 14: 177–89.

Zhang, K., Douglas, B.C. and Leatherman, S.P. (2004) 'Global warming and coastal
erosion', *Climatic Change*, 64: 41–58.

Part I

Environments

2 Impacts of global environmental change on tourism in the polar regions

Margaret E. Johnston

Local effects of global environmental change have been evidenced in both the Arctic and the Antarctic, causing changes, for example, in snow and ice conditions, in vegetation patterns and in animal behaviour. These effects are coincident with changes in opportunities for human activity by local people and visitors. Given that tourism in the polar regions has been largely dependent on scenic landscape attractions, such as snow and ice and access to wildlife populations, there is the potential for major change in this industry. This chapter describes predicted impacts of climate change in these regions and those that are already occurring. It outlines current patterns in polar tourism and identifies several challenges to the tourism industry related to infrastructure, access and attractions. The chapter describes opportunities and challenges created by the local effects of global environmental change and it concludes with the significance of such changes for tourism in the polar regions.

Climate change in the polar regions

On 9 November 2004, the Arctic Climate Impact Assessment was released by the Arctic Council (Friesen 2004), an eight-nation intergovernmental forum that addresses issues in the Arctic. The report predicts major changes to the Arctic resulting from climate change, including decreased sea ice, warmer and shorter winters, thawing permafrost and changes in wildlife populations. The 1400-page report brings together the work of more than 250 scientists (Sallot 2004). The scientist heading the study reported that the average winter temperature in some parts of the Arctic had increased by 3°C over the past 60 years and that Arctic sea ice had diminished by 10 per cent over the past 30 years (Sallot 2004). These changes, evidenced in satellite images and scientific data, demonstrate the vulnerability of the Arctic to climate change and the resulting environmental changes that occur on a variety of scales. Undoubtedly, these environmental changes will influence human activities in global, regional and local contexts. This chapter addresses the issues of environmental change in both the Arctic and the Antarctic and explores the implications of climate change for tourism.

The human dimensions of global environmental change include a focus on how people influence such transformations, the results of those transformations on people, and their adaptations to change (National Research Council 1999). The institutional environment of humans, their economic, political and social contexts, is considered in this perspective, as are the policy implications. The effects of climate change on individuals, communities and societies depend not only on the climatic events themselves, but also on human vulnerability to change and the capacity of people to adapt (National Research Council 1999). It is important to note that on-the-ground impacts and human vulnerability occur on scales very different to those used in climate models. Community and individual responses will reflect local conditions, including endogenous community-level aspects of culture, economy, history and experience with change (Duerden 2004).

Understanding vulnerability and the capacity to adapt are two facets of climate change impact assessments that are derived from the social science literature on risk and hazard. This recent emphasis on vulnerability and the use of integrated regional assessments to focus on the interplay of all direct and indirect impacts along with the adaptations of people and wildlife are advances in how we understand the impacts of climate change. Vulnerability describes the risk a community and individuals face from change, while adaptability refers to their access to economic resources, the capability of public infrastructure to mitigate impacts, and the availability of institutions to manage both sudden onset events and longer term changes. The vulnerability and adaptability of Arctic peoples to sudden and ongoing local changes related to global environmental change has become an important topic for researchers, community members and governments (Berman *et al.* 2004, Ford and Smit 2004), and this work recognises that people are not passive or predictable in their responses (Duerden 2004).

The polar regions are significant in global environmental change because they both are affected by climate change and influence global climate change. Climate change is expected to be greater in the polar regions, with the Arctic particularly affected: more rapid warming is predicted for the Arctic than any other region (International Arctic Science Committee 1994). The Government of Canada reported that analysis of Canadian temperatures from 1895 to 1991 confirmed a similar trend to the global data, but with warming at a higher rate, consistent with the predictions of climate change models for greater warming in high latitudes (Government of Canada 1994). This polar amplification of change is the most important of the three generalisations about climate change. The other two generalisations are that warming will be greater over the land than the sea, and warming will be greater in the winter than in the summer (see Government of Canada 1997).

Models of global warming show temperature increases in the Arctic and some also show increases in some parts of the Antarctic (Wadhams 1991). Further, the Arctic is a critical influence on changes in global climate patterns due to its role in global air circulation and ocean circulation, and in acting as a sink or source of trace gases (International Arctic Science Committee 1994). The Southern Ocean exercises a cooling influence on the world's oceans and may play a role in the stability of the Antarctic ice sheet (Gordon 1991).

Considerable concern has been raised about the contribution of the melting of polar land glaciers to sea-level rises internationally. In Greenland, an increase in temperature is predicted to cause a retreat of the ice margin and, hence, a substantial contribution to sea-level rise as glacial ablation exceeds accumulation (Letréguilly *et al.* 1991). The influence of glacial melting in the Antarctic differs, at least in the shorter term. Research demonstrates that the current expansion of the ice sheet means it is difficult to attribute sea-level rises to melting of the Antarctic ice sheet. Iceberg fluxes reflect disturbance at the front of the ice shelves (Bentley and Giovinetto 1991), and since the weight of floating ice shelves already is accounted for in sea-level, bergs that break off do not contribute to any rise. Indeed, for the next 100 years, the greatest impact of change in the mass balance is predicted to be sea-level lowering because of increased precipitation over the land ice, an offset to the expected rise from temperate glacier melting (Budd and Simmonds 1991). After about 100 years, the effects of increased ice flow from the continent and ice shelf melting would be felt as sea-level rise.

The ice-albedo feedback mechanism is the physical process that appears to be the most important in understanding why the Arctic is expected to warm to a greater extent than the rest of the planet. Fresh snow has a high albedo – high reflection of incoming short-wave solar radiation. When snow cover melts into pools of water, albedo is lowered and thus more radiation is absorbed, causing more melting. This feedback effect will be experienced both for sea ice and land snow cover (International Arctic Science Committee 1994, see also Morison *et al.* 2000). In the Antarctic this effect will be less important because the land is permanently covered with extremely thick ice, and the lowering of the ice-albedo and the resultant feedback effect will only occur with the sea ice, where it will result in retreat and thinning. In the Antarctic, at least 80 per cent of the sea ice in winter is thin, first-year ice, suggesting that this ice will be unstable with warming; however, the complexity of feedback mechanisms related to open water, salinity and depth of warmer water suggest that this thin ice might be resilient (Wadhams 1991). Further, warming in the Peninsula area and cooling in the continent has been confirmed by temperature records from 1966 to 2000 (Doran *et al.* 2002). The Antarctic Peninsula appears to be warming more rapidly than the sub-Antarctic islands and other Antarctic sites, coinciding with a reduction in sea ice extent in the Bellingshausen and Amundsen seas (Jacobs and Comiso 1997; Kiernan and McConnell 2002).

Marine and land ecosystems will be greatly affected by global climate change; indeed, the greatest change is predicted to occur in the high latitudes. These ecosystems are characterised by low species diversity and high adaptation, leaving them vulnerable to environmental change (International Arctic Science Committee 1994). Warming in the Antarctic Peninsula and the sub-Antarctic islands will lead to changes in species, while summer cooling evident in the continent suggests that biologic declines will result (see Doran *et al.* 2002).

Pan-Arctic climate change assessment

The Arctic Climate Impact Assessment (ACIA) is based on scientific observations, indigenous and local knowledge, and the use of scenarios. It describes impacts that are expected to occur in this century and, for each, provides a degree of likelihood that the stated impact will occur, reflecting the level of confidence of the experts (ACIA 2004). The report is organised around ten key interrelated findings. These findings reinforce and extend much of the earlier scientific work on Arctic climate change. These themes are:

1   Arctic climate is now warming rapidly and much larger changes are projected.
2   Arctic warming and its consequences have worldwide implications.
3   Arctic vegetation zones are very likely to shift, causing wide-ranging impacts.
4   Animal species' diversity, ranges and distribution will change.
5   Many coastal communities and facilities face increasing exposure to storms.
6   Reduced sea ice is very likely to increase marine transport and access to resources.
7   Thawing ground will disrupt transportation, buildings and other infrastructure.
8   Indigenous communities are facing major economic and cultural impacts.
9   Elevated ultraviolet radiation levels will affect people, plants, and animals.
10 Multiple influences interact to cause impacts to people and ecosystems.

(ACIA 2004: 10–11)

The report also provides a sub-regional overview that recognises that change is not uniform across the Arctic and that some impacts are more important for some areas than others. For example, in the sub-region from East Greenland to the Scandinavian Arctic and north-west Russia, sea ice retreats will increase access to hydrocarbons and minerals, traditional animal harvests will be less predictable and distributions of plants and animals will shift northward. In the Siberian sub-region, tundra will retreat northward and permafrost will continue to thaw, resulting in infrastructure damage in addition to that already experienced. Increased opportunities for navigation through the Northern Sea Route will bring economic activity and access to oil and gas.

The need to assess these sub-regional effects of global environmental change has become increasingly evident in the polar regions with the recognition of strong regional variation. For example, in Canada, models show greater warming in the interior, greater winter warming, northward movement of the permafrost boundary, and glacier retreat that is more pronounced in the west (Government of Canada 1997). Temperature data confirm this: the western Canadian Arctic has warmed by about 1.5°C over the past 40 years, while the central Canadian Arctic has warmed by 0.5°C over that period (Government of Canada 2001). Precipitation is also regionally differentiated. For example, in Alaska, winter temperature and precipitation variability over a 30-year time period show a warmer, drier climate shift in the interior and a warmer, wetter climate shift along the southern coast (Milkovich 1991).

Yet, not all areas are warming: Hudson Bay is in a transition zone, between annual warming in north-western Canada and cooling in the Baffin Island region (Cohen *et al.* 1994). In the Nordic Arctic, there have been substantial variations in parts of the region over the past century based on data analysis of 20 climate stations in Greenland, Iceland, the Faeroe Islands and Arctic Norway (Førland *et al.* 2002). Increasing temperatures in recent decades might suggest general warming, but these more recent temperatures are still not as high as those experienced in the 1930s and 1950s in parts of the Nordic Arctic. Site differences will also play a role within regions: permafrost thaw is expected to occur more rapidly in the coastal areas and less rapidly in the wetland areas of the Mackenzie Basin (Cohen 1996). The variation between regions continues to be an ongoing challenge for research in this area. A key issue for coming years is making the assessments and predictions of climate change relevant at the regional (and sub-regional) scale (National Research Council 1999).

The ACIA (2004) predictions reflect not only expectations of what will occur, but also changes that are occurring in the physical and biologic world of the Arctic that are now having impacts on people and their activities. Although very little of the literature on the impacts of climate change on human activities delves deeply into effects on tourism, the link to tourism is not a difficult one. All of these changes are linked and all have implications for human activity. This section will explore further several of these changes that have the greatest relevance for tourism.

Changes in sea ice are seen by scientists as a key indicator of the rapid warming of the Arctic (ACIA 2004). A significant decline in the annual extent and thickness of sea ice is evident over the past 50 years, with an acceleration of this decline apparent in recent decades. The extent of sea ice in summer has seen a disproportionate decrease compared to the annual average decrease. The average loss of sea ice in the climate models used by ACIA projected a greater than 50 per cent decline by the end of the century and some models predicted a complete disappearance of summer sea ice.

The scenarios project average annual temperature increases in the latter part of this century to be 3–5°C over land and 7°C over the seas (ACIA 2004). Winter temperature increases will be disproportionately higher: 4–7°C over land and 7–10°C over seas. Increased evaporation with warmer temperatures will lead to increased precipitation for the Arctic as a whole in the order of 20 per cent. Summer rainfall in Scandinavia is expected to decrease and winter precipitation in all areas except southern Greenland is expected to increase. Coastal regions will be most affected by increased precipitation, especially in autumn and winter. The extent of snow cover has already decreased by 10 per cent and it is projected to decrease by another 10–20 per cent by the end of the century, with greatest declines in spring snow cover.

Change in the population and distribution of animal species is another key change expected and evident to some degree in the Arctic (ACIA 2004). Given the major changes to sea ice and the ocean environment, concomitant changes are projected for marine life. Marine mammals are susceptible to changes in climate, as evidenced for numerous species by the effects of El Niño warming in the northern Pacific Ocean, although some, such as the polar bear, might be adaptable to changes in temperature and diet. However, decline in prey populations as a result of sea ice changes and decreased productivity at the ice edge is a threat to polar bears (Ono 1995). Reports from Inuit and scientists also suggest that polar bear territory is being increasingly invaded by grizzly bears, along with other unusual species such as marten, wolverine, robin and Pacific salmon (Struzik 2003).

ACIA (2004) predicts devastating implications for polar bear populations. The health and survival of females and cubs is dependent on success in seal hunting, which requires good spring ice conditions when the bears emerge from their hibernation dens. Polar bears hunt for seals on the sea ice throughout the spring and summer. As sea ice extent and stability continue to decline with warming, individual polar bears and populations will be affected.

> Polar bears are unlikely to survive as a species if there is an almost complete loss of summer sea-ice cover, which is projected to occur before the end of this century by some climate models.
>
> (ACIA 2004: 58)

Impacts on polar bear population health – declines in numbers of adult bears, live births and cub weight – have been documented for the Hudson Bay area, the southern limit of the population. Stirling *et al.* (1999) state that these health declines in the Western Hudson Bay polar bears do not yet threaten this population but that, if trends continue, there will be a detrimental impact on population sustainability.

Some ice-dependent species, such as the ringed seal, spotted seal, harp seal, ivory gull and walrus, will be negatively affected by the decline in sea ice and perhaps will be unable to adapt. Other species, such as the harbour seal and grey seal, might extend their distribution with the reduction in sea ice. Research in the Beaufort Sea indicates that warming and the associated decline in sea ice have already affected the productivity of the ice algae at the base of the marine food chain (ACIA 2004).

Land animals will also be affected, for example as hydrological changes force new migration patterns, flash floods claim lives, and predator–prey relationships are altered through changes in population distribution (ACIA 2004). For example, 'Despite uncertainty about future ecological relationships, the most likely outcome of a warmer climate will be a decline in caribou and musk-oxen' (Gunn 1995). Given an increase in weather fluctuations, there will be less winter forage, changes in migration patterns, energy effects of deep, slushy snow, and increased insect harassment, and these will have a negative effect on caribou and musk-oxen.

Flora and fauna changes have been confirmed in the Antarctic as well. Two scientists reported on the ecological impacts of global warming on sub-Antarctic Heard Island (Pockley 2001). Glaciers on the island have retreated by 12 per cent since 1947 and the sea surface temperature has increased. Changes include the development of lush vegetation, the expansion of bird, seal and insect populations. King penguins were reported to have increased from three breeding pairs in 1947 to 25,000 50 years later, and fur seal numbers to have increased from near extinction to 28,000 adults. According to the British Antarctic Survey, grass has become established in the Antarctic – a direct effect of a warming climate (*The Globe and Mail* 2004).

The decline in Arctic sea ice extent is also causing increased wind-generated wave action and more intense storm surges along the Arctic coastline, resulting in severe erosion that threatens numerous low lying Arctic communities (ACIA 2004). Along with increased air and water temperatures, this continuing shoreline erosion is projected to destabilise permafrost, thereby causing increased thawing of the permafrost and further erosion. The effects of these intensified storm surges have been observed in small communities in Alaska (USA) and the Northwest Territories (Canada), where buildings have been abandoned, and transportation infrastructure and continued habitation are at risk.

Even small increases in temperature have an impact on permafrost, the partially or completely frozen ground that lies beneath an active layer of soil or the sea bed in the Arctic. The active layer freezes in the winter, but thaws every summer. When some of this layer fails to re-freeze, the permafrost degrades – the result can be settling of the soil causing damage to overlying structures. Permafrost degradation has already begun across the region and is projected to occur over 10–20 per cent of the current permafrost extent, including a northward shift in the southern limit of permafrost by as much as several hundred kilometres (ACIA 2004). The thawing of permafrost has caused damage to buildings and transportation infrastructure; such problems are increasingly common in northern Russia, largely due to engineering inadequacies and maintenance failures.

Tourism is only one of many interactions that people have with the environment. As such, it will be affected by how much climate change and related environmental changes affect local and regional resource use, and also by unrelated stresses on the system (Table 2.1). For example, with warming in the Mackenzie Basin, deciduous trees are expected to fare better than coniferous trees, as are younger trees compared to older trees (Rothman and Hebert 1997). This will have an influence on forest fires and forest productivity. Yet, at the current time, the forest sector and forestry-dependent communities internationally are being tested by a variety of other forces including technological change and globalisation. Climate change is an additional force that is experienced within this context.

Patterns in polar tourism

A number of sources describe tourism development and issues in the Arctic and Antarctic, including both site-specific studies and more general examinations (e.g. Cessford and Dingwall 1994; Hall and Johnston 1995; Jacobsen 1994; Johnston and Viken 1997; Jones 1998; Notzke 1999; Kaltenborn 2000; Tracey 2000; Bauer 2001). Rather than repeat the detail that these sources can provide, this section outlines some general aspects of polar tourism and relates the impacts of climate change to these. It is important to reiterate that on-the-ground impacts will be experienced very specifically and this current work cannot do justice to that scale of impact.

Tourism in the Antarctic has increased dramatically in recent years although, in comparison to almost all other destinations, numbers remain low. Statistics available on the website of the International Association of Antarctic Tour Operators

Table 2.1 Some aspects of global environmental change with relevance for Arctic tourism

Effects on infrastructure

- Increased open water leads to increased storm surges and shoreline erosion
- Permafrost melting/land instability leads to construction and engineering problems and structural damage

Effects on access

- Decline in sea-ice extent leads to extended shipping season
- Glacier melting leads to increased iceberg hazards
- Shorter seasonal river ice duration leads to access difficulties related to winter roads
- Earlier and greater spring floods leads to access hazards
- Greater snow accumulation leads to access difficulties
- Northward movement of permafrost line leads to increased access through road construction

Effects on attractions

- Greater snow accumulation leads to new opportunities for snow-based activities
- Shorter snow duration leads to seasonal challenges for some activities
- Warmer summer and winter temperatures lead to extension of seasonal activities
- Warmer summer temperatures lead to increased insect challenge
- Warmer winter temperatures lead to new opportunities
- Ecosystem changes lead to alterations in distribution and abundance of existing animal species
- Ecosystem changes lead to appearance of new species in north
- Environmental changes alter local activity possibilities
- Scenic values altered through environmental changes locally and regionally

(IAATO), an industry coalition, show that the number of visitors who made landings doubled from the 1992/93 season – when there were 6,704 estimated ship and land-based tourists – to 2002/03 – when there were 13,571 tourists. By 2003/04 the number had tripled to nearly 20,000. An additional 1,500 are predicted by IAATO for 2004/05 (IAATO 2004). These numbers refer to visitors who actually physically visited the Antarctic by ship or by air with a landing on the islands along the Peninsula or the continent itself. Cruises without a landing in 2003/04 contributed another 4,939 tourists, while overflight tourism added 2,827.

Most of the tourism in the Antarctic is ship-based expedition cruising. The 2003/04 statistics show about 400 individuals who travelled by air and participated in activities on land, and 185 tourists who arrived via small sailing vessels (IAATO 2004). Attractions in the Antarctic include wildlife (e.g. penguins, seals), scenic landscape, cultural sites and scientific bases (Tracey 2000). The season is limited to about five-months: mid-November to the end of March, and is concentrated along the western side of the Antarctic Peninsula, an area with a concentration of accessible attractions. Landings by small rubber inflatable craft occur at a variety of popular sites that provide suitable access, as well as diversity and reliability of attractions. Based on the IAATO statistics, the top five most-visited sites from 1989 to 2001 appear to be Whalers Bay, Port Lockroy, Half Moon Island, Cuverville Island and Peterman Island, particularly in the most recent seasons. These island sites in the Peninsula region have visits that number in the thousands. At Whalers Bay in 2003/04 there were 9,941 tourists and at Cuverville Island there were 9,901 (IAATO 2004). In comparison, visits on the continent number at most in the hundreds. Tracey (2000) reports that only a small proportion of sites receive high visitor numbers each year; the vast majority are not used on a regular basis and there continues to be extension of available sites through new site use each year.

Accessibility requires ice-free landings and a reasonably shallow beach or rock entrance to enable boats to approach shore safely and allow visitors to step ashore. There are no docking facilities or jetties at these sites in the Antarctic. It is commonly suggested that this ice-free land is about 1 per cent of the total land mass in the Antarctic. Not coincidentally, there is tremendous pressure on the use of this space; tourism competes at many sites with penguin rookeries, flora and seal haul-outs. As these sites become free of seasonal snow in the spring, they are colonised by successive waves of species.

Helicopters are used regularly for landings, more frequently outside the Peninsula region. The Ross Sea and East Antarctica receive much less visitation than the Peninsula; helicopter use extends the options available and has included, in some cases, access to inland regions such as the Dry Valleys (Tracey 2000).

A small number of tourists each year arrive by air. These visitors are the true adventure tourists in the Antarctic – their access is provided by small planes such as Hercules or Twin Otters to blue-ice runways where seasonal or temporary camps are established as support (Tracey 2000). Activities include climbing, cross-country skiing, wildlife viewing and South Pole visits.

In the Arctic there is a greater range and extent of tourist activity, including traditional hunting and fishing, expedition-style and destination cruising, wildlife viewing, northern lights tourism, skiing and snow-machine riding, dog-sledding, and cultural and aboriginal tourism (see Johnston 1995; Jones 1998; Viken and Jorgenson 1998; Notzke 1999; Tracey 2000; Timothy and Olsen 2001). There is a much longer tourism history in the Arctic, with northern mainland Norway and Svalbard featuring centrally in early Arctic travel (Jacobsen 1994; Viken and Jorgenson 1998). In addition to ship travel, there is considerable air travel and road travel – the latter is more prevalent in the European Arctic where good access leads to destinations such as North Cape. In the North American and eastern Russian Arctic roads are sparse.

For a variety of reasons, accurate numbers of tourists are hard to provide for the Arctic, although estimations can be made using several sources. For example, it is estimated that approximately 33,000 leisure travellers visited the Northwest Territories, Canada, in 1999/2000 (Government of Northwest Territories 2003). Yukon Territory received about 32,000 tourists in 2002 and Alaska had nearly 1.5 million pleasure travellers in 2001 (Pagnan 2003). For all three jurisdictions, it must be noted that there has been no attempt made to separate the Arctic visitors from those who travel to the sub-Arctic. This is a logistical difficulty in much of the north and prevents a clear picture of truly Arctic tourism. Very little Russian data exist, although ship tourism to the Murmansk, the White Sea, Archangel and the North Pole is on offer. The Kamchatka Peninsula is another developing destination region (Pagnan 2003). Nunavut, an Arctic territory in Canada, reported 12,000 visitors and Greenland reported 32,000 (Pagnan 2003).

Iceland (277,800), Finland (1.6 million), Sweden and mainland Norway maintain tourism numbers that are substantially higher. Svalbard, the Arctic archipelago off the northern coast of Norway, has had a rapidly growing tourism industry since the early 1990s when Norwegian authorities began to de-emphasise coal production and promote leisure travel (Viken and Jorgenson 1998). It is reported that ship-based travellers number around 40,000 and that an additional 10,000 tourists arrive by air (Pagnan 2003).

Certain sites have attained international renown for tourism and these draw large numbers of visitors. North Cape, Norway, is one of these, where road access has enabled hundreds of thousands of tourists to access 'the end of the world' (Jacobsen 1994). Another well-known site is Churchill, Manitoba, often called the polar bear viewing capital of the world due to the reliable congregations of polar bears in the vicinity. Churchill, at the boundary of the Arctic and sub-Arctic and on the shore of Hudson Bay, boasts a unique combination of physical features, biological attractions and facilities that contribute to hunting, fishing, bird watching, whale watching, northern lights tourism and, of course, polar bear viewing (Lemelin 2005). It is this latter activity that provides Churchill with its international tourism reputation. Polar bears congregate at Churchill in late autumn to await the formation of sea ice. Nowhere else in the Arctic do congregations of polar bears regularly occur of the magnitude experienced at Churchill.

In addition to the well-known destinations, tourism has been developing gradually at numerous other sites in the Arctic and, indeed, is often sought as a new form of income in the mixed economy communities looking for incoming cash flows (Notzke 1999) and in regions where other forms of industry have declined. A number of studies suggest there is local support for tourism growth, provided it meets particular requirements related to community interests (e.g. Nickels *et al.* 1991; Johnston and Viken 1997; Dressler 1999).

Climate change has been seen as having a direct effect on tourism numbers in the Arctic. Increasing tourism is seen as a potential outcome of warming in particular (Kochmer and Johnson 1995). Although this possibility is attractive to many individuals and communities, it brings concomitant concerns about how Arctic tourism can be integrated, for example, with the reality of an animal harvesting lifestyle (Wenzel 1995). Nunavut communities that are close to the natural attractions of protected areas may see increased visitation as tourists seek mountain climbing, hiking and bird watching, but they must address the varied costs of shielding visitors from animal harvesting activities (Wenzel 1995). As attractive as increased tourism might be, it nonetheless represents an economic, social, political and environmental agent of change that must be addressed within the context of a particular community.

Climate change and polar tourism

A number of authors have examined the likely impacts of climate change on tourism. Brotton *et al.* (1997) explored the possible impacts of climate change scenarios for the sub-Arctic Nahanni National Park Reserve and for caribou hunting activities in the region. Their analysis indicates little change for river recreation activities in Nahanni stemming from an increase in temperature and increased precipitation, but possible changes in visitor experiences related to increased forest fires and ecological shifts associated with warmer temperatures (see also Staple and Wall 1996). Although the overall tourist season might be lengthened, this might be offset by closures due to forest fire hazard. Further scenarios suggest a possible decrease in caribou hunting as a result of climate change-related stresses on the animal population. The authors suggest that it may well be that consumptive activities are more likely than non-consumptive activities to be negatively affected because of their requirements for a specific resource rather than a general setting. Brotton *et al.* (1997) note that changes in attractions and in opportunities for recreation will result in further alterations to natural ecosystems through increased pressure and new requirements for infrastructure.

Weller and Lange (1999) report on the magnitude of expected changes on tourism in the Arctic. They note expectations of low impacts on tourism from changes in precipitation, sea ice, hydrology, plant mortality, animal distribution and animal migration. Expectations of medium impacts on tourism are noted from changes in air temperature and animal productivity. These expectations are given for the region as a whole.

Pagnan (2003) reports on discussions with Arctic tourism professionals that provide some sub-regional distinctions. In Alaska, for example, tourism professionals have observed changes in bird and whale migrations, melting ice, permafrost degradation and sparseness of snow. These changes have had immediate and practical implications for the tourism industry related to tourism attractions, experiences and accessibility. The season is being extended for activities such as cruise ship tourism and whale watching. Greenland officials reported disappointment on the part of visitors whose expectations of ice features and ice-dependent wildlife were not met because of a lack of ice along the west coast. Although not all tourism officials could confirm changes in climate that affected tourism, all were willing to have further discussions on the topic. Pagnan concludes that the environmental outcomes of climate change could have dramatic impacts on tourism and that the industry is not ready to respond. He also points to challenges related to infrastructure, access and attractions that must be addressed.

The final section of this chapter describes some of the opportunities and challenges for tourism as a result of environmental changes.

Tourism infrastructure and the polar regions

Infrastructure in the Arctic is particularly vulnerable to climate change. As sea ice continues to decline, open water will increase, causing greater and more frequent storm surges and an acceleration of shoreline erosion. This poses problems for communities and transportation infrastructure located along the coast, including facilities used for tourism. The outcomes of permafrost melting and instability are also of concern. The varied nature of infrastructure across the Arctic makes it difficult to generalise beyond this.

In the Antarctic, there is little dedicated tourism infrastructure. Most tourists arrive via cruise ship; a much smaller number arrive by independent yacht or by air. The blue-ice runways that are used to land the latter group and the temporary camps that support them are seasonal infrastructure. Given the location in the interior of the continent, where warming has not been observed, there might be little to no impact on this infrastructure. In the Peninsula, where warming and related effects are confirmed, scientific bases that might support tourism to some degree could be affected. With very little permafrost throughout the Antarctic and sub-Antarctic, and a few highly exposed sites used for scientific bases, any further examination must be on a site-specific basis.

Tourism access and the polar regions

One of the most important changes evident in the Arctic is the decline in the extent of sea ice. The number of days per year that the Northern Sea Route will be navigable is projected to increase from the current 20–30 to 90–100 days by 2080, with even more navigable days for vessels with ice-breaking capability (ACIA 2004). The Northern Sea Route has been used for tourist voyages. For example, in the summer of 1999 a nuclear icebreaker travelled to the North Pole with tourist passengers, and in 2000 there were two similar voyages (Brigham 2001). The Northwest Passage, through the Canadian Arctic Archipelago, will also be freer of sea ice for more of the summer, leading to increased opportunities for cruise ship tourism. Although the decline in sea ice will provide more open water for navigation, there might be additional hazards to navigation through an increase in iceberg calving and through the instability of what pack ice remains. It is possible that the Northwest Passage could become less predictable for shipping because of high year-to-year variability in sea ice (ACIA 2004). Nonetheless, it seems inevitable that cruise tourism will increase to most destinations, perhaps even on the scale already seen in Svalbard. The primary challenge to such an increase is an infrastructural one for much of the Arctic. Small communities have difficulty now in accommodating the needs of cruise tourists when ships visit their communities. Without substantial capital input, the opportunities will be hard to access.

In the Antarctic Peninsula, financial and physical access have increased in recent years as capacity has been expanded and lower-cost opportunities been made available. With numbers of cruise ship visitors approaching 20,000, there will be increasing competition for access to sites, particularly the top sites that currently receive the highest visitation. How warming will affect these key sites is an important and pressing question for the industry and it can only be answered by detailed site-specific studies. Will warming and increased precipitation affect snow and ice conditions at landings? Will sites become dangerous because of changes in snow and ice? Will there be increased competition among animal and bird species at these sites, and subsequent greater conflict with human users?

Tourism attractions and the polar regions

There will be climate-related changes to attractions in both polar regions. Particular activities can expand or shrink depending on the local outcomes of environmental change. For example, where snow accumulation increases in the Arctic, snow-based activities could grow; where snow accumulation lessens or becomes more variable year-to-year, these activities could decline or require greater investment in snowmaking technology. Seasons might shorten or lengthen; some activities might disappear. Changes in wildlife will have an influence, perhaps leading to new opportunities as species migrate in new patterns or are concentrated in new localities. Predictions of polar bear decline, behaviour changes and the possible disappearance of the species illustrate the degree of gravity for the wildlife viewing component of the industry.

Scenic attractions will change, resulting in site-specific as well as regional challenges or opportunities. Pagnan (2003) points out that the tourism industry in the Arctic relies on traditional perceptions of the Arctic environment and expectations about the experience that relate to ice and snow, mountains and tundra, and wildlife. As the Arctic changes, general perceptions used to sell tourist travel will change. What will it mean for tourists when the inaccessible becomes accessible, and the 'inhospitable' climate appears more hospitable? How important is the idea of an ice and snow-covered, challenging environment to expectations and satisfaction? How will a decline in polar bear populations and other typical Arctic wildlife affect tourism? Shifts and variability in biology and physical geography will affect how attractions are perceived. If tourist numbers continue to grow, related in part to changing environmental conditions, perceptions of crowding may well begin to replace perceptions of solitude. This could have devastating impacts for the Antarctic brand especially.

Adaptation and the future

This chapter has described the predicted and observed environmental changes attributed to climate change, and outlined some of the key challenges for the tourism industry in terms of infrastructure, access and attractions. There is some anecdotal evidence of tourism already being affected by changes in the environment. The ability of individuals and communities to respond adequately to challenges will be tested over the next decades. Opportunities exist for new activities, replacement and diversification, in short, for tourism operators, the communities and tourism officials to moderate the negative and benefit from the positive. Threats exist and there will be site-specific circumstances that result in some individuals and communities being unable to adapt. The vulnerable communities or segments of the industry are those whose local or business conditions do not currently demonstrate the capacity to change or support the flexibility to respond to changes that, in some cases, will be rapid and dramatic.

References

ACIA (Arctic Climate Impact Assessment) (2004) *Impacts of a Warming Arctic*. Cambridge: Cambridge University Press. www.acia.uaf.edu

Bauer, T. (2001) *Tourism in the Antarctic – Opportunities, Constraints and Future Prospects*. London: The Haworth Hospitality Press.

Bentley, C.R. and Giovinetto, M.B. (1991) 'Mass balance of Antarctica and sea level change', in G. Weller, C.L. Wilson and B.A.B. Severin (eds) *International Conference on the Role of the Polar Regions in Global Change*. Proceedings of a conference held 11–15 June 1990 at the University of Alaska Fairbanks, volume II. Fairbanks, AK: University of Alaska Fairbanks, pp.481–8.

Berman, M., Nicolson, C., Kofinas, G., Tetlichi, J. and Martin, S. (2004) 'Adaptation and sustainability in a small Arctic community: results of an agent-based simulation model', *Arctic* 57(4): 401–14.

Brigham, L.W. (2001) 'The Northern Sea Route, 1999–2000', *Polar Record*, 37(203): 329–36.

Brotton, J., Staple, T. and Wall, G. (1997) 'Climate change and tourism in the Mackenzie Basin', in S.J. Cohen (ed.) *Mackenzie Basin Impact Study, Final Report*, Downsview, Ontario: Atmospheric Environment Service, Environment Canada, pp.253–64.

Budd, W.F. and Simmonds, I. (1991) 'The impact of global warming on the Antarctic mass balance', in G. Weller, C.L. Wilson and B.A.B. Severin (eds) *International Conference on the Role of the Polar Regions in Global Change*. Proceedings of a conference held 11–15 June 1990 at the University of Alaska Fairbanks, volume II. Fairbanks, AK: University of Alaska Fairbanks, pp.489–94.

Cessford, G. and Dingwall, P. (1994) 'Tourism on New Zealand's sub-Antarctic islands, *Annals of Tourism Research*, 21(2): 318–32.

Cohen, S.J. (1996) 'What if the climate warms? Implications for the Mackenzie Basin', in J. Oakes and R. Riewe (eds) *Issues in the North*. volume 1, Edmonton: Canadian Circumpolar Institute, pp.199–201.

Cohen, S.J., Agnew, T.A., Headley, A., Louie, P.Y.T., Reycraft, J. and Skinner, W. (1994) *Climate Variability, Climate Change, and Implications for the Future of the Hudson Bay Bioregion*. Ottawa, Ontario: The Hudson Bay Programme, Canadian Arctic Resources Committee.

Doran. P., Priscu, J.C., Berry Lyons, W., Walsh, J.E., Fountain, A.G., McKnight, D., Moorhead, D., Virginia, R.A., Wall, D.H., Clow, G.D., Fritsen, C.H., McKay, C.P. and Parsons, A.N. (2002) 'Antarctic climate cooling and terrestrial ecosystem response', *Nature*, 415: 517- 20.

Dressler, W. (1999) 'Tourism and Sustainability in the Beaufort-Delta Region', unpublished thesis, University of Manitoba, Canada.

Duerden, F. (2004) 'Translating climate change impacts at the community level', *Arctic*, 5(2): 204–12.

Ford, J.D. and Smit, B. (2004) 'A framework for assessing the vulnerability of communities in the Canadian Arctic to risks associated with climate change', *Arctic*, 57(4): 389–400.

Førland, E.J., Hanssen-Bauer, I., Jónsson, T., Kern-Hansen, C., Nordli, P.Ø., Tveito, O.E. and Vaarby Laursen, E. (2002) 'Twentieth-century variations in temperature and precipitation in the Nordic Arctic', *Polar Record*, 38(206): 203–10.

Friesen, J. (2004) 'Arctic melt may open up Northwest Passage', *The Globe and Mail* (Toronto, Canada), 9 November: A3.

Gordon, A.L. (1991) 'The Southern Ocean: Its involvement in global change', in G. Weller, C.L. Wilson and B.A.B. Severin (eds) *International Conference on the Role of the Polar Regions in Global Change*. Proceedings of a conference held 11–15 June 1990 at the University of Alaska Fairbanks, volume I, Fairbanks, AK: University of Alaska Fairbanks, pp.249–55.

Government of Canada (1994). *Canada's National Report on Climate Change: Actions to meet Commitments under the United Nations Framework Convention on Climate Change*. Ottawa, Ontario: Government of Canada.

Government of Canada (1997) *Canada's Second National Report on Climate Change: Actions to meet Commitments under the United Nations Framework Convention on Climate Change*. Ottawa, Ontario: Government of Canada.

Government of Canada (2001) *Canada's Third National Report on Climate Change: Actions to meet Commitments under the United Nations Framework Convention on Climate Change*. Ottawa, Ontario: Government of Canada.

Government of Northwest Territories (2003) *Parks and Tourism: Tourism Research and Statistics*. Yellowknife: Northwest Territories Resources, Wildlife and Economic Development.

Gunn, A. (1995) 'Responses of Arctic ungulates to climate change', in D.L. Peterson and D.R. Johnson (eds) *Human Ecology and Climate Change: People and Resources in the Far North*. Washington, DC: Taylor & Francis, pp.90–104.

Hall, C.M. and Johnston, M.E. (eds) (1995) *Tourism: Tourism in the Arctic and Antarctic Regions*. London: John Wiley and Sons Ltd.

IAATO (2004) Tourism Statistics. Available at: www.iaato.org/> (accessed 21 November 2004).

International Arctic Science Committee (1994) *Scientific Plan for a Regional Research Programme in the Arctic on Global Change*. Proceedings of a workshop at Reykjavik, Iceland 22–25 April 1992, Washington, DC: National Academy Press.

Jacobs, S.S. and Comiso, J.C. (1997) 'Climate variability in the Amundsen and Bellingshausen seas', *Journal of Climate*, 10(4): 697–709.

Jacobsen, J.K.S. (1994) *Arctic Tourism and Global Tourism Research Trends*, special paper no. 37. Thunder Bay, Ontario: Lakehead University, Centre for Northern Studies.

Johnston, M.E. (1995) 'Patterns and issues in Arctic and Subarctic tourism', in C.M. Hall and M.E. Johnston (eds) *Polar Tourism: Tourism in the Arctic and Antarctic Regions*. Chichester: John Wiley and Sons, pp.27–42

Johnston, M.E. and Viken, A. (1997) 'Tourism development in Greenland', *Annals of Tourism Research*, 24(4): 978–82.

Jones, C.S. (1998) 'Predictive tourism models: are they suitable in the polar environment?', *Polar Record*, 34(190): 197–202.

Kaltenborn, B.P. (2000) 'Arctic-alpine environments and tourism: can sustainability be planned? Lessons learnt from Svalbard', *Mountain Research and Development*, 20(1): 28–31.

Kiernan, K. and McConnell, A. (2002) 'Glacier retreat and melt-lake expansion at Stephenson Glacier, Heard Island World Heritage Area', *Polar Record*, 38(207): 297–308.

Kochmer, J.P. and Johnson, D.R. (1995) 'Demography and socioeconomics of northern North America: current status and impacts of climate change', in D.L. Peterson and D.R. Johnson (eds) *Human Ecology and Climate Change: People and Resources in the Far North*. Washington, DC: Taylor & Francis, pp.3–53.

Lemelin, R. H. (2005) 'Wildlife tourism at the edge of chaos: complex interactions between humans and polar bears in Churchill, Manitoba', in F. Berkes, R. Huebert, H. Fast, M. Manseau and A. Diduck (eds) *Breaking Ice: Renewable Resource and Ocean Management in the Canadian North*. Calgary, Alberta: University of Calgary Press (in press).

Letréguilly, A., Reeh, N. and Huybrechts, P. (1991) 'The Greenland ice sheet contribution to sea-level changes during the last 150,000 years', in G. Weller, C.L. Wilson and B.A.B. Severin (eds) *International Conference on the Role of the Polar Regions in Global Change*. Proceedings of a conference held 11–15 June 1990 at the University of Alaska Fairbanks, volume II, Fairbanks, AK: University of Alaska Fairbanks.

Milkovich, M. (1991) 'A winter season synoptic climatology of Alaska: 1956–1986', in G. Weller, C.L. Wilson and B.A.B. Severin (eds) *International Conference on the Role of the Polar Regions in Global Change*. Proceedings of a conference held 11–15 June 1990 at the University of Alaska Fairbanks, volume II, Fairbanks, AK: University of Alaska Fairbanks.

Morison, J., Aagaard, K., and Steele, M. (2000) 'Recent environmental changes in the Arctic: a review', *Arctic*, 53(4): 359–71.

National Research Council (1999) *Dimensions of Global Environmental Change: Research Pathways for the Next Decade*. Washington, DC: National Academy Press.

Nickels, S., Milne, S. and Wenzel, G. (1991) 'Inuit perceptions of tourism development: the case of Clyde River, Baffin Island', *Inuit Studies*, 15(1): 157–69.

Notzke, C. (1999) 'Indigenous tourism development in the Arctic', *Annals of Tourism Research*, 26(1): 55–76.

Ono, K.A. (1995) 'Effects of climate change on marine mammals in the far north', in D.L. Peterson, and D.R. Johnson (eds) *Human Ecology and Climate Change: People and Resources in the Far North*. Washington, DC: Taylor & Francis, pp.105–21.

Pagnan, J. (2003) 'Climate change impacts on Arctic tourism – a preliminary review', *Climate Change and Tourism*. Proceedings of the First International Conference on Climate Change and Tourism, Djerba, Tunisia, April 2003. Madrid: World Tourism Organization.

Pockley, P. (2001) 'Climate change transforms island ecosystem', *Nature*, 410(6829): 616.

Rothman, D. and Hebert, D. (1997) 'The socio-economic implications of climate change in the forest sector of the Mackenzie Basin', in S.J. Cohen (ed) *Mackenzie Basin Impact Study, Final Report*. Downsview, Ontario: Atmospheric Environment Service, Environment Canada, pp.225–41.

Sallot, J. (2004) 'Report to predict big changes in Arctic', *The Globe and Mail* (Toronto, Canada), 1 November: A5.

Staple, T. and Wall, G. (1996) 'Climate change and recreation in Nahanni National Park Reserve', *The Canadian Geographer*, 40(2):109–20.

Stirling, I., Lunn, N.J. and Iocozza, J. (1999) 'Long-term trends in the population ecology of polar bears in Western Hudson Bay in relation to climate change', *Arctic*, 53(3): 294–306.

Struzik, E. (2003) 'Grizzlies on ice', *Canadian Geographic*, 123(6): 36–48.

The Globe and Mail (Toronto) (2004) 'Warmer?', 30 December: A20.

Timothy, D. J. and Olsen, D. H. (2001) 'Challenges and opportunities of marginality in the Arctic: a case of tourism in Greenland', *Tourism Recreation Research*, 49(4): 299–308.

Tracey, P. (2000) 'Managing Antarctic Tourism', unpublished thesis, University of Tasmania, Australia.

Viken, A. and Jorgenson, F. (1998) 'Tourism on Svalbard', *Polar Record*, 34(189): 123–8.

Wadhams, P. (1991) 'Variations in sea ice thickness in the Polar Regions', in G. Weller, C.L. Wilson and B.A.B. Severin (eds) *International Conference on the Role of the Polar Regions in Global Change*. Proceedings of a conference held 11–15 June 1990 at the University of Alaska Fairbanks, volume II, Fairbanks, AK: University of Alaska Fairbanks. pp.4–13.

Weller, G. and Lange, M. (eds) (1999) *Impacts of Climate Change in the Arctic Regions. Report from a Workshop on the Impacts of Global Change*, Tromso, Norway, April 1999. International Arctic Science Committee.

Wenzel, G. (1995) 'Warming the arctic: environmentalism and Canadian Inuit', in D.L. Peterson and D.R. Johnson (eds) *Human Ecology and Climate Change: People and Resources in the Far North*. Washington, DC: Taylor & Francis, pp.169–82.

3 Global environmental change and mountain tourism

Daniel Scott

Introduction

Mountain regions represent approximately one fourth of the Earth's terrestrial surface and contain some of the most diverse and fragile ecosystems. The international community recognised the importance of mountain regions in the global environment–development agenda at the 1992 United Nations Conference on Environment and Development in Rio de Janeiro, when mountain regions were included as a specific chapter in Agenda 21, thereby receiving equal priority with deforestation, desertification and climate change. The importance of mountains in the global ecosystem was also emphasised with the United Nations declaration for the year 2002 as the International Year of Mountains.

The pristine landscapes and biodiversity that make mountain regions important for the global environment are among their principal attractions for tourism. After coasts and islands, mountains are the most important destinations for global tourism, constituting an estimated 15–20 per cent of the global tourism industry (Price *et al.* 1997). Tourism is of great economic importance to many mountain communities and is one of the fastest growing economic sectors for mountain regions of the world.

Mountain ecosystems are also among the most vulnerable to global environmental change. A number of reviews of the effects of global change in mountain regions around the world (Price 1999; Beniston 2000) and the science documenting ongoing environmental change in mountain regions (glacial retreat, melting permafrost, elevation of treeline, changes in species composition, nonnative species introductions, increased geomorphic processes) is progressing steadily.

Climate change is the most important and widespread form of global change affecting mountain regions. The threat of global climate change to mountain regions was recognised in the second assessment of the United Nations' Intergovernmental Panel on Climate Change (IPCC 1995), when an entire chapter was devoted to the impacts of climate change in mountain regions. Although tourism was identified as one of the sectors where important consequences were possible – along with water, biodiversity, agriculture and forestry – virtually no research was available upon which to assess the potential magnitude of climate change impacts for tourism in any mountain region of the world. More recently, the World Tourism Organization (WTO) also recognised the relative vulnerability of tourism in mountain regions to climate change when it identified the impacts of climate change in mountains as one of four main theme areas at the first international conference on climate change and tourism (Djerba, Tunisia) in 2003.

This chapter will concentrate on two of the predominant tourism segments in mountain regions, which are also the better researched areas of environmental change and tourism in mountain regions: skiing and nature-based tourism. By concentrating on these two major segments of mountain tourism, contrasts in the potential implications of environmental change for winter and summer tourism can be examined. The chapter is organised into winter and summer tourism sections accordingly.

The sustainability of winter tourism, and the ski industry in particular, has been repeatedly identified as highly vulnerable to global climate change (Wall 1992; IPCC 2001; WTO 2003). The multinational research literature on climate change and skiing is perhaps the best developed in the tourism sector, with some of the earliest studies conducted in the late 1980s. The section on winter tourism will review the findings of studies from seven nations and consider the relative vulnerability of major ski regions around the world (Europe, North America, Japan and Australia). The methodologies employed in these studies will also be examined in order to discuss their comparability.

Nature-based tourism is a very important component of tourism in mountain regions of the world and is one of the fastest growing tourism market segments globally, increasing at an annual rate of 10–30 per cent according to Carter *et al.* (2001: 266). Nature-based tourism in mountain regions is likely to be affected by climate change in two ways, altered seasonality and changes in the physical landscape. Because of severe winter conditions nature-based tourism is highly seasonal in most mountain regions. Therefore, an extended summer season could provide new opportunities for this tourism market. A study of the nature-based tourism market found that the natural setting was the most critical factor in the determination of a quality tourism product (HLA and ARA 1995). It is increasingly recognised that the value of the mountain landscape for tourism depends not just on the presence and quality of tourism infrastructure, but also on the quality of the mountain landscape. Consequently, if climate change adversely affects the natural setting (for example, loss of glaciers, reduced biodiversity, fire or disease impacted forest landscape, reduced snow cover) at a destination, the quality of the tourism product could be diminished with implications for visitation and local economies.

In many mountain regions of the world, parks and other types of protected areas are key resources for nature-based tourism. The second section of this chapter will compare the results of very recent studies on the implications of climate change and related environmental change on nature-based tourism in some of the internationally renowned national parks in the Rocky Mountains of Canada and the USA. Unfortunately, no comparable research has yet to be conducted in other mountain regions of the world.

Climate change and winter tourism: are billions at risk?

Winter tourism has been repeatedly identified as vulnerable to global climate change due to diminished snow conditions required for the sports (alpine and nordic skiing, and snowmobiling in North America) that dominate the winter tourism market. The alpine ski industry is largely concentrated in the mountainous regions of the world and has received greater research attention because of its large economic value in some regions. While it is difficult to compare the economic size of the ski industry around the world, because of the differences in the way ski resorts are operated – single or multiple owners of ski lifts, mountain restaurants and accommodations, ski schools, retail operations – and the quality of data, direct revenues from the global ski industry approach US$9 billion each year. In the USA, members of the National Ski Areas Association (NSAA) had revenues of over US$3 billion in 2003 (NSAA 2004). In 2003, the ski industry in Canada reported annual revenues of approximately US$647 million (Statistics Canada 2003). Estimates by Lazard (2002), show the ski industry in western Europe and Japan have annual revenues of over US$3 billion and US$1.4 billion respectively. In Australia, the ski industry was worth approximately US$94 million in 2000 (KPMG 2000). How much of this multi-billion dollar winter tourism industry is at risk to climate change?

This section will provide an overview of the range of studies that have examined the implications of climate change for the ski industry. Impacts on skiing supply and demand will be discussed separately and organised according to the proportion of global skier visits (Western Europe 54 per cent, North America 21 per cent, Japan 16 per cent, Australia less than 1 per cent) not the chronology of research or the relative vulnerability of each region. Because of the climatic diversity and large distances between ski areas in regions of North America – for example, the distance between a ski resort in Quebec and California can be more than 4500km – this region will be divided into eastern and western North America.

Implications for skiing supply

The importance of snow conditions for skier satisfaction was emphasised by the research of Carmichael (1996), who found that snow condition was by far the key attribute in tourist image and destination choice for winter sports holidays. Sufficient amounts of snow (either natural or machine-made) and the inter-annual reliability of climatic conditions to provide good snow conditions are also critical factors in determining the economic success of ski resorts and ski tourism.

Measuring the impact of climate change on the reliability of snow conditions, and thus the economic viability of ski areas, has been accomplished in different ways in the research literature. Studies by König and Abegg (1997) and Elsasser and Bürki (2002) have used the concept of 'snow reliability' – which they define in Switzerland as a ski resort having a 100-day ski season (minimum 30–50cm snow depth) seven years out of ten – to assess the potential impact of climate change on the ski industry. In North America and Australia, studies by McBoyle *et al.* (1986), Lamothe and Periard Consultants (1988), Galloway (1988) and Scott *et al.* (2003, 2006) have calculated the change in average ski season, which is defined as the number of days a ski area would be operational by meeting specified climatic criteria – minimum snow depth, suitable snow conditions and temperature range – to assess the potential impact of climate change on the ski industry.

The methodology – snow modelling approach, climate change scenarios used, consideration of snowmaking – and results of these and other climate change impact assessments of the ski industry are discussed below and summarised in Table 3.1.

Western Europe

König and Abegg (1997) examined the impact of three consecutive snow-deficient winters (1987–88 to 1989–90) in Switzerland as potential climate change analogues. They found the number of skiers transported by four ski resorts in the Canton and Grisons regions declined, while skier visits to high elevation glaciers increased. König and Abegg's analysis of the impact of a hypothetical 2°C warming on the snowline in the Swiss Alps indicated that the number of 'snow reliable' ski areas (using a 100-day criteria) dropped from 85 per cent to 63 per cent. By comparison, the DJF climate change scenarios for the central Alps in the ACACIA (2000) project indicated warming of 0.9–2.0°C in the 2020s, 1.4–3.7°C in the 2050s, and 1.7–5.7°C in the 2080s. Elsasser and Bürki (2002) later indicated that the number of snow reliable ski areas in Switzerland could drop to 44 per cent if the snowline were to rise to 1800m above sea level (masl), although the climate change scenario responsible for such a shift is not identified. The snow modelling methodology (from Foehn 1990) used for these studies was not described and cannot be compared to other studies.

In Austria, Breiling and Charamza (1998) developed a statistical model of monthly temperature, precipitation and snow cover depth at climate stations across the country and estimated that changes in snow cover from a hypothetical 2°C warming (with no change in precipitation) could put several major low elevation resorts (including Kitzbühel) at risk. Their study estimated the resulting losses in winter tourism revenue at 10 per cent. With various economic multipliers included, the projected losses approached 30 per cent (or roughly 1.5 per cent of Austrian GDP).

Harrison *et al.* (1999) examined the trend in ski season length at the Cairngorm ski area in Scotland from 1972 to 1996. The ski season was getting shorter on average, but the highest elevation ski lift (1060–1150 masl) indicated no change. Using a spatial statistical model of monthly frequencies of frost and days with snow cover, Harrison *et al.* used the analogue winters of 1985–86 and 1988–89 – respectively rated severely cold and exceptionally mild in their Winter Severity Index – to map the difference in days with snow cover. They observed that the warm analogue had the smallest reduction in

Table 3.1 Comparison of climate change impacts on the ski industry[a]

Region and study	Snowmaking incorporated	Climate change scenarios used	Loss of ski season (~2050s)	Other impacts
Western Europe				
Switzerland – König and Abegg (1997)	No	+2°C, no change P[b]		snow reliable[c] ski areas drop from 195 to 145
Austria – Breiling (1994)	No	+1.5°C, no change P	–15 days	
Eastern North America				
S. Ontario – McBoyle et al. (1986)	No	+3 to +5°C, +9% P[d]	–40 to –100%	
Quebec – Lamothe and Periard (1988)	No	+4.5°C, –15% P[d]	–42 to –87%	
S. Vermont – Badke (1991)	No	+3 to +6°C, +5% to +10% P[d]	–56 to –92%	
S. Michigan – Lipski and McBoyle (1991)	No	+3 to +5°C, +9% to +11% P[d]	–59 to –100%	
S. Ontario – Scott et al. (2005a)	Yes	+3.6 to +8°C, +15% to +19% P[e]	–8 to –46%	f
Quebec – Scott et al. (2005a)	Yes	+4 to +7.8°C, +14% to +35% P[e]	–4 to –32%	f
S. Michigan – Scott et al. (2005a)	Yes	+3.3 to +7°C, +4% to +16% P[e]	–12 to –65%	f
S. Vermont – Scott et al. (2005a) / N. Vermont – Scott et al. (2005b)	Yes	+4 to +6.3°C, +4% to +22% P[e]	–14 to –60% (400 masl) / +1 to –30% (1200 masl)	f

Region and study	Snowmaking incorporated	Climate change scenarios used	Loss of ski season (~2050s)	Other impacts
Japan				
Fukushima *et al.* (2003)	No	+3°C, no change P		−30% skier visits nationally, −50% in southern regions
Australia				
Galloway (1988)	No	+2°C, −20% P	−64 to −81%	
Köenig (1998)	No	+1.3°C, −8% P / +3.4°C, −20% P		snow reliable[g] ski areas drop from 8 to 5 / drop from 8 to 0
Hennessy *et al.* (2003)	Yes	2050s: +0.6 to +3°C, +2% to −24% P[h]		potential volume of snowmaking −27 to −55%

[a] although Scott *et al.* (2003), Hennessy *et al.* (2003) and Scott *et al.* (2006) provide impact projections for the 2020s as well, only the more common 2050s results are included in this table for comparisons.

[b] P = precipitation.

[c] using '100 day' criteria.

[d] IS92a-based equilibrium GCMs.

[e] five SRES-based GCMs.

[f] other modelled outputs not summarised here include probability of operations during key holiday periods, volume of snowmaking required, water use, snowmaking costs.

[g] using '60 day' criteria.

[h] SRES-based CSIRO GC.

snow cover and concluded that ski areas above 1000m would still have sufficient natural snow cover for skiing in warmer winters. Unfortunately, it was not
indicated how much warmer the 1988–89 winter was, so it cannot be compared
against climate change scenarios for this region.

Eastern North America

Like Switzerland, some ski areas in North America have faced challenging
climatic conditions in the late 1980s and more recently the late 1990s. Figure 3.1
demonstrates the impact of inter-annual climate variability on the length of ski
seasons in the three eastern ski regions of the USA from 1975–76 to 2001–02. The
tremendous variability in the 1970s preceded the widespread implementation of
comprehensive snowmaking in these regions.

Some of earliest research on the potential impact of climate change on the ski
industry was conducted in eastern North America (Table 3.1). McBoyle *et al.*
(1986), using the equilibrium IS92a based climate change scenarios of that period
(GFDL and GIS model runs from 1984), estimated that the ski season in southern
Ontario would contract substantially or possibly be eliminated – 40 per cent to 100
per cent reduction. Using similar methods, a study by Lamothe and Periard
Consultants (1988) projected that the number of skiable days in southern Quebec
would decline by 50–70 per cent. Comparable results were also projected for ski
areas in the eastern USA. Lipski and McBoyle (1991) estimated that the ski season
in central Michigan would be reduced by 30–100 per cent and Badke (1991) estimated a 56–92 per cent reduction in average seasons in central Vermont.

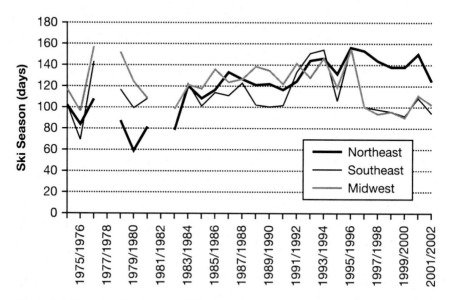

Figure 3.1 Historic ski season variability in the eastern USA

Data sources: National Ski Area Association state of the ski industry reports (1975–2002)

These early studies of climate change and skiing in eastern North America had two fundamental limitations. The first was the criteria used to define a skiable day. The criteria were based on the work of Crowe *et al.* (1973), who defined the minimum snow depth for ski operations as 2.5cm. No downhill ski area in Canada or in the world will operate with such little snow because it is unsafe and will cause damage to both ski equipment and the landscape. The second critical limitation was the omission of snowmaking as a climate adaptation strategy. Snowmaking has been an integral component of the ski industry in eastern North America for more than 20 years, as ski areas in eastern Canada and the mid-west, north-east and south-east regions of the USA have made multi-million dollar investments in snowmaking technology in order to reduce their vulnerability to current climate variability. Today, all ski areas in the north-east, south-east and mid-west ski regions of the USA and the eastern provinces of Canada use snowmaking to some extent (see Chapter 15 for a more detailed discussion of snowmaking as a climate adaptation).

A second generation of climate change assessments on the ski industry of this region emerged, reflecting the integral nature of snowmaking in the ski industry of eastern North America (Scott *et al.* 2003, 2006). These studies benefited from new transient climate change scenarios and downscaling techniques – weather generator parameterised to local climate stations – that allowed for the development of a methodology that used a physically based daily snow cover model at its core. These studies were also the first to incorporate snowmaking into the climate change assessment by integrating a snowmaking module, with climatic thresholds and operational decision rules based on interviews with ski area managers, directly with the snow cover model.

Using a range of climate change scenarios, Scott *et al.* (2003) found that, with current snowmaking capabilities, doubled-atmospheric CO_2 equivalent scenarios (~2050s) projected a 7–32 per cent reduction in average ski season in southern Ontario. The findings demonstrated the importance of considering snowmaking in climate change impact assessments, because the vulnerability of the ski industry was reduced substantially relative to previous studies that projected a 40–100 per cent loss of the ski season in the same study area (McBoyle *et al.* 1986) (Table 3.1). The authors recommended that similar reassessments be completed in areas of eastern North America where previous, and widely cited, climate change studies projected very large impacts on the ski season. Scott *et al.* (2006) examined how current snowmaking capacity affects the climate change vulnerability of ski areas in six locations in Ontario, Quebec, Vermont and Michigan where previous climate change assessments did not incorporate snowmaking. Consistent with the southern Ontario reassessment, the range of season losses projected for the 2050s were much lower (Quebec –4 to –32 per cent, southern Michigan –12 to –65 per cent, southern Vermont –14 to –60 per cent, and northern high elevation Vermont +1 to –30 per cent) than earlier studies that did not account for snowmaking (Table 3.1). At most locations examined in the reassessment, the losses projected under the 'worst case' 2050s scenario approximated the 'best case' from earlier studies.

Western North America

The Rocky Mountains are home to some of North America's most widely known winter tourism destinations. Although snow cover modelling in the mountains of north-western USA projected a 75–125cm reduction in average winter snow depth under two climate change scenarios and an estimated upward shift in the snowline from 900masl to 1250masl (US National Assessment Team 2000), the implications for major ski areas in the region have not yet been comprehensively examined.

Australia

Galloway (1988) used both a statistical model of snow cover and a physically based snow model to examine the impact of a hypothetical 2°C warming and 20 per cent reduction in winter precipitation for Australia's three main ski areas (Perisher, Hotham and Mt Selwyn). It was projected that the mean duration of their snow season would decline from 130, 135, and 81 days respectively to 60, 60, and 15 days (a 64, 66 and 81 per cent loss respectively) (Table 3.1). Unfortunately, the criteria defining the 'snow season' were not defined (i.e. duration with 5cm, 30cm or 50cm), limiting the comparability to other studies. The author also noted that the ski industry had installed snowmaking systems at these locations, but did not attempt to integrate snowmaking into the analysis.

A more recent study in Australia did examine the implications of climate change for snowmaking in Australia's Snowy Mountains. Using a physically based snow model linked with a snowmaking module (similar to the work of Scott *et al*. 2003, 2005 in North America) Hennessy *et al*. (2003) examined the potential impact of two climate change scenarios from the CSIRO Global Climate Model (GCM) at six ski area locations. Their analysis found that the potential volume of machine-made snow could be reduced by 4–10 per cent in the 2020s and 27–55 per cent in the 2050s. Considering the target snow depth required by ski area managers, it was concluded that, with sufficient investment in snowmaking systems, the six ski areas examined would be able to cope with the impact of projected climate change until at least 2020.

Other ski regions

It should be noted that no climate change assessments of the ski industry in Spain, Eastern Europe and China have been conducted; yet these are the regions with the greatest growth in the industry (Lazard 2002).

Implications for skiing demand

Although most climate change impact assessments of the skiing industry have focused on potential changes to the ski season (supply), the potential impact on skiing demand is also very important. Like studies of the impacts of climate change on skiing supply, different methodologies have been used to examine the potential impacts of climate change on skiing demand. The different research approaches and the findings of each study are summarised below.

Europe

Bürki (2000) conducted a survey to investigate how skiers at five Swiss ski areas perceived the threat of climate change and how they would change their skiing patterns if climate change conditions were realised. Relevant to the future of skiing demand, skiers were asked: 'Where and how often would you ski, if you knew the next five winters would have very little natural snow?' The majority (58 per cent) indicated they would ski with the same frequency – 30 per cent at the same resort and 28 per cent at a more snow reliable resort. Almost one-third (32 per cent) of respondents indicated they would ski less often and 4 per cent would stop skiing altogether. With more than one-third of the sampled ski market skiing less or quitting, the implications of climate change for skiing demand in Switzerland are significant. No similar surveys have been conducted in other European nations, so it is uncertain whether these results can be generalised to the European ski market.

Eastern North America

Another method of examining skiers adaptation to future climate change is to examine how they respond to shorter ski seasons during warmer winters. This approach offers advantages over surveys in that it is based on the observed behavioural responses of the entire ski market to real climatic conditions, not stated behavioural responses of a sample of the ski market to hypothetical climatic conditions. Ideally, this analogue approach would examine the difference between skiing demand during winters representative of normal current climatic conditions and winters that might represent what a normal winter is expected to be like under a changed climate, and would be conducted over a relatively short period of time when economic conditions and market competition are very similar.

The winters of 2000–01 and 2001–02 in eastern North American provided the contrast needed for such an analysis. The winter of 2000–01 had temperatures fairly representative of the 1961–90 normal in southern Ontario, southern Quebec and many of the New England states. The winter of 2001–02 was the record warm winter throughout these same regions and approximated the temperatures expected of a mid-range 2050s scenario (i.e. approximately 4.5°C warmer than the DJF 1961–90 normal). Analysis of the difference in skier demand during these contrasting winters, revealed consistently lower demand in the 2050s analogue (2001–02): –11 per cent in the Northeast ski region of the US, –7 per cent in Ontario, and –10 per cent in Quebec (NSAA 2004, Canadian Ski Council, 2004). Although this finding is not surprising, what is somewhat surprising is how small the reduction in demand during this 2050s analogue season was.

One possible explanation for this lower than expected decline in skiing demand is behavioural adaptation by skiers. In a shorter ski season skiers can participate more frequently and still ski as much as they would in a normal year (i.e. go skiing every weekend instead of every two weeks). This type of behavioural adaptation is particularly possible in a ski season that starts later than usual, because skiers know they will be likely to have fewer opportunities that season. This type of behavioural adaptation by skiers can be seen in a measure of asure of the ratio of actual skier visits to the physical capacity for skier visits at a ski area over the ski season – calculated as the daily visitor capacity times number of days of operation. The trend line in Figure 3.2 indicates that utilisation decreases during longer ski seasons. Greater utilisation during shorter ski seasons suggests that behavioural adaptation by skiers is indeed occurring in this region. Notably, similar relationships between season length and utilisation were found in the south-east and mid-west ski regions of the USA.

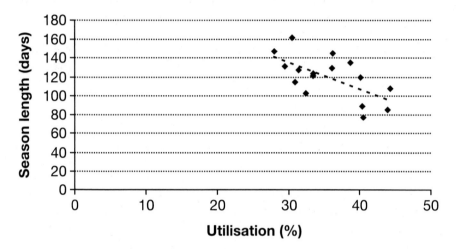

Figure 3.2   Ski area utilisation in the northeast ski region (1974–75 to 1995–96)

Data sources: National Ski Area Association state of the ski industry reports (1975–1996)

The results of this analogue approach stand in contrast to the changes in demand projected by the surveys in Switzerland and Australia, although it must be stated that, because this 2050s analogue winter occurred for only one winter (not five in a row) and was buffered by the presence of snowmaking (not natural snow only), the situations skiers were responding to are not directly comparable. Nonetheless, explaining the differences in the findings of this analysis and those of climate change and skiing demand in Europe, Japan and Australia provide interesting avenues for future research.

Japan

Fukuskima *et al.* (2003) developed a statistical model of snow depth and skier demand in order to examine the potential impact of climate change on the Japanese ski industry. Using a snow model to project future snow cover depth, their assessment of the national ski industry (61 ski areas) estimated that a 3°C warmer scenario would result in an overall 30 per cent decline in skier visits. Ski areas in southern regions were the most vulnerable with skier visits falling by 50 per cent. Conversely, the impact of climate change was projected to be negligible in some northern high altitude ski areas.

Recognising that different relationships between snow conditions and skiing demand may exist in different regions of the world, predicting skier demand based on snow depth at a ski area is a somewhat questionable approach. Most ski areas will not even open until a safe snow base is in place (usually at least 30cm). Therefore, there can be no relationship between visitation and snow depth when there are low amounts of snow. Once a safe operating snow depth is achieved and a ski area opens, there is often a surge of demand in the early part of the new ski season. This initial surge of demand occurs when snow depth levels are still relatively low compared to later in the season. Furthermore, skiers will generally not know whether the snow base is 30cm or 100cm as long as there is full coverage of the ski slope. Only if there are bare patches on the ski slope might skier visitation be affected by snow depth. In other words, the relationship between skier visits and snow depth is thought to be almost binary, with the snow depth either suitable for operations or not. The quality of snow conditions – well groomed, fresh powder snow, icy – is known to have an important impact on skier visitation, but good quality snow conditions are not dependent on snow depth once a safe operational depth has been achieved. More studies using the Fukuskima *et al.* (2003) approach are required to test for similar relationships in other ski regions.

Australia

At the same time that Bürki (2000) conducted his survey of skiers and climate change in Switzerland, König (1998) conducted a very similar survey at three Australian ski areas. Skiers were similarly asked how often and where they would ski if the next five winters had little snow cover. Only 25 per cent of respondents indicated they would continue to ski as often in Australia. Nearly one third (31 per cent) of the sampled ski market would ski less often, but still in Australia. An even greater portion of the sampled ski market would be lost to the Australian ski industry, with 38 per cent of respondents indicating they would substitute destinations and ski overseas (mainly in New Zealand and Canada) and a further 6 per cent of the market would quit skiing. With 44 per cent of the ski market potentially lost and 31 per cent skiing less often, the implications of climate change for Australia's ski industry appear ominous. Whether the remaining skiing demand would be sufficient to sustain the ski industry in Australia remains an important uncertainty. Conversely, the more snow reliable ski resorts in New Zealand and Canada could potentially benefit from the demise of the Australian ski industry.

Summer tourism: opportunities and risks

A number of authors have warned of the potential negative affects of environmental change in sensitive alpine environments for mountain tourism (Wall 1992,

Elsasser and Bürki 2002, Scott 2003). Unfortunately, there has been very little empirical research conducted to explore this issue. The only studies to have examined the potential implications of environmental change for mountain tourism are limited to the Rocky Mountain region of North America and consequently that region was selected as a case study for this section.

Nature-based tourism is an important component of North American tourism, and the national parks in Canada and the USA are central components of this tourism market. Eagles *et al.* (2000) estimated there were over 2.6 billion visitor days in parks and protected areas in Canada and the USA in 1996 (including over 300 million in national parks). A large proportion of park tourism in North America is concentrated in the mountain parks of western Canada and the USA. For example, approximately 65 per cent of visitation to national parks in Canada in 2000–01 occurred in the six national parks in the Rocky Mountains. The economic impact of the three mountain national parks in the Province of Alberta alone is estimated to exceed US$600 million annually (Alberta Economic Development 2000).

Tourism in the many parks in the Rocky Mountains of western North America, displays marked seasonality. Tourism in this region is constrained by climate and the concentration of tourist visitation during warm weather months suggests that a lengthened and improved tourism season could provide opportunities to increase visitation levels in parks in this region. Richardson and Loomis (2004) and Scott and Jones (2005) used regression analysis of historical monthly visitation data (1987–99 and 1996–2001 respectively) to model the current influence of climate on park visitation and project changes under climate change scenarios. Richardson and Loomis (2004) used climate change scenarios from the Canadian and Hadley Centre GCMs forced with IPCC IS92a emission scenarios for the 2020s only. Scott and Jones (2005) examined five SRES-based climate change scenarios for the 2020s, 2050s and 2080s (the same scenarios used in the skiing assessment by Scott *et al.* 2006).

The results of the two studies for the 2020s are very similar (Table 3.2), with Yoho and Banff National Parks projected to have the least increase in visitation (3–6 per cent range), Rocky Mountain and Kootenay National Parks projected to have moderate increases (7–12 per cent range), and Waterton Lakes National Park showing the greatest potential increase (10–19 per cent). These projected increases in visitation would have benefits for local economies and tourism employment near each park. Conversely, increased visitation could also exacerbate existing ecological pressures from visitors and tourism infrastructure, particularly in high visitation parks like Banff (4.5 million annually), Rocky Mountain (3 million annually), Kootenay (1.6 million annually), and Yoho (1.3 million annually).

More astounding to tourism operators and perhaps alarming to park managers, are the projected changes in visitation in the 2050s and 2080s scenarios (Table 3.2). If the findings of Scott and Jones (2005) are suggestive of the longer term effects of climate change on visitation in other alpine parks in the Rocky Mountains region, to say nothing of future increases in demand from population growth, park-based tourism economies would increase dramatically. Such large growth in

Table 3.2 Visitation to parks in the Rocky Mountains under climate change scenarios

Park	2020s	2050s	2080s
Banff National Park (Alberta, Canada)[b]	+4 to +6%	+8 to +23%	+10 to +41%
Kootenay National Park (British Columbia, Canada)[b]	+7 to +10%	+13 to +41%	+15 to +69%
Rocky Mountain National Park (Colorado, USA)[a]	+7 to +12%	—	—
Waterton Lakes National Park (Alberta, Canada)[b]	+10 to +19%	+18 to +65%	+21 to +107%
Yoho National Park (British Columbia, Canada)[b]	+3 to +5%	+5 to +19%	+5 to +27%

[a] Richardson and Loomis (2004); [b] Scott and Jones (2005).

visitation in parks that already report ecological stress from tourism would also necessitate intensive visitor management, including strategies such as demarketing, visitor quotas, and variable pricing for peak and shoulder demand periods.

The research into the potential impact of climate change on park visitation in the Rocky Mountains of North America summarised in Table 3.2 only examined the implications of changes in seasonality (i.e. a longer and improved season for warm weather tourism activities). These projected changes in visitation did not take into account effects of climate change-induced environmental change in these parks.

Climate-induced environmental change has already been documented in the Rocky Mountain region (Luckman and Kavanagh 2000; British Columbia Ministry of Land, Water and Air Protection 2002) and a growing body of literature indicates the magnitude of change will only increase if climate change projections for the twenty-first century are realised. Scott and Suffling (2000) identified a range of potential climate change impacts on ecosystems in the mountain parks of Canada's Western Cordillera. Vegetation modelling suggests that the Rocky Mountain region will experience both latitudinal and elevational ecotone changes, with the potential for species reorganisations and implications for biodiversity. The upslope migration of the tree line has already been documented in Jasper National Park (Alberta, Canada) (Luckman and Kavanagh 2000). Similar impacts are expected in Yellowstone National Park (Wyoming, USA), where vegetation modelling results project that the range of high-elevation species will decrease, some tree species will be regionally extirpated, and new vegetation communities with no current analogue will emerge through the combination of existing species and non-native species (Bartlein *et al.* 1997). Vegetation modelling in Glacier National Park (Montana, USA) projected a 20m per decade upslope advance of forest through 2050, with considerable spatial variation determined by soil conditions and aspect (Hall and Farge 2003). A study of mammal populations in the isolated mountain tops of the Great Basin in the western United States projected that regional average warming of 3°C would cause a loss of 9–62 per cent of species inhabiting each mountain range and the extinction of three to fourteen mammal species in the region (McDonald and Brown 1992).

Like glaciers around the world, those in western North America have been retreating over the past century. Glacier National Park (Montana, USA), which early visitors referred to as the 'little Switzerland of America', has lost 115 of its 150 glaciers over the past century and scientists estimate that the remaining 35 glaciers will disappear over the next 30 years (Hall and Farge 2003). Similar projections have been made for glaciers in Canada's Rocky Mountain parks. Climate records show that the Rocky Mountains of Canada have experienced a 1.5°C increase in average temperatures over the past century, almost three times the global average of 0.6°C. All of the glaciers in this region have shown a strong decline over the same period and glaciers less than 100m thick are expected to disappear over the next 30 to 40 years (Brugman *et al.* 1997).

The quality of the alpine environment is essential for successful tourism in mountain regions. How might the types of climate-induced environmental change detailed above affect tourist perceptions of the landscape and visitation to parks in the Rocky Mountain region? Two research projects examined this question using similar survey methodologies. Richardson and Loomis (2004) surveyed visitors to Rocky Mountain National Park, asking how their visitation patterns (number and length of stay) might change under the specified hypothetical environmental change scenarios provided. The scenarios were partially developed on the basis of climate change studies of potential ecological impacts in the park. The three future scenarios outlined a range of environmental changes, including the climate, recreation access (scenic roads and trails) and crowding, wildlife populations and vegetation compositions in the park. Two of the scenarios were based on the Canadian and Hadley Centre GCMs forced with IPCC IS92a emission scenarios for the 2020s and a third was a warmer hypothetical scenario – it was not indicated if the 'extreme heat' scenario was based on climate change projections for 2050s or 2080).

Scott and Konopec (2005) used a similar approach, asking tourists in Glacier-Waterton Lakes International Peace Park to consider three environmental change scenarios, and to indicate whether they would still visit the park and, if so, more or less frequently. The three scenarios of environmental change were hypothetical, but where possible based on biophysical climate change impact studies in the region. The three scenarios were designed to reflect the types and magnitude of change in the 2020s, 2050s and 2080s in order to examine the potential long-term effect of climate-induced environmental change on park tourism in the region. Several types of environmental changes in the park were outlined in each scenario, including: wildlife populations, number of glaciers, vegetation composition, mammal and rare plant species lost, forest fire occurrence, water temperatures, and fishing catch rate.

Richardson and Loomis (2004) found that between 9 and 11 per cent of respondents would change their visitation behaviour under the two 2020s scenarios. A larger proportion (16 per cent) indicated they would alter their visitation patterns under the 'extreme heat' scenario. The large majority indicated they would not change their visitation patterns based on the scenarios provided. The changes in visitation behaviour in the two 2020s scenarios resulted in a 10–14 per cent increase in annual visitation, while the 'extreme heat' scenario caused a 9 per cent loss in visitation.

Scott and Konopec's (2005) findings for the 2020s were consistent with those of Richardson and Loomis. After considering the environmental changes outlined in scenario 1 (2020s) the vast majority (99 per cent) of respondents indicated they would still visit the park and 9 per cent indicated they would visit more often (Figure 3.3). The vast majority of respondents (97 per cent) indicated they would visit the park if the environmental change in scenario 2 (2050s) occurred, however 14 per cent of those who would still visit indicated they would visit less often. An important threshold of environmental change was reached for many in scenario 3 (2080s), where 19 per cent of respondents indicated they would not visit the park and, of those who indicated they would still visit the park, 37 per cent stated they would visit less often. With 56 per cent of respondents indicating they would not visit the park or would visit less often, substantive environmental changes later in the twenty-first century may reduce total park visitation (Figure 3.3).

This last finding provides an important qualification to the large increases in park visitation projected for the 2080s in Table 3.1. Although seasonality changes may be favourable to increased visitation, environmental change may reduce the attractiveness of the mountain landscape to such an extent that these impacts override the opportunities provided by an improved climate for tourism.

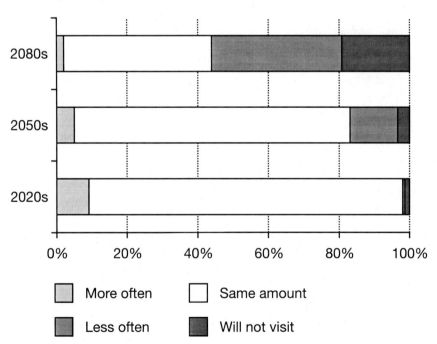

Figure 3.3 Impact of environmental change on visitation to Glacier-Waterton Lakes International Peace Park

Conclusions

This chapter has examined the potential implications of global climate change for tourism in mountain regions, with a focus on those tourism markets that have been reasonably well researched: skiing and nature-based tourism in parks and protected areas. Thus far, the geographical focus of climate change research on mountain tourism has been quite limited, with most studies focusing on the European Alps or areas of North America. Drawing a central conclusion about the implications of climate change for mountain tourism from these disparate studies is difficult. Assessing the timeframe most relevant to the tourism industry (i.e. the 2020s), it seems that a minor climatic warming, similar to that projected by the IPCC for the 2020s or even low-emission 2050s scenarios, could benefit mountain tourism in North America economically. An extended and climatically improved summer tourism season is projected to increase visitation and tourism revenues, while the projected warming is not enough to adversely affect the winter tourism season at high elevations or exceed the coping range of snowmaking in lower elevations. The sustainability of these changes – increased visitor use pressures, increased water and energy use for snowmaking – is uncertain.

Based on the available studies, this does not seem to be the case in the European Alps, however. Without the same investment in snowmaking, the European ski industry does not have the same adaptive capacity as its North American counterpart and warming of only 2°C (a high-impact 2020s scenario or only half of the high-impact scenario for the 2050s) is projected to threaten a significant number of ski areas. The potential benefit of a slightly warmer climate for mountain tourism in North America should not be construed as an argument in favour of global climate change or the abandonment of effort to mitigate climate change, but rather considered support for policies that would achieve a low emission future (i.e. an IPCC 'B2-world').

The implications of climate change for mountain tourism beyond the 2020s is even more problematic. With the substantive caveat that 'all else will remain equal' and that only the climate will change, available studies suggest winter tourism in many locations will be adversely affected by warming projected for the 2050s, while summer nature-based tourism may still benefit. The significant warming projected for the 2080s appears detrimental for both winter tourism and summer nature-based tourism. Of course, 'all else will not remain equal' particularly in the very dynamic global tourism industry. Most mass tourism markets have existed for less than 50 years and how the many major influencing variables – globalisation and economic fluctuations, fuel prices, demographic changes in existing and future demand markets, increased travel safety and health concerns, increased cultural awareness, advances in information and transportation technology, regional and local environmental limitations, such as water supply and pollution – will affect the tourism sector over the next 50 years in unknown ways. Examining how global environmental change in mountain regions may interact with other major factors influencing the tourism sector should be a focus for future research.

Research on global environmental change and mountain tourism is quite limited. Although some notable methodological developments have been made in recent years, a number of recommendations for future research are offered to conclude this chapter.

Winter tourism is one of the better-developed areas of global environmental change and tourism research. Nonetheless, there are several potential ways forward. One of the most obvious conclusions from the overview of the international skiing studies in Table 3.1 is the lack of comparability between studies. The climate change and skiing literature is sufficiently well developed that researchers, government, investors and the ski industry will want to begin to compare studies to assess the relative vulnerability of winter tourism regions. For example, in 2003 the International Olympic Committee indicated that it would include climate change in its considerations of where to hold future winter games. How will they compare the relative vulnerability of different locations to climate change? The research community needs to adopt similar terminology and standard climate change impact indicators to facilitate such comparisons. For example, calculating the average length of ski seasons and the probability of being operational during certain time periods could become the standard for impact assessments for the ski industry. These variables, because they are calculated daily, can then also be converted to regional measures of economic viability such as 'snow reliability' – whether using a 60- or 100-day rule. To facilitate future comparisons and tourism sector meta-analysis, researchers must heed the advice of the IPCC and do a better job of clearly identifying the climate change scenarios used in impact assessments.

Scott *et al.*'s (2003, 2006) reassessment of the impact of climate change on the ski industry in eastern North America, with snowmaking incorporated in the methodology, revealed a much lower vulnerability than previous studies. This research illustrates the critical importance of including adaptation in future climate change assessments in the tourism sector (see Chapters 13 to 16 for further discussion of adaptation). Similar reassessments are needed for the European ski industry. Snowmaking technology is not as widespread in the European Alps and research similar to that done in North America could provide insight into the ability of snowmaking to reduce the risk of climate change and determine in which locations the large investment in snowmaking would be justified economically.

Endeavouring to explain the somewhat contrasting results of the two approaches used to explore potential changes in skiing demand – i.e. skier surveys and skier visits during a climate change analogue year – is another interesting direction for future research in this field. If possible, both methods should be applied in the same location in an effort to compare stated and observed behavioural responses to poor snow conditions.

The findings of the studies of park visitation in the Rocky Mountain region of North America were remarkably consistent and raise many questions for future research. Will certain tourism market segments – for example first time visitors, international visitors, local recreationists, ecotourists – respond differently to environmental changes? Which environmental changes have the most impact on visitation and can adaptation strategies overcome these impacts? How might destination substitution affect regional and international tourism patterns? It is important that the future studies of tourist response to environmental change in mountain regions – or other types of tourist destinations – examine the impacts of the magnitude of environmental change anticipated later in the twenty-first century. Although it was noted that the 2020s are the most relevant to the planning and management decisions in the tourism industry, the North American studies indicate that considering only the results for 2020s would portray an overly optimistic future and would therefore be misleading about the potential long-term threat environmental change poses to tourism in some mountain regions. Of course, the findings from these North American studies cannot be generalised to other mountain regions and similar research on visitor responses to environmental change needs to be conducted in mountain regions around the world, particularly in developing nations where tourism is a vital component of local or regional economies.

The dependence on a climate-sensitive natural resource base to attract visitors places mountain tourism at greater risk to the impacts of global change than many other tourism destinations. Increased collaboration between climate change and tourism research communities, government tourism officials and the private tourism sector is paramount to advancing our understanding of the implications of global environmental change for tourism dependent economies of mountain communities.

Acknowledgements

The author is grateful to the Government of Canada's Climate Change Action Fund for partial financial support of this research. He also thanks the many individuals from the ski industry and Parks Canada whose insights on mountain environments and interest in the potential implications of climate change for tourism made some of the research in this chapter possible.

References

ACACIA (2000) 'Tourism and recreation', in M. Parry (ed.) *Assessment of Potential Effects and Adaptations for Climate Change in Europe*. Norwich: Jackson Environment Institute, University of East Anglia, pp.217–226.

Alberta Economic Development (2000) *The Economic Impact of Visitors to Alberta's Rocky Mountain National Parks in 1998*. Edmonton: Alberta Economic Development.

Badke, C. (1991) 'Climate change and tourism: the effect of global warming on Killington, Vermont', unpublished thesis, Department of Geography, University of Waterloo, Canada.

Bartlein, P., Whitlock, C. and Shafer, S. (1997) 'Future climate in the Yellowstone National Park Region and its potential impact on vegetation,' *Conservation Biology*, 11(3): 782–92.

Beniston, M. (2000) *Environmental Change in Mountains and Uplands*. London: Arnold.

Breiling, M. (1994) 'Climate variability: the impact on the national economy, the alpine environments of Austria and the need for local action', *Proceedings of the Conference on Snow and Climate*, September, Geneva, Switzerland.

Breiling, M. and Charamza, P. (1999) 'The impact of global warming on winter tourism and skiing: a regionalized model for Austrian snow conditions', *Regional Environmental Change*, 1(1): 4–14.

British Columbia Ministry of Land, Water and Air Protection (2002) *Climate Change in British Columbia: Present and Future Trends*. Victoria: Ministry of Land, Water and Air Protection.

Brugman, M., Raistrick, P. and Pietroniro, A. (1997) 'Glacier related impacts of doubling atmospheric carbon dioxide concentrations on British Columbia and Yukon', in E. Taylor and B. Taylor (eds) *Canada Country Study: Climate Impacts and Adaptation – British Columbia and Yukon*. Ottawa, Ontario: Environment Canada.

Bürki, R. (2000) *Klimaaenderung und Tourismus im Alpenraum – Anpassungsprozesse von Touristen und Tourismusverantwortlichen in der Region Ob- und Nidwalden*. PhD dissertation, Zurich: Department of Geography, University of Zurich.

Carmichael, B. (1996) 'Conjoint analysis of downhill skiers used to improve data collection for market segmentation', *Journal of Travel and Tourism Marketing*, 5(3): 187–206.

Carter, R., Baxter, G. and Hockings, M. (2001) 'Resource management in tourism: a new direction?', *Journal of Sustainable Tourism*, 9: 265–80.

Crowe, R., McKay, G. and Baker, W. (1973) *The Tourist and Outdoor Recreation Climate of Ontario – Volume 1: Objectives and Definitions of Seasons*. Toronto, Ontario: Atmospheric Environment Service, Environment Canada.

Eagles, P.F., McLean, D. and Stabler, M.J. (2000) 'Estimating the Tourism Volume and Value in Parks and Protected Areas in Canada and the USA', *George Wright Forum*, 17(3): 62–76.

Elsasser, H. and Bürki, R. (2002) 'Climate change as a threat to tourism in the Alps', *Climate Research*, 20: 253–7.

Foehn, P. (1990) 'Schnee und Lawinen', in *Schnee, Eis und Wasser: die Alpen in einer Wärmeren Atmosphäre*. Internationale Fachtagung, Mitteilungen VAW ETH Zurich No. 108, pp.33–48.

Fukuskima, T., Kureha, M., Ozaki, N., Fukimori, Y. and Harasawa, H. (2003) 'Influences of air temperature change on leisure industries: case study on ski activities', *Mitigation and Adaptation Strategies for Climate Change*, 7: 173–89.

Galloway, R. (1988) 'The potential impact of climate changes on Australian ski fields', in G. Pearman (ed.) *Greenhouse: Planning for Climatic Change*. Melbourne, Australia: CSIRO, pp.428–37.

Hall, M. and Farge, D. (2003) 'Modeled climate-induced glacier change in Glacier National Park, 1850–2100', *BioScience*, 53(2): 131–40.

Harrison S., Winterbottom S. and Sheppard, C. (1999) 'The potential effects of climate change on the Scottish tourist industry', *Tourism Management*, 20: 203–11

Hennessy, K., Whetton, P., Smith, I., Batholds, J., Hutchinson, M. and Sharples, J. (2003) *The Impact of Climate Change on Snow Conditions in Mainland Australia*. Aspendale, Australia: CSIRO Atmospheric Research.

HLA Consultants and ARA Consulting Group Inc. (1995) *Ecotourism-Nature-Adventure-Culture: Alberta and BC Market Demand Assessment*. Vancouver, British Columbia: Department of Canadian Heritage.

IPCC (Intergovernmental Panel on Climate Change) (1995) *IPCC Second Assessment – Climate Change 1995*. Geneva: United Nations Intergovernmental Panel on Climate Change.

IPCC (2001) *Climate Change 2001: Impacts, Adaptation and Vulnerability*. Third Assessment Report. Geneva: United Nations Intergovernmental Panel on Climate Change.

König, U. (1998) *Tourism in a Warmer World: Implications of Climate Change due to Enhanced Greenhouse Effect for the Ski Industry in the Australian Alps*. Wirtschaftsgeographie und Raumplanung, Vol.28, Zurich: University of Zurich.

König, U. and Abegg, B. (1997) 'Impacts of climate change on tourism in the Swiss Alps', *Journal of Sustainable Tourism*, 5(1): 46–58

KPMG (2000) Victorian Alpine Resorts – Economic Significance Study 2000. Alpine Resorts Co-ordinating Council, online. Available at: www.dse.vic.gov.au/dse/nrenrt.nsf/childdocs/C6F5A4A5BA082FC64A256A650023A21F0DB6816 FF97D14464A256B8A001A4768?open#3 (accessed 11 July 2005).

Lamothe and Periard Consultants (1988) *Implications of Climate Change for Downhill Skiing in Quebec*. Ottawa, Ontario: Environment Canada, Climate Change Digest 88–03.

Lazard, A. (2002) 'Ski winter: world flat', *Ski Area Management*, September: 24–7.

Lipski, S. and McBoyle, G. (1991) 'The impact of global warming on downhill skiing in Michigan', *East Lakes Geographer*, 26: 37–51.

Luckman, B. and Kavanagh, T. (2000) 'Impact of climate fluctuations on mountain environments in the Canadian Rockies', *Ambio*, 29: 371–80.

McBoyle, G., Wall, G., Harrison, K. and Quinlan, C. (1986) 'Recreation and climate change: a Canadian case study', *Ontario Geography*, 23: 51–68.

McDonald, K. and Brown, J. (1992) 'Using montane mammals to model extinctions due to global change', *Conservation Biology*, 6(3): 409–15.

National Ski Areas Association (2004) Available online at http://www.nsaa.org (accessed 1 September 2004).

Price, M. (1999) *Global Change in the Mountains*. New York: Parthenon.

Price, M., Moss, L. and Williams, P. (1997) 'Tourism and amenity migration', in B. Messerli and D. Ives (eds), *Mountains of the World. A Global Priority*. New York: Parthenon, pp.249–80.

Richardson, R. and Loomis, J. (2004) 'Adaptive recreation planning and climate change: a contingent visitation approach', *Ecological Economics*, 50: 83–99.

Scott, D. (2003) 'Climate Change and Tourism and the mountain regions of North America', in *Climate Change and Tourism*. Proceedings of the First International Conference on Climate Change and Tourism, Djerba, Tunisia, 9–11 April. pp.1–9.

Scott, D. and Jones, B. (2005) 'Climate change, seasonality and visitation in Canada's national park system'. *Tourism Management* (in review).

Scott, D. and Konopec, J. (2005) 'Tourist response to environmental change scenarios in Glacier-Waterton International Peace Park'. *Global Environmental Change* (in review).

Scott, D. and Suffling, R. (2000) *Climate Change and Canada's National Parks*. Toronto, Ontario: Environment Canada.

Scott D., McBoyle, G. and Mills, B. (2003) 'Climate change and the skiing industry in Southern Ontario (Canada): Exploring the importance of snowmaking as a technical adaptation', *Climate Research*, 23; 171–81.

Scott, D., McBoyle, G., Mills, B. and Minogue, A. (2005 – in review) 'Implications of climate change for the Vermont ski industry,' *Applied Geographer*.

Scott, D., McBoyle, G., Mills, B. and Minogue, A. (2006) 'Climate change and the sustainability of ski-based tourism in eastern North America: a reassessment', *Journal of Sustainable Tourism* (in press).

Statistics Canada (2003) *The Daily* 12 May, online. Available at: http://dissemination.statcan.ca/Daily/English/050512/d050512d.htm (accessed 11 July 2005).

United States National Assessment Team (2000) *Climate Change Impacts on the United States: the Potential Consequences of Climate Variability and Change*. New York: Cambridge University Press.

Wall, G. (1992) 'Tourism alternatives in an era of global climate change', in V. Smith and W. Eadington (eds) *Tourism Alternatives*. Philadelphia, PA: University of Pennsylvania, pp.194–236

WTO (World Tourism Organization) (2003) *Climate Change and Tourism*. Proceedings of the First International Conference on Climate Change and Tourism, Djerba 9–11 April. Madrid: World Tourism Organization.

4 Lakes and streams

Brenda E. Jones, Daniel Scott and Stefan Gössling

Introduction

On 31 January 2003, the US National Park Service released a statement indicating that it has closed another boat ramp on Lake Powell, the most popular recreation lake in the state of Utah. The ramp closure, the third of six ramps to be closed since 2002, was in response to very low water levels (24.3 metres below fill level). The pessimistic state of affairs prompted one environmental reporter to ponder, 'Lake Powell, where have you gone?'

(Hollenhorst 2003: 1)

Water – recreation and tourism have an affinity for it. An extensive number of streams, rivers and natural and engineered (for example, reservoirs) lakes are critical resources for the recreation and tourism industry around the world. These water resources are the foundation of such industries as sport fishing, recreational boating, white-water rafting and diving, and indirectly support other important land-based industries including golf (irrigation) and skiing (snowmaking). These recreation activities are sensitive to natural and human-induced changes in the availability and quality of the water resources (e.g, climate change, water-level management, over-exploitation). This chapter will examine how global environmental change may affect water resources in the lakes and streams that recreation and tourism is dependent on. It is beyond the scope of this chapter to examine the breadth of impacts brought about by global environmental change. Rather, the potential impacts on water-based recreation and tourism are examined in relation to four main themes: water levels, water properties (i.e. thermal conditions, water quality), biodiversity and water supply. Case studies are drawn primarily from North America and Europe where most empirical research has been conducted to date. The case studies are from diverse regions, in North America (for example, Great Lakes, Rocky Mountains and New England) and Europe (Sweden and the United Kingdom), and are representative of the types of impacts projected to be experienced by the broader water-related recreation and tourism industries.

Water levels

In North America, the Great Lakes region has long been a popular destination for tourism and recreation – especially recreational boating and fishing – and thus serves as an important example of the implications of fluctuating water levels. Approximately six million recreational boats are registered in the region, and over one half of the 1,413 marinas in the bordering eight states and one Canadian province are located along the shores of the five Great Lakes (Thorp and Stone 2000). The recreation industry in the region is negatively impacted by extremes in both high and low water levels, particularly the latter.

A 1992 survey of marina operators and recreational boaters on the Canadian side of the lakes revealed that most had incurred some degree of financial impact as a result of fluctuating water levels since they opened for business – between five and 30 years before the survey – (Bergmann-Baker *et al.* 1993). During periods of low water levels, over two thirds of survey respondents experienced difficulty accessing docks (i.e. dock too high out of the water) and boat launch ramps (i.e. ramp no longer extended to the water). A smaller proportion experienced reductions in the length of their boating season and structural damage to wooden piers and docks. In response to these problems, marina officials undertook a number of adaptation strategies including dredging channels (55 per cent), adjusting docks (45 per cent), putting restrictions on the size and location of boats (44 per cent), and closing boat slips (27 per cent).

Below average water levels in the Great Lakes during 1999–2002 once again revealed the sensitivity of marinas and the recreational boating industry to climate variability. A 2001 survey of marinas on Lake Ontario and the upper St. Lawrence River revealed that fluctuating water levels had a 'major' or 'devastating' impact on the majority of respondents during the previous five years (McCullough Associates and Diane Mackie Associates 2002). Low water levels on Lake Huron precipitated the Canadian Government's creation of a US\$9.9 million Great Lakes Water-Level Emergency Response Programme to aid marina owners and operators with emergency dredging costs (CCN 2000). Common adverse impacts among respondents were the loss of access to boat slips midseason, reductions in the ability to use equipment (i.e. refuelling hoses no longer reached boats), and the need to move boats to other marinas (i.e. with sufficient channel depths), all of which contributed to lost revenues for operators and customer dissatisfaction among boaters and anglers (Connelly *et al.* 2002).

Most climate change scenarios project reductions in average Great Lakes water levels, with reductions of at least 1m on Lakes Michigan, Huron and Erie by the middle of the twenty-first century (Mortsch *et al.* 2000). The frequency and duration of low water levels in the region are projected to increase (Mortsch *et al.* 2003), thus there is a high likelihood that marinas and recreational boaters will experience similar conditions to those experienced during 1999–2002 on a regular basis. Further, projected water level reductions are likely to contribute to reduced navigability in some channels (for example, from newly exposed sand bars, plant growth), changes in the location of well-established launch points for boats, and even restrictions on the size and weight of boats (for example, large draft sail boats) allowed to operate in certain water bodies.

Lower average water levels in the Great Lakes region will also have important implications for the nature of shoreline environments and their associated tourism potential. Many large freshwater wetlands that serve as important recreation destinations for anglers, hunters and bird watchers are vulnerable to climate change because of projected shifts in the location of shorelines. In Point Pelee National Park (Ontario, Canada), for example, lower average lake levels could cause the vitally important wetlands to dry. Since the wetland is protected by sand spits, it is likely to shift towards a meadow environment (Wall 1998), severely affecting the waterfowl population. Point Pelee is ranked among the top 10 best locations for watching birds in North America, and its waterfowl population attracts approximately 60,000 tourists annually to the park, contributing US$4 million to the local economy (American Birding Association 2003). Changes in the quality of open water wetlands will also contribute to declines in habitat (waterfowl nesting, fish spawning grounds) for recreationally valued species (Koonce *et al.* 1996; Mortsch 1998) and reductions in the biodiversity of wetland environments.

In the USA, low water levels are restricting tourism and recreation in western regions of the country and therefore serve as an important analogue of the potential impacts of climate change on tourism. Drought conditions in Colorado during the spring and summer of 2002 impacted the state's sport fishing and rafting industries. Anglers were restricted from fishing in many state rivers because the fish populations were highly stressed by low water levels and higher water temperatures. Despite the precautions, and to the dismay of anglers, higher water temperatures eventually led to large fish kills at several popular sport fishing spots on the Colorado River (Kenworthy 2002). The river rafting season was also substantially shortened. Low water levels contributed to business reductions of 40 per cent at some river rafting outfitter companies and, statewide, economic losses in the rafting industry exceeded US$50 million (Associated Press 2002a, 2002b).

The prolonged drought in western regions of the USA also negatively affected reservoirs, a major tourism and recreation resource in the country. Lake Mead is the largest functional reservoir in the western USA and is an important recreation destination in southern Nevada, used by nearly 10 million people annually (National Park Service 1999). Water levels in the reservoir have dropped nearly 30m since 1999 due to reduced flows in the Colorado River (National Park Service 2003), and were at record low levels in 2004 (Zimmer 2004). Boaters have been exposed to new navigation hazards including rocks and shifting sand bars, resulting in costly repairs to boats and safety concerns for boaters who stray from deeper waters. A number of launch ramps have been closed because they no longer extend to the water line, and new boat launch sites are exposing boaters to uneven surfaces. The National Park Service estimates that every 6m reduction in Lake Mead's surface water level costs US$6 million to mitigate (Allen 2003).

There are no known published case studies of the impact of changing water levels on freshwater lakes and streams in Europe. However, models predict substantial changes to European precipitation patterns under climate change (Xu

2000; SWECLIM 2002). Increases in precipitation, most of which is projected to occur in winter, will contribute to increased lake inflows, lake levels and runoff, the latter leading to greater frequency of riparian flooding (see Palmer and Räisänen 2002). During the summer, drier conditions, exacerbated by greater evaporation, will reduce lake inflows and lake levels. Higher temperatures and decreasing water levels in summer may also affect thermal stratification, evaporation and species composition of lakes (Hulme *et al.* 2003).

Changing water levels will also affect wetlands and floodplains of importance for tourism. For example, a 1992 review of 344 Ramsar sites showed that 84 per cent were either threatened or experiencing ecological changes through drainage for agriculture and urban development, pollution and siltation (Dugan and Jones 1993, cited in Revenga *et al.* 2000). Human modification of rivers and lakes has substantially increased in recent decades. For example, there are now more than 40,000 large dams (15m and higher) worldwide, most of them built within the last 35 years (Revenga *et al.* 2000). Dams change water levels in streams, and they can both enhance or decrease recreational opportunities. For example, reservoirs created by dams might be used for sailing or swimming, but dams might also interfere with the migratory routes of various fish species, contribute to the destruction of riparian habitat and breeding grounds, affect coastal areas and deltas through sedimentation and nutrient loads, cause changes in water temperature and chemical composition of rivers, or lead to declining water levels (Revenga *et al.* 2000). All of these factors might affect recreational activities directly or indirectly.

Water properties

A direct change in the physical characteristics of freshwater resources can indirectly affect the water-based recreation and tourism industry. In this section, the impacts on recreation and tourism are examined with respect to changes in thermal conditions and water quality.

The multi-billion dollar North American freshwater sport fishing industry (~US$75 billion, American Sportfishing Association 2002) would be impacted by changes in thermal conditions of water bodies induced by climate change. As lakes and streams warm, temperature-induced habitat loss and range shifts are expected to contribute to significant losses in recreationally valued fish populations, especially cold water species. A study by the US Environmental Protection Agency (1995) suggested that the thermal habitat for many cold water sport fish desired by anglers (for example, rainbow, brook and brown trout) would be reduced 50–100 per cent in the Great Lakes region under a doubling of atmospheric CO_2 scenario (~2050s). Under similar climate change scenarios, other studies have shown that warmer water temperatures would eliminate the sport trout fishery from most North Carolina streams (Ahn *et al.* 2000), and reduce habitat for several popular cool water sport fish in eastern Canada, including walleye, northern pike and whitefish (Minns and Moore 1992).

Similar research on the thermal habitat for salmonid species in the Rocky Mountain region of the USA found that the projected 4°C summer warming in the region would reduce habitat area by an estimated 62 per cent (Keleher and Rahel 1996). Note, however, that the response of salmon populations to climate change is not homogeneous (Levin 2003), increasing the degree of uncertainty in predicting their future distribution and abundance. The warming of water bodies would generally seem to limit the physiological tolerance of many cold water sport fish species, resulting in these species migrating to more climatically suitable waters if possible. It is projected that the range of cool water (for example, walleye, perch), and particularly warm water fish species (for example, bass), will expand northward and alter the composition of preferred catches in many lakes and streams of North America (Casselman *et al*. 2002). Magnuson (1998) suggested that the northern limit of some cold water species in the USA could migrate north as much as 500km under climate change.

The cumulative impact and regional vulnerability of freshwater sport fishing in North America to climate change has yet to be assessed, but it is possible that changes in fish species will have mixed impacts on the industry. The US Environmental Protection Agency (1995) estimated that annual losses to the US sport fishing industry from climate change would be US$320 million by the 2050s. Using an economic model, Pendelton and Mendelsohn (1998) estimated that the impact of climate change on the sports fisheries of the north-east region of the USA would range between a US$4.6 million loss and a US$20.5 million benefit, depending on the climate change scenario. Population shifts in important sport fish, especially along the southern margins of species range, are likely to negatively impact the local economy of communities that depend on particular sport fish to attract anglers. Outfitters and charter companies in Canada, however, would be likely to benefit from an increase in sport fish tourism as more American anglers would likely travel north to seek the species they have traditionally desired. In 2000, non-resident anglers (mainly Americans) spent about US$850 million sport fishing in Canada (Canada Department of Fisheries and Oceans 2002). Over the long term, however, it is possible that losses in cold water sport fisheries in many areas of North America could be offset by gains in cool water and warm water fisheries, especially if anglers adapt their preferences to changes in locally dominant species.

Changes in thermal conditions of lakes and streams would also limit the tourism potential for winter sport fishing. In the Great Lakes region of North America, Lake Simcoe (southern Ontario) is an important destination for ice fishers; the local community typically receives more winter sport fishers than summer ones and the winter sport fishery is estimated to be valued at US$17.8 million (Scott *et al*. 2002). Ice conditions, especially the timing of ice cover and ice thickness, are critical determinants of the ice fishing season. The 1997–98 ice fishing season on Lake Simcoe highlighted the future impacts the industry may experience under climate change. The winter of 1997–98 was the second warmest on record in the Great Lakes region (3.7°C above the 1961–90 normal), and the above normal temperatures contributed to a 52 per cent reduction in the fishing season on Lake Simcoe (Scott *et al*. 2002). The winter of 1997–98 was an analogue for winter in the 2050s (according to mid-range climate change scenarios); ice fishing seasons are expected to be reduced by 50 per cent in the 2050s.

Reductions in the duration of ice cover on reservoirs, lakes, streams and even canals would also limit ice-skating opportunities (Stefan *et al.* 1998, Williams *et al.* 2004). In North America, a study of spring ice-out dates on 29 lakes in New England (USA) between 1850 and 2000 found that the duration of ice cover has been reduced by nine and sixteen days in the northern/mountainous and southern regions of New England, respectively (Hodgkins *et al.* 2002). A model of 143 freshwater lakes in North America predicted that a 1°C increase in average air temperature would result in ice-in dates occurring five days later and ice-out dates six days earlier (Williams *et al.* 2004). Changes in the duration of ice cover under climate change will limit ice-skating opportunities on the world's longest outdoor skating rink (7.8 kilometres) – the Rideau Canal Skateway (Ottawa, Ontario). Skating on the Rideau Canal Skateway is a 30-year-old tradition in Canada; the skating experience is a primary attraction to the 1.5 million people that visit Ottawa's annual winter festival (Winterlude) in February, contributing over Can$100 million (approximately US$82 million) to the local economy (Ekos Research Associates 2000).

Under the three climate change scenarios examined, the skating season is expected to diminish (Table 4.1). The skating season is projected to decline from a current average of 61 days to between 43 and 52 days in the 2020s. By the middle of the century (2050s), it is projected that the skating season could be only three to four weeks. In the 2080s the season is projected to be reduced even further and virtually eliminated under the warmest scenario. Analysis from the same study concluded that the Rideau Canal Skateway would also open later. As early as the 2050s, it is projected that the canal would not open to skaters until the winter festival begins, which is four weeks later than at present.

In Europe, suitable climatic conditions for ice-skating and other ice-related activities are also likely to decline substantially (see SWECLIM 2002). Ice-skating has a long history as an important recreational winter activity in many European countries. For example, the Swedish *Stockholms Skridskoseglarklubb* (Stockholm's Ice-skating and Ice-sailing Club) was founded more than 100 years ago and has 10,000 members (SSSK 2004). The Great Dutch Ice-Skating Marathon is an example of ice-related events attracting large numbers of tourists that would be vulnerable under climate change:

Table 4.1 Projected season length of the Rideau Canal Skateway under climate change

	1961–90 (days)	2020s (days)	2050s (days)	2080s (days)
Current average season length for ice-skating	61			
Climate change scenarios				
NCARPCM B21		52	49	42
ECHAM4 A21		46	34	25
CCSRNIES A11		43	20	8

Source: Scott *et al.* 2005b

Known as the *Elfstedentocht* in Dutch, the one-day tour is an obsession for its 16,000 participants and the millions more who follow it worldwide. The event is held in The Netherland's northern province of Friesland but only in those years when the ice freezes over the 124-mile track of lakes and canals that makes up the route. The last tour took place January 4, 1997. The fabled marathon was officially organized as a contest nearly 90 years ago by the Friesian Skating Association though its roots go back generations before that. This century, the race has taken place just 15 times; yet, it's become the biggest phenomenon in Dutch sports.

(The Holland Ring 2004)

Thermal changes in lakes and streams will also influence water quality, which could limit the attractiveness of many water resources for tourism. Water quality influences the solubility of dissolved oxygen, the metabolism and respiration of plants and animals, and the toxicity of pollutants (Stefan *et al.* 1998: 547). Water quality is a subjective concept that depends on socially defined levels of pollution – water quality requirements for water-related recreation and tourism segments are usually high. In Europe, the water quality at many popular recreation lakes has been negatively affected by the influx of nutrients, which has contributed to eutrophication and littoral algae production (Cronberg 1999; Scheidleder *et al.* 1999). Nutrient input-related problems caused by agricultural runoff are reported in most European countries (Scheidleder *et al.* 1999; Revenga *et al.* 2000; see also Kosk 2001; McGarrigle and Champ 1999), which has had a negative effect on recreational use of water bodies (Table 4.2). For example, of 171 freshwater bathing areas in Spain, only 42 per cent met stricter – that is recommended – water quality standards in 2003. In Belgium, 47 per cent of 70 freshwater bathing areas complied with stricter quality standards and 84 per cent with mandatory standards. In Italy, 58 per cent of 775 freshwater bathing areas met recommended quality standards and 71 per cent mandatory standards. Due to very poor water quality, bathing was prohibited in almost 28 per cent of all freshwater bathing areas in Italy.

Increases in nutrient loads entering lakes and streams will also negatively affect recreational fishing. For example, some of western Ireland's recreational fishing lakes, described as 'among the finest, natural, wild brown trout fisheries in Europe' (McGarrigle and Champ 1999: 455), are threatened by the intensification of agricultural production and a concomitant influx of phosphorus. In Lough Conn, a formerly oligotrophic/mesotrophic lake, nutrient inputs have resulted in an increase in littoral algae production and the disappearance of the arctic charr (*Salvelinus alpinus*), a popular sport fish (McGarrigle and Champ 1999). The threat to population densities of popular angling fish species from nutrient inputs have also been reported in England (Eliott *et al.* 1996).

During the summer, changes in water quality induced by warming water conditions could also be a limiting factor in beach recreation. As water warms, its oxygen-carrying capacity is diminished, which can contribute to enhanced algae growth and other water pollution (Poff *et al.* 2002). Bacterial contamination can degrade the aesthetics of beaches and pose a health risk to swimmers. Under

Table 4.2 Fresh water bathing areas in Europe, 2003

Country	Fresh water bathing areas (total)	Respecting stricter quality standards (%)	Respecting mandatory standards (%)
Belgium	70	47	84
Denmark	113	88	97
Germany	1572	80	95
Greece	4	75	100
Spain	171	42	96
France	1405	59	94
Ireland	9	100	100
Italy	775	58	71
Luxembourg	20	40	80
Netherlands	561	64	98
Austria	266	80	97
Portugal	55	11	96
Finland	292	70	98
Sweden	404	83	99
UK	11	46	100

Source: European Union 2004

Quality parameters: total coliforms, faecal coliforms, mineral oils, surface-active substances, phenols

climate change, there is a higher likelihood that beach closures or restricted use could become more common in many popular tourism areas.

Algae growth might also affect increasingly popular recreation sports such as diving. Particularly clear lakes are usually well known by diving tour operators and divers. Although no research has been carried out in the context of lakes, reduced visibility brought about by micro-algae has been documented as an important parameter negatively affecting diving experiences in the tropics (see Gössling *et al.* 2005 for a case study in Mauritius).

Biodiversity

An estimated 12 per cent of all animal species live in fresh water (Abramovitz 1996: 7), highlighting the importance of freshwater ecosystems for biodiversity. Freshwater ecosystems are endangered through pollution, overexploitation and invasive species. For example, more than 20 per cent of the world's 10,000 described freshwater fish species might have become extinct, threatened, or endangered in recent decades (Revenga *et al.* 2000).

Floodplains and riverine wetland ecosystems are also of great importance for biodiversity (Tockner and Stanford 2002). For example, wetland-dependent mega-fauna of importance for tourism in Asia includes different kinds of monkey such as the proboscis monkey (*Nasalis larvatus*), which lives in forested riverine wetlands and sleeps in tall trees along riverbanks, leaf monkeys (*Presbytis* spp.) or crab-eating macaques (*Macaca fascicularis*). Swamp forest is an important habitat for orangutans (*Pongo pygmaeus*) in central Kalimantan/Borneo (Dudgeon 2000). In Europe, 20 per cent of regularly occurring bird species are dependent on inland wetlands and, of all bird species that are categorised as endangered, vulnerable, rare or declining, 30 per cent are inland wetland-dependent species (Revenga *et al.* 2000). Birds are of great importance for tourism and recreation all over the world. In the UK, the red grouse (*Lagopus lagopus*) is the most significant commercial wild game bird. Estimates of gross revenues received by grouse moor owners range from €4.8 million to €14 million, with participants paying between €68 and €115 per pair shot. Grouse shooting may sustain around 2,500 jobs in Britain (IUCN UK 2002).

Bird watching in the UK might be of even greater importance. In 1998/99, the Royal Society for the Protection of Birds' reserves attracted over one million visitors, generating an annual expenditure of €16.8 million. Particularly in rural areas, bird watching can be of great economic significance. For example, in the Shetland Isles, bird watchers bring an estimated €1.4 million to the local economy supporting 40 full-time equivalent jobs. In Scotland, geese are a major winter bird-watching attraction, with at least 44,000 visitors to key goose reserves each year contributing €2.4 million to local economies. The value contributed to the economy by bird watchers in the UK is estimated to be in the order of €322 million, including the purchase of equipment, bird food, membership subscriptions, travel, books and magazines (IUCN UK 2002).

The accidental introduction of non-native species into lakes and streams by human activities is an important global environmental change issue for tourism and recreation in many parts of the world. One of the most well-known examples, illustrating the consequences of invasive species introductions in freshwater systems, is Lake Victoria which is bounded by Uganda, Tanzania and Kenya. The lake contained more than 350 fish species in the cichlid family, of which 90 per cent were endemic. After introducing the Nile perch and Nile tilapia in the 1950s, more than half of the native species became extinct (Revenga *et al.* 2000). Similar consequences of the colonisation of water bodies by non-indigenous or 'invasive' species can be observed in the Great Lakes region of North America, illustrating the potential range of impacts to recreation and tourism.

Two of the most important invasive fish species in the Great Lakes are the sea lamprey and the round goby. Sea lampreys are native to the Atlantic Ocean and are believed to have entered the lake system through shipping canals in the early 1900s. They are well established in Lakes Michigan, Huron and Superior, with smaller populations in Lakes Ontario and Erie. Sea lampreys are parasitic eel-like vertebrates characterised by a round, sucker-like mouth, which it uses to attach itself to other fish. It feeds on the bodily fluids of other native freshwater fish, resulting in deep wounds/scarring and often the death of its prey. The round goby, a member of the *Gobiidae* family, is a more recent arrival to the Great Lakes, having only been detected since the early 1990s. Brought to the Great Lakes by foreign ships (from Eurasia), the main populations of round gobies are located in

Lake Erie and Lake St. Clair, but isolated pockets have been found in Lakes Huron and Michigan (Jude 1997).

Sea lampreys and round gobies threaten the native species that support the large sport fishing industry in the Great Lakes region. Sea lampreys are non-selective in their feeding, attacking many large and valuable sport fish including salmon, trout and perch. A report prepared for the Great Lakes Fisheries Commission indicated that one sea lamprey kills over 40kg of fish during its adulthood, which lasts no more than two years (Ontario Federation of Anglers and Hunters 1998). The sea lamprey contributed to the elimination of lake trout, a valuable sport fish, from Lakes Michigan and Huron during the 1960s (Schneider *et al.* 1996), and severely hindered the establishment of a salmon population introduced to the Great Lakes specifically for sport fishing (Fuller *et al.* 2004). Round gobies pose a threat because they feed on fry and eggs of native fish, a behaviour that reduces populations of native fish (Jude 1997). In addition, both species are rapid reproducers (round goby will spawn five times in one mating season), and are aggressive by nature, typically colonising the spawning grounds of native sport fish.

The accidental introduction and subsequent proliferation of non-native species threatens the valuable sport fishing industry in the Great Lakes. The feeding and reproductive actions of sea lampreys and round gobies have important implications for anglers. Population reductions in popular game fish and the inability to re-establish current populations in some areas could lead to demand shifts in sport fishing destinations, impacting the livelihoods of communities that depend on the industry. Monetary investments to rehabilitate native spawning grounds may become futile in the future, thus serving to further reduce populations of popular sport fish. It is possible that anglers will demand the introduction of new sport fish to the Great Lakes that are immune to these two invasive species, thus contributing to the creation of new hybrid sport fish varieties. Uncertainty exists regarding the effect the introduction of hybrid sport fish might have on the composition of existing sport fish populations. In addition, round gobies appear to be tolerant of poor water quality and warm water. Under climate change, projected changes in water conditions (for example, warmer water, decreased water quality) could permit existing populations of gobies to become more prolific.

Zebra mussels are another invasive species in Great Lakes' waters. Native to the Caspian Sea, zebra mussels were first identified in the Great Lakes in the late 1980s. Control of this invasive species is important because it influences a range of activities in the region's recreation and tourism industry. For example, zebra mussels can negatively affect beach recreation. Littering of beaches by dead zebra mussels that wash ashore and accumulate can diminish the aesthetics of beaches. In one event, zebra mussel shells were 0.6m thick and covered an area 4.6m wide along a 400m stretch of public beach on Lake Erie (Dane County Lakes and Watershed Commission 2003). Decaying zebra mussels also produce a foul odour that can further diminish enjoyment of public beaches and swimming areas. If the population growth and range expansion of zebra mussels continues uninhibited in the Great Lakes region, the presence of large onshore expanses of zebra mussels could also precipitate restrictions in beach use or lead to beach closures because the shells pose an additional safety threat because the razor-sharp shells can cut exposed skin.

Zebra mussels also negatively affect recreational boating. The mussels easily attach themselves to and accumulate on the hulls of boats, which can to lead to reductions in handling capability. A common problem for many recreational boaters is the accumulation of zebra mussels in water intake pipes – clogged pipes increase the risk of onboard fires due to overheated engines (Michigan Sea Grant College Program 1992). Accumulation of mussels on boats and the use of boats on different lakes also contribute to the spread of this invasive species. It is likely that difficult decisions will have to be made in the future to limit the spread of zebra mussels, decisions that may restrict where water access is granted. Wisconsin, for example, recently amended a state law to require all anglers to remove zebra mussels from any boat they put into and remove from state waters – failure to do so is punishable by fines and anglers can be prohibited from using state waters (State of Wisconsin 2002). It is possible that if other jurisdictions enact similar rules to prevent the spread of zebra mussels, recreation and tourism will be negatively affected. Boating and sport fishing in some areas may decline, and shifts to new regions where there are fewer access restrictions and less time is required for inspections and cleaning may occur.

The marine diving industry in the Great Lakes region has also been affected by the invasion and spread of zebra mussels, but the nature of the impact is mixed. There are between 6,000 and 10,000 shipwrecks in Great Lakes' waters (2000 in Lake Michigan alone) (Cigelske 2004; Migliore 2004). Many related water reserves and marine parks, including Fathom Five National Marine Park (Canada) in Lake Huron, are popular diving and water-tour destinations, and are economically important for many local communities. For example, it is estimated that the Great Lakes Shipwreck Museum in Paradise, Michigan, receives over 90,000 tourists annually, contributing approximately US$14 million in direct spending to that local economy (Migliore 2004).

Stakeholders in the marine tourism industry have mixed feelings about the impact of zebra mussels. On the positive side, diving and surface viewing conditions have been enhanced at many shipwreck sites because the water-filtering action of zebra mussels has contributed to better water clarity. In parts of Lake Erie, the mussels have actually improved clarity 77 per cent (to 6.1m) (Claiborne 2000). Continued improvements in water quality could enhance diving experiences (i.e. see shipwrecks better), provide an opportunity to expand the number of sites divers can access and reduce the need for experienced guides, or even increase surface tour operations (for example, in glass-bottom boats). On the negative side, zebra mussels are considered a threat to marine tourism. Zebra mussels are colonising many popular shipwrecks, blurring ship details and contributing to their rapid disintegration (Claiborne 2000). Marine businesses that depend on the quality of shipwrecks are likely to be negatively impacted. It is also possible that demand will shift to reserves and marine parks with less abundant zebra mussel populations, or demand will increase for deep water dive sites where the species has not yet colonised the shipwrecks.

The invasion of non-native fish species is also a problem in many European lakes popular with anglers. The English Lake District, for instance, currently contains relatively few native fish species. Illegal introductions of exotic fish species can threaten native sport fish through competition (for example, by roach, *Rutilus rutilus*), predation (for example, zander, *Sander lucioperca*; ruffe, *Gymnocephalus cernuus*), or habitat modification (common carp, *Cyprinus carpio*). This affects recreational angling of native species including sea trout (*Salmo trutta*), Atlantic salmon (*Salmo salar*), perch (*Perca fluviatilis*) and pike (*Esox lucius*) (Winfield and Durie 2004). Recreational angling is of great importance for the local tourist industry. In England and Wales, the total value of inland fisheries is estimated at £3032 million (IUCN UK 2002).

Water supply

Mark Twain once said: 'Whiskey is for drinking, water is for fighting over.' The availability of fresh water is currently an important issue for tourism operations in some locations. Water availability will increasingly be a critical issue for the sustainability of tourism as demand for water from other users (for example, industry, agriculture, cities) increases and climate change affects the reliability of water supplies (see Chapter 10, this volume).

In North America, the supply of fresh water may become a critical limiting factor in tourism and recreation, particularly in western regions. A recent US government report acknowledged that water supplies in states west of the 100th meridian were insufficient to meet current municipal, agricultural, recreational and environmental water demands (US Department of the Interior 2003). Projections of future water demand, based on population and economic growth, indicated the probability of conflicts over water supplies in many areas of the western USA. A number of areas in the western USA were rated as being at a 'high', 'substantial' or 'moderate' risk of water conflicts by 2025, with several important tourism destinations located within the high and substantial risk regions.

A number of important tourism destinations are located in areas where future conflicts over water supply are highly likely if not inevitable. Las Vegas (Nevada) is one high-profile tourism destination located in a 'high risk' zone (as defined by the aforementioned US government report) for water conflict. Las Vegas is considered one of the world's largest per capita users of water, and a large portion of the water used to support Las Vegas' tourism industry is drawn from the Lake Mead reservoir. As discussed earlier, water levels in the reservoir have dropped nearly 30m since 1999 due to reduced flows in the Colorado River (National Park Service 2003) due to the region's prolonged drought. In order to ensure water availability for tourism in Las Vegas over the next 20 years, it is very likely that difficult decisions will have to be made with respect to the regulation of existing water-intensive tourism operations (e.g. golf courses, hotels with large fountains) and possible restrictions on future tourism development in the area.

By comparison, Phoenix (Arizona) is located in a region with 'substantial' risk for water-use conflicts over the next 20 years. As an important golf and winter-getaway tourism destination, Phoenix also has a high demand for water. Municipal and federal governments and agricultural agencies possess historic water-use rights in this region and, in periods of low water availability, these historic water rights will be given priority over more recent allocations granted to the recreation sector (for example, golf). Under such a scenario, these golf oases in the desert would be unsustainable. The courses are likely to sustain long-lasting damage to turf areas very quickly with longer-term implications for tourism demand.

It is important to acknowledge that the US government's projections of water-use conflicts were based on forecasts of population and economic growth to the year 2025. The projections did not consider the potential impact that climate change would have on current water supplies and related issues. Disregarding the climate change issue is negligent given that climate change is projected to exacerbate reductions in water supplies in the region (US National Assessment 2000).

In terms of individual recreation industries, irrigation by the golf sector represents a significant demand on water in some areas of the USA. In the USA, approximately 500 billion gallons of water are drawn from lakes, streams and aquifers annually to keep fairways green (Walsh 2004) and new golf course development is booming in areas where water availability problems are already common (Florida, California, Arizona, Nevada and Texas). No known published study has examined the potential impacts of climate change on long-term irrigation needs, but water demands from the golf industry will increase, with important consequences in major golf destinations that have high course concentrations. Competitive relationships between major golf destinations (for example, Phoenix and Myrtle Beach) could also change as a result of water availability, with potential losses in areas with inadequate irrigation to sustain optimal playing conditions. Over the even longer term, limitations on water availability could eventually influence the overall design of golf courses. In the future, new golf courses may incorporate narrower fairways and permit browner roughs as a means to reduce irrigation requirements (Selcraig 1993; Walsh 2004).

Snowmaking also places a large demand on water supplies, and the associated costs typically encompass a share of operating expenses at many ski areas in North America, especially in the east (see Scott, this volume). Snowmaking permits ski areas to open earlier and close later, particularly when natural conditions are less than optimal (for example, not enough snow, mid-winter thaws). A recent study by Scott *et al.* (2005a) assessed the amount of snowmaking required to maintain ski seasons in six areas of eastern North America under several climate change scenarios. The need for snowmaking was projected to increase between 8 per cent and 66 per cent by the 2020s and between 18 per cent and 161 per cent by the 2050s.

Changes in water availability in many regions of North America will undoubtedly lead to increased competition among users, including tourism and recreation. In the western USA, nine federal lawsuits have been filed in a dispute over the priority given to water used from the entire length of the Missouri River (Boldt 2003). Upper Missouri states want the federal court to give priority to water-based tourism and recreation (for example, sport fishing), which is valued at an estimated US$66 million annually in these states. Lower Missouri states want priority given to shipping, even though its value is one tenth that of tourism. In one specific lawsuit, North Dakota Game and Fish is suing the US Army Corps of Engineers over low water levels in reservoirs that it says contributes to fish kills, thus hindering the sport fishery (Gunderson 2003). Decisions regarding the value placed on different water uses (for example, tourism, agriculture, industry) and the priority given to individual industries will become increasingly difficult in many areas of North America under climate change. The Missouri case is only one example, but it is likely that the judicial system will be asked to resolve many such water-use conflicts in the future.

In Europe, unsustainable groundwater withdrawals threaten the tourism potential of natural environments along the shorelines of lakes and streams. Table 4.3 summarises cases of groundwater exploitation from European countries for which data were available. In 33 of the 126 cases of overexploitation, wetlands were endangered. In 53 cases, saltwater intrusion resulted, particularly along the coastlines of Spain and Turkey (Scheidleder *et al.* 1999). Over-abstraction in the Republic of Moldova is causing saltwater intrusion through the rise of highly mineralized water from deeper aquifers. The main cause for over-abstraction in the Mediterranean countries is public water demand, including tourism (WWF 2001). Tourism thus contributes to over-abstraction in these areas, and is likely to suffer both directly – as the amount of available fresh water decreases – and indirectly – if there is less wetland area providing bird watching opportunities – from these developments.

Table 4.3 Endangered wetlands and saltwater intrusion

Country	Number of over-exploited areas	Overexploitation leading to	
		saltwater intrusion	*endangered wetlands*
Cyprus	7	6	1
Denmark	14	10	5
Estonia	3	1	0
Hungary	4	0	2
Latvia	3	1	3
Rep. of Moldova	17	14	0
Poland	18	3	13
Portugal	3	3	0
Romania	3	0	0
Spain	45	11	3
Turkey	9	4	6
Total	126	53	33

Source: Scheidleder *et al.* 1999

Conclusion

The chapter has shown that a wide variety of recreational activities depend on streams, lakes, reservoirs, canals and wetlands. These include bathing and swimming, recreational boating, sport fishing, golf, ice-skating and bird watching. Natural and human-induced changes are likely to affect water levels, water properties, biodiversity and water supply, and hence the very foundations that various recreation and tourism segments depend on. Increasing water temperatures, nutrient inputs and other pollution, as well as changing precipitation patterns and associated changes in river discharge and lake water levels in particular will have consequences for recreational activities. Water abstraction, land-use changes, the introduction of alien species and weather extremes may be additional stressors for freshwater systems. The North American, European and African cases examined in this chapter suggest that global environmental changes to lakes and streams will, on balance, have a negative effect on recreation and tourism, with many tourism industries and the communities that depend on them potentially experiencing substantial economic losses as water resources become more scarce and contested in the future.

References

Abramovitz, J.N. (1996) 'Imperiled Waters, Impoverished Future: The Decline of Freshwater Ecosystems' *Worldwatch Paper 128*, Washington DC: Worldwatch Institute.

Ahn, S., De Steiguer, J., Palmquest, R. and Holmes, T. (2000) 'Economic analysis of the potential impact of climate change on recreational trout fishing in the southern Appalachians: an application of a nested multinomial logit model', *Climatic Change*, 45: 493–509.

Allen, J. (2003) *Drought Lowers Lake Mead*, NASA: Earth Observatory, online. Available at www.earthobservatory.nasa.gov/Study/Lake Mead/ (accessed 21 October 2004).

American Birding Association (2003) *Economics of Birding: The Growth of Birding and the Economic Value of Birders*, online. Available at www.americanbirding.org/programs/consecond4.htm (accessed 5 August 2003).

American Sportfishing Association (2002) *Sportfishing In America: Values of Our Traditional Pastime*, Alexandria, VA: American Sportfishing Association.

Associated Press (2002a) 'Royal Gorge tourism hurt by fires, drought', *The Associated Press*, 3 September.

Associated Press (2002b) 'Rough year for rafters', *The Associated Press*, 3 September.

Bergmann-Baker, U., Brotton, J. and Wall, G. (1993) *Non-Riparian Recreational Boater Study: Canadian Section, Implications Of Water Level Scenarios*, Detroit, MI: International Joint Commission.

Boldt, M. (2003) 'Upstream: states feel slighted by Missouri management', *Grand Forks Herald*, 9 August.

Canada Department of Fisheries and Oceans (2002) '2000 Survey of Recreational Fishing in Canada. Canada Department of Fisheries and Oceans', online. Available at www.ncr.dfo.ca/communic/statistics/RECFISH/new2002/sum2000_e.htm (accessed 3 April 2003).

Casselman, J., Brown, D., Hoyle, J. and Eckert, T. (2002) 'Effects of climate and global warming on year-class strength and relative abundance of smallmouth bass in eastern Lake Ontario', *American Fisheries Society Symposium*, 31: 73–90.

CCN (2000) 'Dhaliwal moves ahead with $15m in federal funding for emergency dredging in the Great Lakes', *NewsCan*, 17 July.

Cigelske, T. (2004) 'Great Lakes: great diving, great shipwrecks', *Associated Press*, 2 January.

Claiborne, W. (2000) 'A threat to underwater history', *The Washington Post*, 22 August: A3.

Connelly, N., Guerrero, K. and Brown, T. (2002) *New York State Inventory of Great Lakes' Marinas and Yacht Clubs – 2002*, HDRU Publication No. 02–4, Ithaca, NY: Department of Natural Resources, Cornell University.

Cronberg, G. (1999) 'Qualitative and quantitative investigations of phytoplankton in Lake Ringsjön, Scania, Sweden', *Hydrobiologia*, 404: 27–40.

Dane County Lakes and Watershed Commission (2003) *Invasive Work Group Report On Zebra Mussels*, Madison, Wisconsin: Dane County Lakes and Watershed Commission.

Dudgeon, D. (2000) 'The ecology of tropical Asian rivers and streams in relation to biodiversity conservation', *Annual Review of Ecology and Systematics*, 31(1): 239–63.

Dugan, P.J. and Jones, T. (1993) 'Ecological Change in Wetlands: A Global Overview' in M. Moser, R.C. Prentice and J. van Vessem (eds) *Waterfowl and Wetland Conservation in the 1990s: A Global Perspective. Proceedings of an IWRB Symposium*, St. Petersburg Beach, Florida/USA, 12–19 November 1992, IWRB Special Publication No. 26. Slimbridge, UK: The International Waterfowl and Wetlands Research Bureau (IWRB), pp.34–8.

Ekos Research Associates (2000) *Evaluation of the 2000 Winterlude Festival*, report prepared for the National Capital Commission. Ottawa, Ontario: Ekos Research Associates and Conference Board of Canada.

Eliott, J.M., Fletcher, J.M., Eliott, J.A., Cubby, P.R. and Baroudy, E. (1996) 'Changes in the population density of pelagic salmonids in relation to changes in lake enrichment in Windermere (northwest England)' *Ecology of Freshwater Fish*, 5: 153–62.

European Union (2004) *Water bathing quality*, online. Available at http://europa.eu.int/water/water-bathing/index_en.html (accessed 1 January 2004).

Fuller, P., Nico, L. and Maynard, E. (2004) *Petromyzon marinus*, Nonindigenous Aquatic Species Database, Gainesville, FL: United States Geological Survey.

Gössling, S., Helmersson, J., Liljenberg, J. and Quarm, S. (2005) 'Diving tourism and global environmental change. A case study in Mauritius.' *Tourism Management*, submitted.

Gunderson, D. (2003) 'Water wars: recreation on the Missouri River', *Minnesota Public Radio News*, 2 July.

Hodgkins, G.A., James II, I.C. and Huntington, T.G. (2002) 'Historical changes in lake ice-out dates as indicators of climate change in New England, 1850–2000', *International Journal of Climatology*, 22: 1819–27.

Hollenhorst, J. (2003) 'Low water levels at Lake Powell put officials on alert', *KSL.com*, 31 January.

Hulme, M., Conway, D. and Lu, X. (2003) 'Climate Change: An Overview and Its Impact on the Living Lakes', report prepared for the 8th Living Lakes Conference *Climate change and governance: managing impacts on lakes*, Zuckerman Institute for Connective Environmental Research, University of East Anglia, Norwich, UK, 7–12 September.

IUCN UK (2002) *Use of Wild Living Resources in the UK*, a review, UK Committee of IUCN – the World Conservation Union, online. Available at www.iucn-uk.org/PDF/wild_living.pdf (accessed 3 January 2004).

Jude, D. (1997) 'Round gobies: cyberfish of the third millennium', *Great Lakes Research Review*, 3(1): 27–34.

Keleher, C. and Rahel, F. (1996) 'Thermal limits to salmonid distributions in the Rocky Mountain Region and potential habitat loss due to global warming', *Transactions of American Fisheries Society*, 125: 1–13.

Kenworthy, T. (2002) 'Drought in West being embraced by some', *USA Today*, 29 May.

Koonce, J., Busch, W. and Czapia, T. (1996) 'Restoration of Lake Erie: contribution of water quality and natural resource management', *Canadian Journal of Fish and Aquatic Sciences*, 53(1): 105–12.

Kosk, A. (2001) 'Management issues of the Lake Peipsi/Chudskoe region. Lakes & Reservoirs', *Research and Management*, 6: 231–5.

Levin, P.S. (2003) 'Regional differences in responses of Chinook salmon populations to large-scale climatic patterns', *Journal of Biogeography*, 30: 711–17.

McCullough Associates and Diane Mackie and Associates (2002) *Ontario Marina Impact Survey, Final Report*, report prepared for the International Lake Ontario–St. Lawrence River Study, Ottawa, Ontario: McCullough Associates and Diane Mackie and Associates.

McGarrigle, M.L. and Champ, W.S.T. (1999) 'Keeping pristine lakes clean: Loughs Conn and Mask, western Ireland', *Hydrobiologia*, 395/396: 455–69.

Magnuson, J. (1998) 'Regional climate change and fresh water ecology', paper presented at the *Upper Great Lakes Regional Climate Change Impacts Workshop*, University of Michigan, May.

Michigan Sea Grant College Program (1992) *Zebra Mussels in the Great Lakes*, East Lancing, MI: Michigan State University.

Migliore, G. (2004) 'Theft of shipwreck artefacts may damage Michigan maritime tourism', *Capital News Service*, 1 October.

Minns, C. and Moore, J. (1992) 'Predicting the impact of climate change on the spatial pattern of freshwater fish yield capability in eastern Canadian lakes', *Climatic Change*, 22: 327–46.

Mortsch, L. (1998) 'Assessing the impact of climate change on the Great Lakes shoreline wetlands', *Climatic Change*, 40: 391–416.

Mortsch, L., Alden, M. and Scheraga, J. (2003) 'Climate change and water quality in the Great Lakes region – risks, opportunities and responses', in *Climate Change and Water Quality in the Great Lakes Basin*, Detroit, MI: Great Lakes Quality Board, International Joint Commission.

Mortsch, L., Hengeveld, H., Lister, M., Logfren, B., Quinn, F., Slivitzky, M. and Wenger, L. (2000) 'Climate change impacts on the hydrology of the Great Lakes–St. Lawrence system', *Canadian Water Resources Association Journal*, 25(2): 153–79.

National Park Service (1999) *Lake Mead National Recreation Area, Business Plan Executive Summary*. Washington, DC: National Park Service, US Department of the Interior.

National Park Service (2003) *Lake Mead*, National Park Service, online. Available at www.nps.gov/lame/whylow.html (accessed 10 August 2004).

Ontario Federation of Hunters and Anglers (1998) *Sea Lamprey, the Battle Continues*, report prepared for the Great Lakes Fishery Commission. Toronto, Ontario: Ontario Federation of Hunters and Anglers.

Palmer, T.N. and Räisänen, J. (2002) 'Quantifying the risk of extreme seasonal precipitation events in a changing climate', *Nature*, 415: 512–4.

Pendelton, L. and Mendelsohn, R. (1998) 'Estimating the economic impact of climate change on the freshwater sportfisheries of the Northeastern US', *Land Economics*, 74(4): 483–96.

Poff, N., Brinson, M. and Day, B. (2002) *Aquatic Ecosystems and Global Climate Change: Potential Impacts on Inland Freshwater and Coastal Wetland Ecosystems in the United States*. Arlington, VA: Pew Centre on Global Climate Change.

Revenga, C., Brunner, J., Henninger, N., Payne, R. and Kassem, K. (2000) *Pilot analysis of global ecosystems: freshwater systems*. Washington, DC: World Resources Institute.

Scheidleder, A., Grath, J., Winkler, G., Stärk, U., Koreimann, C. and Gmeiner, C. (1999) *Groundwater quality and quantity in Europe*, technical Report 22. Copenhagen: European Environment Agency.

Schneider, C., Owens, R., Bergstedt, R. and O'Gorman, R. (1996) 'Predation by sea lamprey (*Petromyzon marinus*) on lake trout (*Salvelinus namaycush*) in southern Lake Ontario, 1982–1992', *Canadian Journal of Fisheries and Aquatic Sciences*, 53(9): 1921–32.

Scott, D., Jones, B., Mills, B., McBoyle, G., Lemieux, C., Svenson, S. and Wall, G. (2002) *The Vulnerability of Winter Recreation to Climate Change in Ontario's Lakelands Tourism Region*, Department of Geography Publication Series 18. Waterloo, Ontario: University of Waterloo.

Scott, D., McBoyle, G., Mills, B., Minogue, A. (2005a) 'Climate change and the sustainability of ski-based tourism in eastern North America: a reassessment', *Journal of Sustainable Tourism* (in press).

Scott, D., Jones, B., and Abi Khaled, H. (2005b). *The Vulnerability of Tourism in the National Capital Region to Climate Change*, Technical Report to the Government of Canada's Climate Change Action Fund. Waterloo, Ontario: University of Waterloo.

Selcraig, B. (1993) 'Green Fees, Whose eagles, which birdies? Nature pays a price for our love affair with golf', *Sierra*, 78:70–7, 86–8.

SSSK (Stockholms Skridskoseglarklubb) (2004) online. Available at www.sssk.se/index.htm (accessed 31 December 2004).

State of Wisconsin (2002) *Navigable Waters, Harbors and Navigation*, Wisconsin Statutes Database, Wisconsin, online. Available at www.legis.state.wi.us/statutes/Stat0030.pdf (accessed 14 October 2004).

Stefan, H.G., Fang, X. and Hondzo, M. (1998) 'Simulated climate change effects on year-round water temperatures in temperate zone lakes', *Climatic Change*, 40: 547–76.

SWECLIM (Swedish Regional Climate Modelling Programme) (2002) online. Available at www.smhi.se/sweclim (accessed 27 February 2005).

The Holland Ring (2004) *Elfstedentoch*, online. Available at www.thehollandring.com/11stedentocht.shtml (accessed 03 January 2005).

Thorp, S. and Stone, J. (2000) *Recreational Boating and the Great Lakes – St. Lawrence Region, A Feature Report*. Ann Arbor, MI: Great Lakes Commission.

Tockner, K. and Stanford, J.A. (2002) 'Riverine flood plains: present state and future trends', *Environmental Conservation*, 29(3): 308–30.

US Department of the Interior (2003) *Water 2025: Preventing Crises and Conflict in the West*. Washington, DC: US Department of the Interior.

US Environmental Protection Agency (1995) *Ecological Impacts from Climate Change: An Economic Analysis of Freshwater Recreational Fishing*, Report No. 220–R–95–004. Washington, DC: US Environmental Protection Agency.

US National Assessment (2000) *Climate Change Impacts on the United States: The Potential Consequences of Climate Variability and Change – Northeast*. Washington, DC: US Global Change Research Program.

Wall, G. (1998) 'Implications of global climate change for tourism and recreation in wetland areas', *Climatic Change*, 40: 371–89.

Walsh, J. (2004) 'War over water', *Golf Course News*, October.

Williams, G., Layman, K.L. and Stefan, H.G. (2004) 'Dependence of lake ice covers on climatic, geographic and bathymetric variables', *Cold Regions Science and Technology*, 40: 145–64.

Winfield, I.J. and Durie, N.C. (2004) 'Fish introductions and their management in the English Lake District', *Fisheries Management and Ecology*, 11: 195–201.

World Wide Fund for Nature (WWF) (2001) *Tourism threats in the Mediterranean*, WWF Background information, Gland: WWF Switzerland.

Xu, C.-Y. (2000) 'Modelling the effects of climate change on water resources in central Sweden', *Water Resources Management*, 14: 177–89.

Zimmer, M. (2004) 'Water restrictions impact golf courses in the desert southwest', *Golf Course News*, September.

5 Tourism and forest ecosystems

Stefan Gössling and Thomas Hickler

Introduction

Forest ecosystems are of major importance for recreation and tourism and attract millions of visitors every day, particularly in the mid-latitudes. Global environmental change will affect forest ecosystems through land-use changes, changing temperatures and precipitation patterns, increasing CO_2 concentrations, and nitrogen deposition (Sala *et al.* 2000). This might, in the medium-term future, have consequences for tourism and recreation. Forests, in particular in the tropics, also host a majority of the world's biodiversity (Myers *et al.* 2000), with many individual species being of great importance for tourism. As biodiversity is under serious stress through global climate change (e.g. Thomas *et al.* 2004), loss of species might also affect forest tourism. The chapter provides an assessment of global environmental change affecting forest ecosystems and outlines the consequences for tourism and recreation.

Tourism in forest ecosystems

Worldwide, forests provide attractive scenery for recreation and tourism. Mid-latitude forests in Europe and the United States, for instance, have for centuries been recognised for their recreational benefits (e.g. Carhart 1920). In countries like Austria, Switzerland, Sweden, the UK, Japan, Australia or New Zealand, they are an important element of the landscapes that attract millions of domestic and international tourists each year (see Bostedt and Mattsson 1995; McShane and McShane-Caluzi 1997; Hall and Higham 2000; Kearsley 2000; Knight 2000). In mid-latitude forests, an increasing number of consumptive and non-consumptive activities are carried out by urban populations in search of recreational spaces. Activities include, for example, walking, hiking, mountain-biking, cross-country skiing, dog-sledding, fishing, hunting, bird watching, or mushroom and berry collection (Table 5.1). Such is the level of participation in forest-based recreational activities that some authors have started to describe forests as 'sites of mass recreation' (Knight 2000: 341, in the context of Japanese forests).

In many countries, the growing importance of forests for recreation and tourism has even been an argument for promoting forest expansion (e.g. Broadhurst and

Table 5.1 Forest-based activities

• Animal and plant observation	• Orienteering
• All-terrain vehicle driving	• Outdoor education
• Bird watching	• Painting
• Caving	• Photography
• Canoeing/kayaking	• Picking mushrooms, berries, flowers and herbs
• Collection of other forest products	
• Cross-country skiing	• Picnicking
• Dirt bike driving	• River and lake fishing
• Dog-sledding	• Scenic drives along mountain ridges or through forests
• Fishing (lakes, rivers)	
• Hiking	• Snowmobiling
• Horse riding	• Snowshoeing
• Hunting	• Survival training
• Mountain biking	• Technical rock climbing
• Mushroom and berry collection	• Walking
• Nature observation	• Watching animals
• Nature photography	• White water rafting

Source: Bostedt and Mattsson 1995, McShane and McShane-Caluzi 1997, Kearsley 2000, Knight 2000, Vail and Hultkrantz 2000

Harrop 2000; Ghimire 1994). Even northern boreal forests, which are less renowned for recreation, have gained importance as hiking and adventure tourism destinations, as well as tropical rainforests, which are becoming increasingly popular as tourist destinations (Lindberg *et al.* 1998; Gössling 1999). Some of the most well-known national parks in the world, such as the Monte Verde Cloud Forest in Costa Rica or the Volcanoes National Park in Rwanda, are located in the tropics. Even though the overall number of visitors being attracted by rainforests is small, rising tourist numbers and increasing interest in nature-based vacations have turned even lesser known reserves and parks in the tropics into tourist magnets. The attractiveness of protected areas in the tropics will often be attributed to certain 'charismatic' animals and a diversity of observable, spectacular, dangerous or colourful species is a precondition for rainforest tourism (Roth and Merz 1997). However, the very term 'tropical rainforest' seems to increasingly have the potential to attract visitors because, like no other ecosystem, tropical rainforests stand for the western conception of pristine wilderness, exotic fauna and flora, and the diversity of life.

Charismatic forest fauna includes large mammals, such as any kind of monkey, jaguar, leopard or tiger, or in the northern latitudes, bear, beaver, reindeer or moose. Other appreciated animals are colourful or peculiar birds, such as hummingbirds, chameleons, lizards, iguanas, toads, frogs, snakes, ants, bats, butterflies and bugs. In Madagascar, for example, the giraffe-necked weevil (*Trachelophorus giraffa*), a tiny insect with an un-proportionally long neck, is a major attraction in the Ranomafana National Park. Species such as these also have the power to create and maintain narratives of evolution and diversity which explain part of the attractiveness of tropical rainforests.

In the northern latitudes, rather fewer species might be attractive for tourism, but these species seem disproportionally important. The moose in Northern Europe, for example, is not only a symbol of boreal forests, but a symbol of Scandinavia. Without this species – which appears in many tales and on countless souvenir items – northern forests and Scandinavia itself might lose much of their mystical power and tourist appeal. It should be noted, though, that certain species might also have the potential to scare away tourists. Ticks, for example, have become increasingly common in European forests, carrying and transmitting potentially life-threatening diseases such as tick-borne encephalitis or Lyme disease. This can be seen as a result of climate change (Gustafson and Lindgren 2001) and might make people more reluctant to visit forests. For some areas, there might even be explicit warnings (e.g. USDA Forest Service 2004).

Plants species of importance for tourism usually have aesthetic, nutritional or medical properties. In Europe, for instance, flowering plants covering forest floors attract large visitor numbers in spring, while in autumn, mushroom collection has great touristic appeal. In Japan, visitors are also known to collect herbs (Knight 2000). The structure of forest ecosystems is also of importance for tourism, as indicated by the attractiveness of certain forest types such as the Red Wood forests in the USA or the Alerce forests in Chile. Mangrove trails are popular tourist sights in the tropics, and savannah shrubs are essential elements of some African landscapes and wildlife-tourism experiences. Arboreta and botanical gardens are also important sights, often located in proximity to or within cities.

The reasons for the growing interest in forest tourism are multiple. As Urry (1995) remarks, tourism is increasingly built on the marketing of nature and the natural, which have become central elements of travel. Nature has, in many contexts, become a playground for adventure and experience-seeking tourists (see Gössling 2005; Gyimóthy and Mykletun 2004). Overall, tourists seem more environmentally aware and there is a general trend towards more educative and challenging vacations (see Urry 1995, Lindberg *et al.* 1998). This development seems to be self-reinforcing, because environmental consciousness comes into existence through education, increased media attention and the comparison of the character of the physical and built environment of different places through travelling (Urry 1995). The conclusion would be that the relationship of environmental awareness and travel is a self-reinforcing one, because a heightened environmental consciousness will lead to more travel, while more travel will in turn lead to increased environmental awareness (see Gössling 2002). Due to their attractiveness for recreation, forests play an important role in this process.

In industrialised countries, forests also have important educative, spiritual and religious roles, and they might often function as links between urbanised and industrialised societies and the natural environment. This might go along with processes of mystification and romanticisation of forests (see e.g. Knight 2000 for Japanese forests). On a more proximate level, easier access to remote destinations, better information through the internet and greater travel experience of a growing number of tourists might also explain the observed growth in nature-based tourism, particularly in the tropics (see Lindberg *et al*. 1998). It is likely that tourism and recreation in forest areas will increase in the future. This development will, in industrialised countries, be a result of the wish to recover from daily urban life, and in developing countries due to a growing interest in nature tourism by both domestic and international tourists (see Font and Tribe 2000; Ghimire 2001).

Forest tourism has a substantial economic value. On a global scale, nature tourism – i.e. tourism based on natural areas – might generate as much as 7 per cent of all international travel expenditure (Lindberg *et al*. 1998). A proportional share of the turnover from international tourism can therefore be attributed to the existence of ecosystems, often forests. In Costa Rica, for instance, more than 50 per cent of all tourists visited at least one protected area (usually forests) during their stay in 1988, and about 40 per cent stated that protected areas were important or primary reasons for choosing the country as destination (Boo 1990). Similar is true for Ecuador, with 65 per cent of all tourists stating that protected areas are an important or primary reason for their travel choice, and with 75 per cent having visited at least one protected area (Boo 1990). Broadhurst and Harrop (2000) cite the UK Day Visits Survey (SCPR 1998), which suggests that some 350 million day visits are made each year to woodlands in the UK alone. On average, day visitors spend £3.20, resulting in an estimated expenditure on woodland recreation of more than £1 billion. Direct revenues from forest-based tourism can be substantial with, for example, entrance fees of up to US$250 per individual per day being paid for visiting gorillas in Rwanda and Uganda (Wilkie and Carpenter 1999).

Not all forest values are captured in markets (see Broadhurst and Harrop 2000). Forests support ecological functions essential to humanity such as the maintenance of the global carbon cycle (Costanza *et al*. 1997; Daily 1997). Forests also have a range of other values that are outside the market, usually resulting in underestimations of their value in cost–benefit analyses and decision making, which in turn often has led to clear-cutting, particularly in the tropics (Myers *et al*. 2000). This is of importance because tourism is often giving economic value to otherwise 'useless' natural areas (see Gössling 1999; Hall and Higham 2000). The privately owned 21,750ha Haliburton Forest and Wildlife Reserve in Ontario, Canada, for example, earns 67 per cent of its total revenue from tourism-related activities (Sandberg and Midgley 2000). Even in industrialised countries, where access to forests is usually free of charge and generates only little value to forest owners (Font and Tribe 2000), forest tourism can generate income. For example, guesthouses, hotels and restaurants in proximity to well-known national parks and other protected areas will usually profit from tourism and, in many countries, guided tours, or experience-packages such as beaver or moose safaris, or souvenir selling industries have developed. The latter include wildlife farming for meat or the cultivation of medicinal herbs (Knight 2000). Forest tourism may also develop rather unusual forms. For example, Knight (2000) reports that some municipalities in Japan have developed 'tourist forestry', where tourists plant, weed, prune or thin timber plantations.

From an environmental point of view, forest tourism can be problematic because it usually concentrates on rather limited, 'attractive' areas. Problems of heavy use involve the disturbance of wildlife, trampling of vegetation, forest fires caused by camp fires or cigarette butts, erosion of soil and impacts of cars through off-road driving and emissions of different trace gases (Knight 2000).

Forests and climate change

The distribution of major forests types, such as tropical forests and boreal forests, is strongly controlled by climatic variables, in particular temperature and moisture (e.g. Prentice *et al*. 1992). Cold winters, short growing seasons and drought are environmental factors that limit the distribution of forests (Woodward 1987; Prentice *et al*. 1992). Winter temperatures are, in particular in the north, projected to increase by several degrees. Parts of the Arctic could be up to 10°C warmer by 2100 (IPCC 2001). Such changes in temperature would enhance forest productivity in many northern areas, and trees would be able to expand their ranges to the north and to higher altitudes. The observed warming of the northern hemisphere between 1982 and 1999 has already led to increased forest growth (Lucht *et al*. 2002). Furthermore, temperate trees would become more competitive in nowadays-boreal forests, but it is largely uncertain at what rate competitive replacement of different forest types may occur (Solomon 1997). Climatic change is projected to occur so fast that most forests will rather become non-adapted to the new climate than change toward a new composition (Davis and Shaw 2001). Non-adapted trees may suffer from decreased vigour and increased sensitivity to, for example, storms and pest outbreaks (e.g. Bradshaw *et al*. 2000).

Changes in rainfall patterns could also have a substantial impact on forests. Prolonged droughts can severely damage or kill large forest tracts, either directly, or indirectly through increased fire intensities (Bachelet *et al*. 2003). Dynamic Global Vegetation Models (DGVMs) have been developed to simulate large-scale climate impacts on vegetation and ecosystems (Cramer *et al*. 2001). For specific climate change scenarios, DGVMs have projected substantial forest dieback in the tropics (Cox *et al*. 2000; White *et al*. 2000; Schaphoff *et al*. 2005), as well as in south-eastern USA (Bachelet *et al*. 2003) and boreal forests (Joos *et al*. 2001). However, whether large-scale forest dieback is projected by the models depends on CO_2 emission scenarios (White *et al*. 2000) and on simulated changes in precipitation patterns (Bachelet *et al*. 2003; Schaphoff *et al*. 2005), the latter differing substantially between climate models (IPCC 2001; Bachelet *et al*. 2003; Schaphoff *et al*. 2005). Therefore, it is highly uncertain if such strong impacts will occur.

The most severe impacts are simulated to occur after 2050. However, even though general trends in climate change may be unlikely to cause large-scale forest dieback in the near future, it is uncertain to what extent forests will be disturbed by the general climatic trends in combination with more extreme weather events, including higher frequencies of windstorms (Leckenbusch and Ulbrich 2004), and prolonged drought periods (Semmler and Jacob 2004), which could promote forest fires.

Nitrogen and CO_2 fertilization

Further important aspects of global environmental change are increasing concentrations of CO_2 in the atmosphere and increased nitrogen depositions. Both CO_2 (e.g. Farquhar *et al.* 1980; Norby *et al.* 1999) and nitrogen (e.g. McGuire *et al.* 1992) are limiting resources for plants, and anthropogenic emissions are thought to have increased forest productivity (Spiecker *et al.* 1996). Nitrogen depositions have been rapidly increasing during the last 50 years (Vitousek *et al.* 1997; Holland *et al.* 1999) – in many areas, in particular in Europe, to levels at which trees suffer from negative side-effects, such as acidification, leaching of cations, and nutrient imbalances (Schulze 1989; Aber *et al.* 1995; Vitousek *et al.* 1997). In coniferous evergreen forests in Europe, nitrogen depositions have caused forest decline and increased tree mortality (Schulze 1989), and deposition rates will generally further increase in the future (Galloway 2001).

In the case of CO_2, concentrations of 580 to 970 parts per million (ppm) by year 2100 (IPCC 2001) could lead to substantial fertilisation effects (Cramer *et al.* 2001; Long *et al.* 2004). Over the time span of a couple of years, elevating atmospheric CO_2 by 200ppm has been documented to increase productivity of young trees by about 25 per cent (e.g. Norby *et al.* 1999; Hamilton *et al.* 2002; Nowak *et al.* 2004). Elevated CO_2 could have particularly beneficial effects in dry areas, because many plants respond to elevated CO_2 by decreased leaf conductance, which decreases water losses through transpiration (Drake *et al.* 1997; Long *et al.* 2004). However, it is highly uncertain to what extent other resources, such as nitrogen, phosphorus and cations will constrain CO_2 fertilisation effects (Finzi *et al.* 2002; Hungate *et al.* 2003). Likewise, it is not known how mature forests, where, for example, competition for light can strongly limit growth, will respond to elevated CO_2 (Norby *et al.* 1999).

Land-use changes

Between 20 and 30 per cent of natural forests have been replaced by anthropogenic land-use types, such as agriculture and pastures, and forest losses have been most severe in temperate and warm non-tropical forests (Klein Goldewijk 2001). Today, northern and mid-latitude forests are in a rather stable condition (Dixon *et al.* 1994). In Europe, the total forested area is projected to increase because of abandonment of agricultural areas (Kankaanpää and Carter 2004), but tropical forest are under serious threat. Clearing of tropical humid forests eliminates about one million km^2 every five to 10 years, with burning and selective logging severely damaging several times the cleared area (Pimm and Raven 2000).

Biodiversity

Substantial extinction of species may occur in the near future because of global environmental change (Sala *et al.* 2000). Within a wide range of taxa, such as plants, birds, reptiles and butterflies, it is estimated that climate change alone will cause 15–37 per cent of the species to become committed to extinction by 2050 (Thomas *et al.* 2004; Bakkenes *et al.* 2002). The estimates vary in relation to the climate change scenario and assumptions concerning species' ability to disperse to new habitats if the climate in their old habitat is no longer suitable (Thomas *et al.* 2004). Note that these scenarios only consider the suitability of the climate in a given location; this excludes changes in available habitat that occur as a result of human land-use change, or changes in habitat quality caused by nitrogen deposition or elevated CO_2. In many areas, species diversity could become more severely affected by human land-use change and nitrogen deposition than by climate change (Sala *et al.* 2000; Thomas *et al.* 2004).

Nitrogen deposition generally decreases biodiversity by favouring the most responsive species, which often outcompete rare species confined to nutrient-poor habitats (Mooney *et al.* 1999, Sala *et al.* 2000). In the Netherlands, for example, which has the highest nitrogen deposition rates in the world, nitrogen deposition causes the conversion of heathlands to species-poor grasslands and forests (Aerts and Berendse 1988).

Elevated atmospheric CO_2 will also strongly affect biodiversity (Sala *et al.* 2000). Some species will respond more strongly to elevated CO_2 than others, which will affect the competitive balance between species currently coexisting in ecosystems (Mooney *et al.* 1999). As elevated CO_2 changes the chemical composition of produced biomass, which is consumed at higher trophic levels, not only primary producers but the whole food-web will be affected (e.g. Percy *et al.* 2002). Nevertheless, predicting the outcome of these effects on species composition in ecosystems remains difficult (Mooney *et al.* 1999; Poorter and Navas 2003). Because current CO_2 concentrations never have been exceeded during the past 26 million years (Long *et al.* 2004) – the period during which current plant life evolved – no past analogue exists for future conditions and its consequences for ecosystem composition and functioning. In general, however, rare habitat specialists are likely to suffer more from global change than widespread generalists (Warren *et al.* 2001; Travis 2003), and it is usually the former, often endemic species, which are more important for tourism.

Conclusion

The chapter has outlined the importance of forest ecosystems for tourism and recreation. It is clear that forests are likely to see substantial changes in their extension, structure and species composition. This might in particular be true for the northern latitudes, where the largest temperature increases are projected. Forest growth will probably be enhanced in many areas in the north, while elevated CO_2 could have particularly beneficial effects on forest growth in dry areas. However, climate change will occur so fast that many forests will not be adapted to the new climate, and during the second part of the century this could cause large-scale forest dieback. Furthermore, forests will generally be more likely to suffer from extreme climate events and, in most areas, forests will host a smaller number of species.

Mid-latitude forests are most important in terms of volume, because they provide the scenery for recreation in large parts of Europe, North America, Australia, New Zealand and Japan. Access to forest ecosystems is free of charge in industrialised countries (e.g. Broadhurst and Harrop 2000) and some nations, such as Sweden, even explicitly grant a Right of Public Access (Bostedt and Mattsson 1995). The economic consequences of losing species and forest landscapes might thus rather be indirect in these countries, including forgone opportunities for recreation and tourism. However, indirect tourism-related economic losses could be substantial because, for example, Switzerland would lose much of its attractiveness without its forest landscapes. Similar is true for Sweden, where the loss of charismatic species such as moose could be of great importance. Forest loss will also have cultural and societal consequences. This is because knowledge about ecosystems, production processes and the physical environment are considered to be central to sustainable development (Borgström-Hansson and Wagernagel 1999), but are increasingly lost in industrialised societies. Because forests are popular for walks, nature observation and outdoor activities, including extractive activities such as picking mushrooms or berries, hunting or fishing, forest-related activities might often be the last 'authentic' links that people in industrialised countries have to the natural environment. Forests thus have important educative, and even spiritual and religious functions (see e.g. Knight 2000, for Japanese forests). Hence, reduced access to forest ecosystems might affect the perception of and knowledge of the environment in industrialised societies. In this context it is worth mentioning that forest protected areas in many western European countries have the explicit task of mediating nature-related knowledge, compensating for the loss of 'traditional' knowledge.

In the tropics, forest national parks and nature reserves are often essential sources of income (see Langholz 1996; Gössling 1999) and their revenue-generating properties are one of the strongest arguments for their conservation. Should tourism to these areas decline, important income might be lost. Note, however, that this relationship has two sides, because tourism itself is an important contributor to climate change (Gössling et al. 2005). Loss of revenue from protected areas might become even more important in areas where a substantial part of local livelihoods depends on forest products such as meat, timber or herbs, which might also decline. Obviously, the loss of forests will have severe consequences for people in the tropics, with some 200–300 million people being directly dependent on these ecosystems for their livelihoods (Myers et al. 2000).

In conclusion, global change will considerably alter forest structure, functioning and biodiversity, including the possibility of large-scale forest damage or dieback, but it is currently not possible to predict how forest-based tourism in particular will be affected. Scientists can only provide scenarios of global change impacts, i.e. a wide range of possible developments. The future use of forests for tourism will also depend on the tourists' aesthetic perception of landscapes and forest types, their perception of damage caused by weather extremes, the loss of charismatic species and, potentially, their risk perception if disease-carrying vectors become more abundant.

References

Aber, J. D., Magill, A., McNulty, S.G., Boone, R.D., Nadelhoffer, K.J., Downs, M. and Hallett, R. (1995) 'Forest biogeochemistry and primary production altered by nitrogen saturation', *Water, Air, and Soil Pollution*, 85: 1665–70.

Aerts, R. and Berendse, F. (1988) 'The effect of increased nutrient availability on vegetation dynamics in wet heathlands', *Vegetatio*, 76: 63–9.

Bachelet, D., Neilson, R.P., Hickler, T., Drapek, R.J., Lenihan, J.M., Sykes, M.T., Smith, B., Sitch, S. and Thonicke, K. (2003) 'Simulating past and future dynamics of natural ecosystems in the United States', *Global Biochemical Cycles*, 17: 1045.

Bakkenes, M., Alkemade, J.R.M., Ihle, F., Leemans, R. and Latour, J.B. (2002) 'Assessing effects of forecasted climate change on the diversity and distribution of European higher plants for 2050', *Global Change Biology*, 8: 390–407.

Boo, E. (1990) *Ecotourism: The potentials and pitfalls*, vol. 1 and 2. Washington, DC: WWF.

Borgström-Hansson, C. and Wackernagel, M. (1999) 'Rediscovering place and accounting space: how to re-embed the human economy', *Ecological Economics*, 29(5): 203–13.

Bostedt, G. and Mattsson, L. (1995) 'The value of forests for tourism in Sweden', *Annals of Tourism Research*, 22(3): 671–80.

Bradshaw, R.H.W., Holmqvist, B., Cowling, S.A. and Sykes, M.T. (2000) 'The effects of climate change on the distribution and management of *Picea abies* in southern Scandinavia', *Canadian Journal of Forestry Research*, 30: 1992–8.

Broadhurst, R. and Harrop, P. (2000) 'Forest Tourism: Putting Policy into Practice in the Forestry Commission', in X. Font and J. Tribe (eds) *Forest Tourism and Recreation. Case Studies in Environmental Management*. New York: CABI Publishing, pp.183–99.

Carhart, A.H. (1920) 'Recreation in the forests', *American Forests*, 26: 268–72.

Costanza, R., d'Arge, R., de Groot, R., Farber, S., Grasso, M., Hannon, B., Limburg, K., Naeem, S., O'Neill, R.V., Paruelo, J., Raskin, R.G., Sutton, P. and Van den Belt, M. (1997) 'The value of the world's ecosystem services and natural capital', *Nature*, 387: 253–60.

Cox, P.M., Betts, R.E., Jones, C.D., Spall, S.A. and Totterdell, I.J. (2000) 'Acceleration of global warming due to carbon-cycle feedbacks in a coupled climate model', *Nature*, 408: 185–7.

Cramer, W., Bondeau, A., Woodward, F.I., Prentice, I.C., Betts, R.E., Brovkin, V., Cox, P.M., Fisher, V., Foley, J.A., Friend, A.D., Kucharik, C., Lomas, M.R., Ramankutty, N., Sitch, S., Smith, B., White, A. and Young-Molling, C. (2001) 'Global response of terrestrial ecosystem structure and function to CO_2 and climate change: results from six dynamic global vegetation models', *Global Change Biology*, 7: 357–73.

Daily, G.C. (ed.) (1997) *Nature's Services. Societal Dependence on Natural Ecosystems*. Washington, DC: Island Press.

Davis, M. and Shaw, R.G. (2001) 'Range shifts and adaptive responses to quarternary climate change', *Science*, 292: 673–8.

Dixon, R.K., Brown, S., Houghton, R.A., Solomon, A.M., Trexler, M.C. and Wisniewski, J. (1994) 'Carbon pools and flux of global forest ecosystems', *Science*, 263: 185–90.

Drake, B.G., Gonzales-Meler, M.A. and Long, S.P. (1997) 'More efficient plants: a consequence of rising atmospheric CO_2?', *Annual Reviews of Plant Physiology and Plant Molecular Biology*, 48: 609–39.

Farquhar, G.D., Von Caemmerer, S. and Berry, J.A. (1980) 'A biochemical model of photosynthetic CO_2 assimilation in leaves of C_3 plants', *Planta*, 149: 78–90.

Finzi, A.C., DeLucia, E.H., Hamilton, J.G., Richter, D.D. and Schlesinger, W.H. (2002) 'The nitrogen budget of a pine forest under free air CO_2 enrichment', *Oecologia*, 132: 567–78.

Font, X. and Tribe, J. (eds) (2000) *Forest Tourism and Recreation. Case Studies in Environmental Management*. New York: CABI Publishing.

Galloway, J.N. (2001) 'Acidification of the world: natural and anthropogenic', *Water, Air, and Soil Pollution* ,130: 17–24.

Ghimire, K.B. (1994) 'Parks and people: livelihood issues in national parks management in Thailand and Madagascar', *Development and Change*, 25: 195–229.

Ghimire, K.B. (2001) *The Native Tourist: Mass Tourism within Developing Countries*. London: Earthscan Publications.

Gössling, S. (1999) 'Ecotourism – a means to safeguard biodiversity and ecosystem functions?', *Ecological Economics*, 29: 303–20.

Gössling, S. (2002) 'Human-environmental relations with tourism', *Annals of Tourism Research*, 29(4): 539–56.

Gössling, S. (2005) 'Ecotourism as experience-industry', in M. Kylänen (ed.) *Articles on Experiences 2*. Lapland Centre of Expertise for the Experience Industry, pp.28–39.

Gössling, S., Peeters, P., Ceron, J.-P., Dubois, G., Pattersson, T. and Richardson, R. (2005) 'The eco-efficiency of tourism', *Ecological Economics* (in press).

Gustafson, R. and Lindgren, E. (2001) 'Tick-borne encephalitis in Sweden', *The Lancet*, 358 (9275): 16–8.

Gyimóthy, S. and Mykletun, R.J. (2004) 'Play in adventure tourism. The case of Arctic trekking', *Annals of Tourism Research*, 31(4): 855–78.

Hall, C.M. and Higham, J. (2000) 'Wilderness Management in the Forests of New Zealand: Historical Development and Contemporary Issues in Environmental Management', in X. Font and J. Tribe (eds) *Forest Tourism and Recreation. Case Studies in Environmental Management*. New York: CABI Publishing, pp.143–60.

Hamilton, J.G., DeLucia, E.D., George, K., Naidu, S.L., Finzi, A.C. and Schlesinger, W.H. (2002) 'Forest carbon balance under elevated CO_2', *Oecologia*, 131: 250–60.

Holland, E.A., Dentener, F.J., Braswell, B.H. and Sulzman, J.M. (1999) 'Contemporary and pre-industrial global reactive nitrogen budget', *Biogeochemistry*, 46: 7–43.

Hungate, B.A., Dukes, J.S., Shaw, M.R., Luo, Y. and Field, C.B. (2003) 'Nitrogen and climate change', *Science*, 302: 1512–3.

IPCC (Intergovernmental Panel of Climate Change) (2001) *Climate change 2001: the Scientific Basis*, contribution of the working group I to the third assessment report of the Intergovernmental Panel of Climate Change. Cambridge: Cambridge University Press.

Joos, F., Prentice, I.C., Sitch, S., Meyer, R., Hooss, G., Plattner, G.K., Gerber, S. and Hasselmann, K. (2001) 'Global warming feedbacks on terrestrial carbon uptake under the Intergovernmental Panel of Climate Change (IPCC) emission scenarios', *Global Biochemical Cycles*, 15: 891–907.

Kankaanpää, S. and Carter, T. (2004) *Construction of European Forest Land Use Scenarios for the 21st Century*. Helsinki: Finnish Environment Institute.

Kearsley, G. (2000) 'Balancing tourism and wilderness qualities in New Zealand's native forests', in X. Font and J. Tribe (eds.) *Forest Tourism and Recreation. Case studies in environmental management*. New York: CABI, pp.75–92.

Klein Goldewijk, K. (2001) 'Estimating global land use change over the past 300 years: the HYDE database', *Global Biogeochemical Cycles*, 15: 417–33.

Knight, J. (2000). 'From timber to tourism: recommoditizing the Japanese forest', *Development and Change*, 31: 341–359.

Langholz, J. (1996) 'Economics, objectives, and success of private nature reserves in Sub-Saharan Africa and Latin America', *Conservation Biology*, 10(1): 271–80.

Leckenbusch, G.C. and Ulbrich, U. (2004) 'On the relationship between cyclones and extreme windstorm events over Europe under climate change', *Global and Planetary Change*, 44(1–4): 181–93.

Lindberg, K., Furze, B., Staff, M. and Black, R. (eds) (1998) *Ecotourism in the Asia-Pacific Region: Issues and Outlook*. Rome, Bangkok: FAO/USDA Forest Service/The Ecotourism Society.

Long, S. P., Ainsworth, E.A., Rogers, A. and Ort, D.R. (2004) 'Rising atmospheric carbon dioxide: plants FACE the future', *Annual Review of Plant Biology*, 55: 591–628.

Lucht, W., Prentice, I.C., Myneni, R.B., Sitch, S., Friedlingstein, P., Cramer, W., Bousquet, P., Buermann, W. and Smith, B. (2002) 'Climatic control of the high-latitude vegetation greening trend and pinatubo effect', *Science*, 296: 1687–9.

McGuire, A.D., Melillo, J.M., Joyce, L.A., Kicklighter, D.W., Grace, A.L., Moore III, B. and Vorosmarty, C.J. (1992) 'Interactions between carbon and nitrogen dynamics in estimating net primary productivity for potential vegetation in North America', *Global Biogeochemical Cycles*, 6: 101–24.

McShane, T.O. and McShane-Caluzi, E. (1997) 'Swiss forest use and biodiversity conservation', in C. Freese (ed.) *Harvesting Wild Species: Implications for Biodiversity Conservation*. Washington, DC: Island Press, pp.132–66.

Mooney, H.A., Canadell, J., Chapin III, F.S., Ehrlinger, J.R., Körner, C., McMurtrie, R.E., Parton, W.J., Pitelka, L.F. and Schulze, E.D. (1999) 'Ecosystem physiology responses to global change', in B.H. Walker, W.L. Steffen, J. Canadell and J.S.I. Ingram (eds) *Implications of Global Change for Natural and Managed Ecosystems: A Synthesis of GCTE and Related Research*. Cambridge: Cambridge University Press, pp.141–89.

Myers, N., Mittermeier, R.A., Mittermeier, C.G., da Fonseca, G.A.B. and Kent, J. (2000) 'Biodiversity hotspots for conservation priorities', *Nature*, 403: 853–8.

Norby, R.J., Wullschleger, S.D., Gunderson, C.A., Johnson, D.W. and Ceulemans, R. (1999) 'Tree responses to rising CO_2 in field experiments: implications for the future forest plant', *Cell and Environment*, 22: 683–714.

Nowak, R.S., Ellsworth, D.S. and Smith, S.D. (2004) 'Functional responses of plants to elevated atmospheric CO_2: do photosynthetic and productivity data from FACE experiments support early predictions?', *New Phytologist*, 162: 253–80.

Percy, K.E., Awmack, C.S., Lindroth, R.L., Kubiske, M.E., Kopper, B.J., Isebrands, J.G., Pregitzer, K.S., Hendrey, G.R., Dickson, R.E., Zak, D.R., Oksanen, E., Sober, J., Harrington, R. and Karnosky, D.F. (2002) 'Altered performance of forest pests under atmospheres enriched by CO_2 and O_3', *Nature*, 420: 403–7.

Pimm, S.L. and Raven, P. (2000) 'Extinction by numbers', *Nature*, 403: 843–5.

Poorter, H. and Navas, M.L. (2003) 'Plant growth and competition at elevated CO_2: on winners, losers and functional groups', *New Phytologist*, 157: 175–98.

Prentice, C., Cramer, W., Harrison, S., Leemans, R., Monserud, R.A. and Solomon, A.M. (1992) 'A global biome model based on plant physiology and dominance, soil properties and climate', *Journal of Biogeography*, 19: 117–34.

Roth, H. and Merz, G. (1997) *Wildlife Resources. A Global Account of Economic Use*. Berlin, Heidelberg, New York: Springer Verlag.

Sala, O.E., Chapin III, F.S., Armesto, J.J., Berlow, E., Bloomfield, J., Dirzo, R., Huber-Sanwald, E., Huenneke, L.F., Jackson, R.B., Kinzig, A., Leemans, R., Lodge, D.M., Mooney, H.A., Oesterheld, M., Poff, N.L., Sykes, M.T., Walker, B.H., Walker, M. and Wall, D.H. (2000) 'Global biodiversity scenarios for the year 2100', *Science*, 287:1770–4.

Sandberg, L.A. and Midgley, C. (2000) 'Recreation, Forestry and Environmental Management: The Haliburton Forest and Wildlife Reserve, Ontario, Canada', in X. Font and J. Tribe (eds) *Forest Tourism and Recreation. Case Studies in Environmental Management*. New York: CABI, pp.201–24.

Schaphoff, S., Lucht, W., Gerten, D., Sitch, S., Cramer, W. and Prentice, I.C. (2005) *Terrestrial Biosphere Carbon Storage under Alternative Climate Projections*, submitted.

Schulze, E.D. (1989) 'Air pollution and forest decline in a spruce (*Picea abies*) forest', *Science*, 244: 776–83.

SCPR (Social and Community Planning Research) (1998) *UK Leisure Day Visits: Summary of the 1996 Survey Findings*. Cheltenham: Countryside Commission *et al.*

Semmler, T. and Jacob, D. (2004) 'Modeling extreme precipitation events – a climate change simulation for Europe', *Global and Planetary Change*, 44(1–4):119–27.

Solomon, A.M. (1997) 'Natural migration rates of trees: global terrestrial carbon cycle implications', in B. Huntley, W. Cramer, A.V. Morgan, H.C. Prentice and J.R.M. Allen (eds) *Past and Future Rapid Environmental Changes*. Berlin: Springer Verlag, pp.455–68.

Spiecker, H. K., Mielikäinen, M., Köhl, M. and Skovsgaard, J.P. (1996) *Growth Trends in European forests*, European Forest Institute research report no. 5. Berlin-Heidelberg: Springer Verlag.

Thomas, C.D., Cameron, A., Green, R.E., Bakkenes, M., Beaumont, L.J., Collingham, Y.C., Erasmus, B.F.N., de Siqueira, M.F., Grainger, A., Hannah, L., Hughes, L., Huntley, B., Van Jaarsveld, A.S., Midgley, G.F., Miles, L., Ortega-Huerta, M.A., Townsend Peterson, A., Phillips, O.L. and Williams, S.E. (2004) 'Extinction risk from climate change', *Nature*, 427: 145–8.

Travis, J. (2003) *Climate Change and Habitat Destruction: A Deadly Anthropogenic Cocktail*. Proceedings of the British Royal Society B 270: 467–73.

Urry, J. (1995) *Consuming Places*. London: Routledge.

USDA Forest Service (2004) *Warning: We have Ticks in the Black Hills*, online. Available at www.fs.fed.us/r2/blackhills/publications/information/tick_warning.shtml (accessed 8 December 2004).

Vail, D. and Hultkrantz, L. (2000) 'Property rights and sustainable nature tourism: adaptation and mal-adaptation in Dalarna (Sweden) and Maine (USA)', *Ecological Economics*, 35(2): 223–42.

Vitousek, P.M., Aber, J.D., Howarth, R.W., Likens, G.E., Matson, P.A., Schindler, D.W., Schlesinger, W.H. and Tilman, D.G. (1997) 'Human alteration of the global nitrogen cycle: sources and consequences', *Ecological Application*, 7: 737–50.

Warren, M.S., Hill, J. K., Thomas, J.A., Asher, J., Fox, R., Huntley, B., Roy, D.B., Telfer, M. G., Jeffcoate, S., Harding, P., Jeffcoate, G., Willis, S.G., Greatorex-Davies, J.N., Moss, D. and Thomas, C.D. (2001) 'Rapid responses of British butterflies to opposing forces of climate and habitat change', *Nature*, 414: 65–9.

White, A., Cannell, M.G.R. and Friend, A. (2000) 'CO_2 stabilization, climate change and the terrestrial carbon sink', *Global Change Biology*, 6: 817–33.

Wilkie, D.S. and Carpenter, J.F. (1999) 'Can nature tourism help finance protected areas in the Congo Basin?', *Oryx*, 33(4): 332–8.

Woodward, F.I. (1987) *Climate and Plant Distribution*. Cambridge: Cambridge University Press.

6 The coastal and marine environment

Stephen J. Craig-Smith, Richard Tapper and Xavier Font

INTRODUCTION

Since the very beginnings of tourism as a significant human activity the coastal environment has been a major drawcard (Feifer 1986). The marine environment has engendered a sense of wonderment, awe, reverence and fear since humankind evolved, and much of that early mystique has remained to the present. The coastal environment has been an essential element in tourism since Roman times (Ostia outside Rome was one of the earliest coastal resorts) and the coast is as important today as it has been at any time in its history. Demands and expectations may have changed but the coast has retained its human fascination.

Early demand for coasts, islands and reefs from traditional tourists has focused on areas displaying specific geo-environmental characteristics. Traditional tourists (taken here to be mass tourists with demand characteristics commonly displayed over the past 100–200 years) have a high propensity for passive pursuits generally summed up as sun, sea, sand and sex, so the areas and types of coastline developed for tourism activities displayed remarkably uniform characteristics in most parts of the world. The presence of a sandy beach, relatively calm wave conditions for safe bathing, a relatively pollution-free environment and a relatively warm summer climate were important considerations. It is no coincidence to find the early bathing resorts in Europe, many parts of North America, Australia and even in parts of Southern Africa and South America displaying most of these characteristics. With the advent of cheap air travel the geographical spread of demand for the world's coastlines expanded but for many years the type of coastline in tourist demand displayed very similar characteristics. People travelled further afield but expected much the same product on arrival. The mass resorts in Spain, on the north coast of Africa, in many parts of the Caribbean and even in South East Asia have only superficial differences from the traditional resorts of Northern Europe or North America. Accessibility – albeit further distant geographically – good climate, safe bathing conditions, a reasonably clean environment and supportive tourism infrastructure are as significant in these newer resorts as they were with the traditional ones.

More recently tourism has begun to change and post-modern tourists are demanding different characteristics from the coastal environment. Post-modern tourism is still not as common as traditional tourism but it is growing in importance. While it is important to remember that even post-modern tourists have not totally neglected the traditional resorts, they are increasingly seeking new experiences and as a consequence seeking out new coastal, island and reef areas to exploit. The post-modern tourist is generally less interested in passive pursuits and is more likely to be engaged in more active and or educational activities such as surfing, diving, sailing, walking, bird spotting, nature viewing or general exploring. Some, if not all of this wider portfolio of vacation activities is less dependent on a warm climate (most Antarctic visitation is currently confined to the coastal edge) or safe bathing conditions, so not only are more sections of the world's coastlines coming under coastal tourism development but a greater range of coastal environments are being affected.

The coastline is a particularly unique element of the landscape being the edge of the land where it meets the marine environment. It is exceedingly universal and is found in all countries which border seas or oceans and is even present in modified form on the edge of large inland water bodies such as the Great Lakes. However, the coastline in most places is relatively narrow, comprising that band of the Earth where the marine and terrestrial environments meet. This often results in high energy interaction and mega-episodic effects. Furthermore, this narrowness means that much of the world's coastline is ecologically vulnerable because there is limited scope for migration – there may be room to move laterally along the coast but there is very limited scope for plant and animal life to go either far inland or far offshore. This limited scope for eco-migration displays similar characteristics to the issues in mountain areas described in Chapter 3.

Of course, it is not just tourism which is attracted to coastlines. For a wide variety of reasons the world's coastlines are a magnet for a miscellany of human activities from power generation (the presence of large quantities of water for cooling purposes), agriculture (the presence of flat and often good quality land enjoying the ameliorating effects of the oceans on climatic extremes), industry (convenience of export and transshipment of goods and raw materials), urbanisation (attraction of the coast for lifestyle, moderated climate and employment) and increasingly in recent times retirement development with its associated support structure and recreational activities (Timmerman and White 1997). This multiple and competing use of the coastline is significant for environmental change because much of the environmental change is driven directly be these multiple coastal activities. Add to these locally generated changes more general global change affecting the whole of the planet and it becomes immediately obvious that many coastlines are extremely vulnerable (Kuijper 2003). For tourism to survive and prosper in such locations, long-term environmental monitoring and systematic data collection on environment change is not just desirable, it is essential.

To cover all the coastlines of the world and examine all the permutations and combinations of environmental change associated with them is completely beyond the scope of one chapter so set out below is a framework designed to be representative of the more significant and frequently occurring issues. It is not intended to be totally comprehensive or encyclopaedic in nature.

Environmental change

At any one time, a natural system adapts to the environmental processes operating at that time and coastlines are no exception to this rule (Hansom 2001). If, for whatever reason, the current environmental processes change so also will the natural environment; this is the natural order of things which has operated for the past 4500 million years. Environmental change is brought about by many causes and this chapter attempts to examine many of them, but perhaps the most significant change at present is the rise in world atmospheric temperature. For this reason, global climate change is accorded first priority and other aspects of environmental change are examined later.

In the geological past world climatic conditions have been very different from those of today, at times being much warmer than now and at other times much colder. Given the significant climate variations observed within the geological column one might wonder why there is so much concern about current climate change. There are two reasons why current change is viewed with such alarm. First, the rate of current change is occurring at what we believe is a much faster pace than at any previous time in history. Second, current climate change is caused by human and not natural causes. Furthermore, because of the nature of human interference with the chemical composition of the atmosphere, the resultant changes are global and not regional in scale – unlike most other human activity effects such as sand dredging or uncontrolled pollution discharge which may have significant local impact but rarely have global implications.

Because the nature and causes of global climatic change have been discussed elsewhere in this book, this chapter focuses only on the likely effects such warming may have on the coastal environments of the world and what influence those likely effects will have on coastlines and their subsequent human use. Changes due to global climate change are discussed first and other environmental changes of a more local nature are discussed second.

Sea-level rise

Sea levels have been rising for the last 10,000 years, ever since the last glacial ice sheets of the world started to melt, but the current rapid temperature rise is having an accelerating affect. Sea-level rise is due to two different but related factors. As the world's ice sheets melt, thousands of cubic kilometres of water are returned to the oceans thereby increasing their volume which inevitably leads to sea-level rise. But, in addition, as the existing water in the oceans warms up, so it expands and further compounds the problem. Worldwide, the oceans are rising at around 1–22mm per year but this could accelerate in the future. For a variety of geographic factors sea-level rise is not uniform and some parts of the world have witnessed greater rises than others (Doornkamp 1998). The National Tidal Facility at Flinders University in South Australia, for instance, has reported sea level rises in the South Pacific in the order of 25mm per year – over 10 times the global trend. Satellite data show sea-level rises in the region between Papua New Guinea and

Fiji to be nearer 20–30mm per year. Globally, if current trends continue sea level is predicted to rise by another 5 to 12 centimetres some time around 2020.

On upland coasts sea level rise may not be too serious but in many parts of the world tourism development has occurred on lowland coasts as near the ocean as possible. Many of the coral atoll islands of the Pacific, for instance, with average land levels only a matter of one or two metres above high tide level, are threatened with total extinction (Burns 2000). To maintain some existing coastal resorts in the face of rising sea levels may prove prohibitively expensive and require protective measures unacceptable on environmental grounds.

Changing levels of cloudiness

Although not in the same category as sea-level rise, changing levels of cloudiness can be significant for coastal tourism. Because global warming is affecting pressure and wind patterns, the world's rainfall pattern and cloud cover distribution are also changing. Levels of rainfall and daytime temperatures are also intimately linked with levels of cloud cover. The impact of the level of cloud cover will depend on the current state of cloud cover in the area being considered. In cool and cold regions of the world increased cloud cover could act as a deterrent to future tourism because cloud cover reduces the daytime temperatures and creates the impression of gloominess which most traditional tourists dislike (Viner and Agnew 2000; Wall 1992). In hot regions of the world, however, increased cloud may actually be better. Australia for instance has one of the highest incidences of skin melanoma in the world and increased cloudiness in Northern Australia could be a good thing. Increased cloud cover could have an ameliorating affect on maximum daytime temperatures and lessen the effect of increased global temperatures.

Less cloud cover will also have varying effects on tourism. Less cloud cover in cooler climates could be a good thing with much of northern Europe and Canada standing to benefit greatly from such a phenomenon. Southern European coastal destinations are popular with visitors especially from northern Europe. Giles and Perry (1998) suggest that a temperature increase of only 1 or 2°C might remove the need for northern Europeans to travel south for good weather coastal holidays. In warm and hot climates, however, less cover could be a bad thing by increasing daytime temperatures to levels no longer comfortable for the majority of tourists. Increased temperatures on parts of the Great Barrier Reef off northern Australia and on reefs off the Maldives have led to a killing off of some corals which results in what is called coral bleaching. The colour in coral reefs is caused by the presence of small organisms and once these die off the coral reefs lose their colour (Agnew and Viner 2001).

In areas which currently experience a wet regime, such as the west coast of Scotland or the west coast of the South Island of New Zealand, less rain could be a good thing for tourism where the current number of wet days acts as a deterrent to tourism activity. In areas currently deficient in rainfall, however, such as parts of Australia and the Pacific Islands, lower levels of rainfall in the future could be nothing less than catastrophic. In these areas, current water shortages are leading to unsustainable levels of water use for irrigation purposes (Fiji, for instance) and a further lowering of the water table would have very serious consequences, not just for tourism but for entire island communities in the Caribbean and Pacific. The benefits of a heavier rainfall regime will also depend on geographical location. Heavier rainfall can lead to an increased incidence of coastal flooding which is a bad thing, but it could lead to less fossil water use and water importation dependence in drier areas. Probably the only tourism activity which would directly benefit from a greater incidence of rain would be recreational fishing.

Increasing storminess

Temperature warming is contributing to a world increase in cyclones, tropical lows and storms in general. It has been suggested that tropical cyclones could increase in intensity by between 10 and 20 per cent with a doubling of CO_2 in the Earth's atmosphere (Ker 1999). During the second half of the twentieth century the average number of cyclones in the southern Pacific has been seven per year but this is likely to increase to eight in the early decades of this century. Such storms not only bring with them strong destructive winds but the winds raise sea level by pushing surface seawater in front of them and contribute to widespread flooding as well as structural damage. For many coastal tourism structures and facilities the threat of storms is greater than the threat of rising sea level but the two go hand in hand.

The implications of increased storminess for coastal tourism are many and serious. Storms are not just an inconvenience, they can be life threatening. Storms can disrupt transport communications, cause widespread coastal flooding and coastal erosion, and cause considerable structural damage. It is a well-known meteorological phenomenon that coastal cyclones and storms are worst where they cross the coast and that they tend to die out once they cross a land mass. Protection of beaches and coastal infrastructure against storms is technically possible but the cost may well render the project uneconomic and, as with sea-level rise, protection works may prove unsightly and environmentally unacceptable.

Changing glacial activity

In recent years there has been an increasing interest in actively glacial coastlines. This reflects changing demands of tourists in coastal areas as outlined at the beginning of this chapter. With the advent of more and larger coastal cruise ships, active glacial coasts have come under increasing tourist interest. The coasts of northern Canada, Alaska, southern Chile and Antarctica itself have seen a dramatic rise in tourist activity. Further and faster glacial melt will result in less and less of this environment being available for the tourist gaze. While non-active glacial coasts, such as those on Scotland's western seaboard, undoubtedly attract tourists the active glacial coasts have no close substitute. A further reduction in active glacial coastlines will result in cruise ships having to travel closer to the polar regions to view the best glacial scenery. Because most cruises originate in the warmer regions of the world, greater distances will have to be travelled and, therefore, increases in cost and time will have to be expended to see much of this type of environment.

Management of impacts in coastal areas

So far this chapter has concentrated on the global aspects of environmental change. Given the significance of global climate change this is not surprising, but it would be wrong to assume that the only environmental change is as a result of global warming, especially where coasts are concerned. While other changes are more local in character they can nevertheless be significant in certain areas. With the extensive tourism that coastal areas support, and the general trend for expansion of tourism, it is important that sustainability factors are fully incorporated into tourism planning, development and management at all levels, by both the public and private sectors (Tol *et al*.1998; Kuijper 2003). In particular, the public sector has a key role providing and enforcing planning systems for reviewing and approving proposed developments, and for preparation of land development plans that take into account sustainability issues and which incorporate an integrated approach to tourism and coastal and marine management. A range of tools can be used to help implement sustainability within frameworks based on the above factors, which are summarised in Table 6.1. The selection of the appropriate tools to use in any circumstance will depend on the nature and severity of the issues to be addressed.

Information on the main sources of coastal and marine pollution for states with Caribbean coastlines (Hoagland *et al.* 1995) identifies sewage (28 states), oil (22) and mining and industry (20). Construction erosion (9), solid wastes (6) and fertilisers and pesticides (6) also had significant impacts in some states. Tourism is affected by environmental impacts and pollution from other sectors, as well as causing impacts itself. Generally, adverse impacts from the tourism sector tend to be limited to zones just a few kilometres from where tourism takes place, and the evidence available from the mid-1990s confirms that the environmental effects of tourism activities in coastal areas occurred within national boundaries and territorial

Table 6.1 Regulatory instruments

Regulations	Economic approaches	Soft tools
Laws and regulation	Taxes, subsidies and grants	Community programmes, national and local networks
Special status designation	Tradable rights and permits	Tourism ecolabelling
	Deposit-refund schemes	Environmental management systems
	Product and service charges	Certification/award schemes
		Guidelines, treaties and agreements
		Citizenship and education

waters (UNEP 1994, 1997). In the study for UNEP's Caribbean Environmental Programme (CEP), the major categories of effects from tourism were identified as:

- change in sediment loads
- displacement of traditional uses and users
- groundwater depletion and contamination
- physical changes and habitat damage
- solid waste disposal
- toxic chemicals and nutrification from surface runoff
- visual impacts.

Although tourism impacts are generally localised, they are hugely significant both economically – because they occur in the areas which are of high value for tourism – and environmentally – because many of the features that attract tourism are also of major importance for environmental and conservation reasons. Besides wider environmental costs, adverse impacts caused by tourism undermine both public and private investments from local to national levels. Such adverse impacts from tourism can be avoided through proper and effective planning and development.

A further consideration in small islands is that any impacts have proportionately greater effects than in larger land and coastal regions, simply due to their physical features. For example, a coastal development which results in changes to coastal erosion and sediment deposition that extend over a few kilometres may have limited effects if present on a large landmass, but in a small island may affect a substantial part of the island and its coastline, with consequent economic and environmental losses. Four sources of local impacts are considered in more detail. Each sub-section includes a table summarising the main impacts, along with examples of management tools, and regulatory frameworks and options that are available and in use for addressing such impacts.

- Coastal erosion
- Habitat degradation
- Pollution
- Waste handling and management.

Coastal erosion

The coast is a dynamic environment, with erosion as well as build up of material along coasts and shorelines part of a natural cycle of change. Examples of activities sometimes associated with tourism that can affect the rate, pattern and extent of these physical processes are coastal development, beach protection schemes, schemes for replacement of sand from beaches where it is being otherwise eroded, and sand extraction from beaches.

In some cases protection and stabilisation may be required as a consequence of initial inappropriate site choices for coastal infrastructure or properties, or a failure to understand the environmental or physical impacts associated with certain types of coastal features or protection measures. In other cases the structures themselves may be the cause of the problem when these are located within the area of influence of coastal processes. Structures to consider that are mostly associated with tourism are likely to be hotel developments and recreational facilities for marine activities that are built on the shoreline or extend across the intertidal zone. The effects of such structures on coastal processes may be apparent in the immediate vicinity or many kilometres distant. Making planning decisions about such structures within a broader Integrated Coastal Management programme, such as is the case of Barbados, is much more likely to result in sustainable development.

Reefs and mangroves are two marine habitats that provide natural protection against coastal erosion by reducing the impact of wave action on the shore. Mangroves act to stabilise the coastline and aid the accumulation of sediment. Coral reefs act as a natural breakwater, reducing the force of waves that reach the shoreline. Natural damage to these structures, for example, by hurricanes, or damage caused by human activities such as the blasting of channels through reefs or clearing mangroves can affect coastal processes and make coastlines more vulnerable to changes in beach profiles.

Where new structures are required they must be carefully designed to limit any adverse environmental impact, regularly monitored to ensure subsequent achievement of intended function, maintained when integrity or function is impaired and replaced (or removed) when changed circumstances or conditions justify this. For example, beach sand removal is another key impact along coastal areas. The development of resorts and marinas along the coast as a result of tourism development can bring about significant local environmental change. The beach is the natural defence of the land against the power of the sea. In periods of storms, wave energy is expended along the shore line by removing beach sediment offshore. In intervening periods of calm weather that sediment is returned to the beach. This onshore–offshore movement of beach sediment is a natural process and in the long run is by far the most cost efficient. All too frequently, however, beach material is removed for building purposes on the grounds that it will naturally build up again once removal has taken place. Regrettably this is not always the case and the beaches simply disappear. This problem is compounded when coastal development is protected by the construction of sea walls which might protect the area landward of the sea wall for a short time but in the long run simply starves the beach system of one of its natural sources of sand. Marina construction and channel dredging also change the coastal environment by impeding the natural long shore movement of beach sand (Table 6.2).

Habitat degradation

Tourism-related activities may be a primary or a contributory cause of the degradation of coastal and marine habitats, including coral reefs, seagrass beds and mangroves. Degradation of all these habitats has the potential to affect tourism, especially for visitors who choose destinations for their environmental quality and interest. It should however be noted that a wide range of activities cause these

Table 6.2 Coastal erosion

Issues	Examples of management tools	Regulatory frameworks and options
Beach erosion	Zoning to avoid development in vulnerable areas Modelling studies to improve understanding Monitoring programmes to detect changes	National development plans Environmental Impact Assessment
Sediment budget changes (i.e. changes in the balance between build up and erosion of sediments on a beach/coastline) and beach/dune mobility	Potential effects of coastal works and other activities assessed at a scale appropriate to coastal processes (i.e. not just local)	Regional planning
Loss of tourism infrastructure	Long-term planning horizons e.g. 50 years Set-back limits	Strategic Environmental Assessment
Beach and water quality and safety of beach users	Monitoring changes in beach conditions 'Blue Flag' initiative	

types of degradation, only some of which are associated with tourism. They include physical impacts from coastal development, marine and shore-based activities, and water quality issues such as the discharge of effluents. Many of the impacts are associated with other activities (e.g. fisheries, dredging, effluent discharges, coastal development) and management options are tabulated in the other sections.

One of the most obvious signs of the degradation of coral reefs is physical damage, where sections of coral are broken off or abraded, for example, by anchoring on reefs. Smothering of reef areas by algae, and areas of diseased and dead coral are also very obvious signs of degradation although the causes are less visible. They include siltation – where fine particles of sediment suspended in the water are deposited onto reefs and their living corals – or nutrient enrichment – for example, from disposal of sewage at sea, or from agricultural fertilisers that are washed by rainfall and irrigation, off the land and into rivers and coastal waters – which can lead to a more gradual deterioration of the reef structure and associated groups of plants and animals. Additional impacts on the associated communities range from activities such as the collection of reef fish for the aquarium trade through to commercial fisheries. These effects need to be viewed in combination with natural damage, such as that caused by storms, and regional or global effects such as elevated sea temperatures linked to global warming and El Niño conditions.

Seagrass beds and mangroves are highly productive areas acting as nursery grounds and stabilising sediments on the shoreline and sea bottom. In the case of mangroves, deliberate clearance has taken place in many locations, whereas damage to seagrass is more likely to be inadvertent and the indirect result of other activities. In both cases habitats support considerable biodiversity, and damage to them, sometimes as a result of tourism activities, can reduce the overall attractiveness of areas to tourists, as well as causing a loss of environmental services and impacting on fisheries.

Poor water quality is one of the factors that has resulted in the degradation of saline lagoons, estuaries and inlets where water movement and exchange is limited. Pollution can accumulate in the sediments and in conditions of limited water exchange the concentration of pollution may increase. Estuaries can also focus pollution from the watershed.

Further impacts in coastal areas are the result of deforestation. Human removal of large tracts of natural vegetation for agriculture and other economic ventures is having a significant effect on many segments of the world's coastlines. Large scale vegetation removal renders large tracts of the Earth's surface vulnerable to rainfall runoff erosion. Quite apart from the environmental devastation such activity causes inland, the resultant increased sediment load brought down by the rivers to the coastal environment can have catastrophic consequences on the coastal ecology. Suspended sediments in coastal waters make them less attractive to tourists and can kill off much of the existing coastal biomass.

Similar to deforestation, there are impacts arising from coastal vegetation clearance. Coastal development, especially on lowland and tropical coasts, is renowned for its cavalier attitudes to coastal vegetation removal. The desire for all tourists to be able to see the ocean leads developers to remove coastal vegetation, especially mangroves which tend to hug the shoreline and impede sea views. Mangroves are an ideal fish breeding environment and, once removed, entire local fish stocks are affected. Quite apart form this, mangrove removal can lead to local beach erosion and increased mud sediment load being released to the sea (Table 6.3).

Pollution

Sewage, grey water discharges and litter are those most likely forms of pollution to be associated with tourism activity (see *Waste handling and management*). Other types of pollution, while not normally arising from tourism activities, can deter visitors because of aesthetic problems or worries about health and contamination of food and water quality. In each case there are effective ways to deal with the waste causing such pollution to avoid damage to wildlife and the marine environment.

There are many different forms of oil with varying associated risk of pollution and difficulties with clean up. Lighter types of oil (petrol, marine diesel) evaporate rapidly at sea and are quick to degrade, but can cause damage in confined situations such as lagoons and inlets or if they become trapped in sediments. The action of waves can produce an oil/water emulsion which is difficult to clear up while heavier types of oil may eventually form tar balls which sink to the seabed or get washed up on beaches.

Table 6.3 Habitat degradation

Issues	Examples of management tools	Regulatory frameworks and options
Habitat damage and loss through infrastructure development, construction, upgrade and maintenance (e.g. ports)	Building contracts	Environmental Impact Assessment
Impacts of cruise operation	Internal operating standards	Environmental Management Systems Regulatory operating standards
Impacts from marine recreation	Permanent buoys for anchoring in sensitive areas	Zoning

Pesticides in watercourses and coastal waters are most likely to come from diffuse sources on land, which are then carried to the coast in runoff and down watercourses. They include insecticides, herbicides, fungicides and rodenticides. The best known is DDT which is toxic in its own right but is also only broken down slowly, and then only to other persistent derivatives such as DDE which is also toxic to a variety of organisms. Pesticide pollution linked to tourism is mostly associated with management of the grounds of hotels and sports facilities – for example, golf courses.

Use of fertilisers in agriculture and grounds management can also give rise to water pollution by increasing the concentration of nutrients – particularly of nitrates and phosphates – in the water from runoff and leaching of fertilisers. Such increases in nutrient content adversely affect water quality both directly, and by stimulating the growth of algae leading to 'algal blooms'. Algal blooms are often highly damaging to aquatic ecosystems, and reduce water quality so that it is not suitable for swimming: in some cases, they may also excrete toxins. Toxic substances such as heavy metals (for example, mercury, cadmium, lead and copper) may be present in industrial effluents and can be accumulated to harmful levels in marine organisms, as can organic pollutants such as PCBs which are persistent and fat soluble (Table 6.4).

Desalination plants may be an issue in some areas with the discharges depending on the type of plant. The resulting effluent may have an increased salinity – which will also increase the acidity of the water – temperature and organic content, compared to intake water. There is also the possibility of increased concentrations of copper and other metals in the effluent from the use of descaling and antifoam agents. Resulting effects on plants and animals living on the sea floor and on reefs in the vicinity of these inputs may be localised, limited to species that are unable or less able to move from the affected areas, and apparent on a variety of time scales.

Table 6.4 Pollution

Issues	Examples of management tools	Regulatory frameworks and options
Poor water quality	Discharge consents with conditions Treatment plants Catchment management plans	Regulation of discharges Red lists of substances not to be discharged Water quality standards Environmental Management Systems
Acute effects of pollution from accidental spillages and releases	Oil spill contingency plans	National procedures
Chronic effects of pollution from general use of pollution-causing materials	Monitoring programmes Phase-out programmes for specific pollutants Fertiliser management programmes	Red lists
Aesthetic damage	Beach cleaning programmes	

Waste handling and management

Sewage effluent that is discharged into coastal waters may have been subject to various levels or treatment (for example, settlement, filtration, aeration, UV treatment) or be discharged as raw sewage. The inorganic nutrients in sewage, such as nitrogen and phosphorus, increase the nutrient load to inshore waters with consequences such as depletion of oxygen and algal blooms. This is particularly serious in nutrient-poor reef environments. The release of pathogenic micro-organisms in the sewage can be a public health hazard to bathers or through contamination of local fisheries.

Solid waste such as sewage sludge – from treatment plants – and channel dredgings are another potential source of pollution with detrimental effects caused by smothering of the sea floor with sediments, creating anaerobic conditions (where there is no oxygen present in the water), altering the nature of the seabed and increasing the murkiness of water during disposal and possibly for a time afterwards both at and around any disposal site. Litter is another form of solid waste and may be seaborne (e.g. from vessels, lost fishing gear and other marine debris), disposed on the shoreline, or brought down by streams or blown in from nearby landfill sites. Apart from being unsightly, marine litter can kill seabirds and turtles by entanglement and ingestion of plastics.

Managing tourism activities

The impacts of tourism on marine coastal areas are the result of myriad agents for change at the macro level. However, impacts happen at the micro level and it is

Table 6.5 Water handling management

Issues	Examples of management tools	Regulatory frameworks and options
Poor water quality	Discharge consents with conditions Treatment plants Waste disposal systems Grey water and solid waste recycling	Regulation of discharges Water quality standards Environmental Management Systems
High volumes of waste for disposal	Waste separation, reuse and recycling	Regulations for waste management Charges for solid waste disposal Incentives for composting and waste recycling Development of waste recycling infrastructure
Chronic effects from waste materials	Monitoring programmes	Restricting or banning use of specified chemicals and other toxic materials
Aesthetic damage from waste materials	Beach cleaning programmes	

only the accumulation of these micro impacts that creates the macro conditions. The next section of this chapter reviews the impacts and management tools available for public and private sectors in and around coastal areas to contribute to a better quality environment. Tourism activities will not be managed to tackle climate change alone, but a range of impacts, mainly first concentrating on those that have short-term, local impacts that can be linked to specific forms of production and consumption. For this reason the impacts and management choices reviewed here take a firm-based, not climate change specific approach. Five types of tourism activities are reviewed: cruise ships, local sourcing of produce, marine-based activities, recreational areas and commercial fishing.

Cruise ships

Cruise ships make an important contribution to the tourist industry, for example, in the Caribbean and South Pacific The key issues to consider are those associated with the supporting land-based infrastructure, and the operation of the vessels at sea. On land, the location, facilities, and operation of the port can have both direct and indirect impacts on coastal and marine habitats. These cover a spectrum from the total loss of habitats during port construction (for example, mangrove, seagrass, dunes) to more gradual changes as a result of chronic pollution from day-to-day operations (for example, through discharge of oil/fuel wastes, anti-fouling paints). The implications go beyond the loss of biodiversity *per se* because habitats such as these may be spawning grounds or nursery areas for fish, or make a valuable contribution to wildlife tourism and the associated plants that grow on sediments in coastal waters, and keep them in place. Without these plants sediments may be moved around by water currents, damaging beaches and reefs, and silting up channels used by boats. The effects of pollution may be aesthetic, with oil and tar balls washed up on nearby beaches, or have a detrimental effect on local fisheries, for example, by tainting produce taken for human consumption.

Because most of the ports used by cruise ships are well established, it is the upgrading and maintenance of the port facilities that should probably be given most attention at the present time. The trend towards increasing vessel size, for example, has implications for the need for capital and maintenance dredging to retain access to berths. This will not only have a direct impact on the marine communities in the channels, but also further afield, as sediment plumes that form during dredging operations make the water murky, partially blocking out light, and smother plants and animals that live on the sea floor, damaging or destroying them. In addition, the disposal of sediments dredged from channels to keep them open for boats and shipping can have negative effects on bottom-dwelling plants and animals. These effects can also be increased if the sediments are contaminated with pollutants. The location of anchorages outside the port area will also need to be chosen with care as will the likely effect on any extension of jetties and breakwaters (Table 6.6). These issues are discussed further in the section on habitat degradation.

Another set of issues is associated with the passage of cruise ships within and beyond the national territorial waters and Exclusive Economic Zones. They include the need for responsible disposal of waste (for example, sewage, garbage) which can affect water quality as well as having harmful effects on wildlife, such as the entanglement of turtles and seabirds, and care over the exchange of ballast water because of the possible detrimental effects on native species should alien species carried in ballast water become established. The International Convention for the Prevention of Pollution from Ships (MARPOL) includes the regulation of the disposal of wastes from ships. For example, the wider Caribbean, including the Gulf of Mexico, is a Special Area under Annex V of MARPOL (Regulations for the prevention of pollution by garbage). This prohibits the dumping and disposal of any wastes overboard in the region except food waste, which should be macerated and discharged as far as practicable from land but in any case not less than three nautical miles from land. Where solid and liquid wastes are disposed of in shore reception facilities, the problem is simply transferred onshore where incineration and burial at landfill sites are some of the options that may be used. However, such options are impractical or damaging on small islands where space for such facilities is very limited. Shipping accidents, with the potential for some associated marine pollution, are also always a risk.

Local sourcing of produce from coastal and marine ecosystems

Enjoying the local produce is part of the tourist experience but this can put significant additional pressure on marine resources that are already marketed locally, or

Table 6.6 Cruise ships

Issues	Examples of management tools	Regulatory frameworks and options
Port facilities, dredging	Zoning through land use planning	National development plans
		Environmental Management Systems for Ports
		Environmental Impact Assessment
Accidental discharges	Ships routing measures e.g. VTSS, use of pilots	Oil spill contingency planning
		SOLAS and MARPOL convention
Deliberate discharges	Policing, waste reception facilities	MARPOL convention
Introduction of alien species in ballast water	Tanking facilities, offshore exchange	MARPOL convention
Contamination from anti-fouling paint	Containment of runoff around dry docks	Int. Convention on the Control of Harmful Anti-fouling Systems on Ships

create new markets on a scale that cannot be met by sustainable fisheries. Reef fish and crustaceans, such as groupers, snappers and lobsters, are often popular and where there is a limited local supply, or considerable demand, there is a danger of overfishing. This will have the direct effect of depletion of the target species as well as knock-on effects on the structure of the marine communities from which they have been taken.

Another local sourcing issue of potential concern is the sale and collection of marine curios. Turtle shells, coral, seafans and the colourful shells of marine molluscs are some of the marine souvenirs bought or collected by visitors. Some of these may be threatened or endangered species – and therefore subject to international controls on the collection, sale and trade – while the removal of others, although not prohibited, may lead to a more gradual depletion and change in the structure of marine communities (Table 6.7).

Marine-based activities

Marine-based activities are part of the tourist experience at many coastal destinations. A number of issues need to be considered in relation to how and where these activities are carried out. Diving and snorkelling are very popular and in many places the conditions are ideal for beginners. This creates potential problems because it is the inexperienced groups that are most likely to cause damage to reefs by trampling on corals, holding on to reef structures and stirring up sediment. Large numbers of experienced divers and snorkellers repeatedly using the same

Table 6.7 Local sourcing of products

Issues	Examples of management tools	Regulatory frameworks and options
Collection/sale of marine curios	Educational material for tourists	Nationally protected species lists
	Awareness/training in relation to collection of permitted curios	CITES Convention on Biological Diversity
	Alternative income sources for traders	
Depletion of reef fish	Zoning, closures, quotas, size limits	National fisheries regulations
	Local management of fisheries	Regional fisheries management
	Sourcing information for restaurants and hotels	

area may also have such an effect. Feeding of fish during snorkelling and diving is permitted and even encouraged in some places, however this can affect fish behaviour by drawing fish away from other reefs and making some species more competitive and aggressive. The ecological balance of species on the reef may also be altered because of a secondary effect on the fish and invertebrates that form the normal diet of these species. Fish health may also be affected by feeding with inappropriate or contaminated food.

Spearfishing is also an issue because of the ability of those carrying out such activity to target particular species and preferentially take larger fish from reefs. This will not only deplete the area of such fish but also affect the balance of the reef communities and the behaviour of fish such as groupers, which are often the target of such activity, and which subsequently become much more cautious in the presence of divers and snorkellers.

The potential negative effects of water skiing, jet skiing and windsurfing are mainly tramping effects in launch areas and, in the case of the former two activities, noise disturbance. This can be a nuisance to nearby residents, other beach users and may disturb and displace shore-feeding birds.

In the case of recreational craft there is a potential for conflict with other users, particularly when close to the shore where swimmers, divers and other marine-based activities are taking place. Anchoring may also cause some damage, particularly in reef areas or on seagrass beds, especially if such areas used by large numbers of craft or on a regular basis, and the wake of motorised craft can lead to shoreline erosion.

Sea angling is another popular marine-based activity with both casual and competition fishing enjoyed in many coastal locations. There is growing concern about the capture of some of the large species found in open water, such as marlin and tuna, which are targeted by offshore anglers as well as commercial fishermen.

Table 6.8 Marine-based activities

Issues	Examples of management tools	Regulatory frameworks and options
Habitat damage (reefs, seagrass beds)	Public education programmes Codes of conduct Snorkel/diver training Good site selection for training Zoning of activities	CITES Marine Protected Areas Internationally recognised diver training programmes (e.g. PADI, BSAC)
Anchor damage	Mooring buoys, notification of unsuitable anchoring areas on charts, public information, wardening	Marine management plans with associated regulations
Fish community changes	Prohibition or regulation of spearfishing and fish feeding Angling codes of practice	National/local fisheries regulations
Shoreline erosion	Speed limits	Monitoring programmes
User conflicts	Zoning schemes, wardening	Community management schemes

Depletion of these fish species has resulted in angling associations introducing codes of practice, which include no landing policies so that fish are released alive, back into the wild (Table 6.8).

Recreational areas

The coastal environment is the focus for most of the tourism in coastal areas, with extensive use of beaches as recreational areas. Recreational impacts may arise from a range of recreational activities, for example, shoreline camping, sunbeds and umbrellas, beach sports, reef and coastal trail walking, boating, water skiing and jetskis, fishing, and scuba diving. Issues to consider are the opening up of access to such areas, associated physical development on the shoreline and any effects from the presence of large numbers of people in potentially sensitive coastal and marine environments. These range from disturbance and direct habitat damage to a more gradual deterioration in the quality of the environment (Table 6.9).

Commercial fishing

The sustainable management of commercial fisheries is a difficult challenge for all maritime nations. The links between tourism and fisheries are most often a very minor consideration in any such management, but there are two areas that are

Table 6.9 Recreational areas

Issues	Examples of management tools	Regulatory frameworks and options
Disturbance of turtle nesting areas	Seasonal restrictions on use of areas	Convention on Biological Diversity
	Development control	Land use planning regime
	Mitigation measures (e.g. beach lighting requirements)	
Habitat fragmentation and damage	Networks of protected areas	Planning policy guidance
	Wildlife corridors	
Beach cleaning	Codes of practice	

worth discussing in the current context. The first of these is the effect of unsustainable fisheries on tourism. While not as significant as any effects on the fishing industry itself, the depletion, disruption and destruction of fish stocks and damage to fish habitat will make the promotion of such areas as snorkelling and diving destinations untenable in the long run. Any associated loss of charismatic species, such as turtles and sharks, will further degrade the attraction of visiting these locations compared to destinations where it remains possible to have close encounters with such species.

Negative effects can also take place in the other direction with tourism displacing locals from traditional fishing grounds. This may be because of the physical obstacle of coastal development blocking access to traditional beach launch points, establishment of marine protected areas to attract tourists but where fishing is banned, and incompatibilities in use of the water space, for example jetskiing in areas where mobile gear fisheries are taking place. Effective resolution of these issues in a way that can enhance both fisheries and tourism is possible, as occurred in the Soufrière, St Lucia, where a participatory process ensured the success of a marine management scheme bringing benefits to both fisheries and tourism (Table 6.10).

Conclusions

The world's coastlines have been in the past, and continue at the present time to be, significant tourist attractions and while the traditional tourism demands on this environment might be changing, the coasts' overall attraction remains strong. Unlike some other forms of tourism such as urban or shopping tourism, coastal tourism relies heavily on the natural environment and any change in that environment is bound to have significant consequences for future tourism use. As has been demonstrated in the general account and in the specific case studies within this chapter, future tourism use will be affected in many ways. An increase in temperature and a decrease in rainfall may be beneficial in some coastal environments but

Table 6.10 Commercial fishing

Issues	Examples of management tools	Regulatory frameworks and options
Impacts on marine biodiversity	Fisheries regulations including gear, area and size restrictions	Fisheries management regimes (national and regional)
	Species specific protection measures	Convention on Biological Diversity
		Marine protected areas
Conflicts of use	Zoning schemes	ICZM, marine planning zones
	Codes of conduct	
	Voluntary agreements	
	Community projects	
	Participatory planning	

could be catastrophic in others. An increase in sea-level rise would be of marginal significance in the upland fiord coasts of Norway or New Zealand but could be potentially lethal to low lying atoll regions of the world.

The significance for tourism is great. Some parts of the world's coasts could become more attractive to tourism and might well benefit from a rise in temperature. Coastlines in the higher latitudes may well benefit with more visitation and a longer visitor season but other places, generally in lower latitudes, could well suffer. There could well be a global shift in coastal tourism use to higher latitudes.

The other likely consequence of environment change will be engineering adaptation to existing tourism infrastructure. Stronger structures capable of withstanding higher wind speeds, artificial temperature control by the introduction of more air conditioning systems, and coastline stabilisation by means of sea wall construction are but a few of the possible engineering methods available. There are, however, many downsides to engineering solutions. Stronger structures, air conditioning and sea wall construction are all expensive solutions which may well render many existing coastal tourism operations uneconomic in the future and increased interference with the natural environment flies in the face of current tourism trends for sustainability, authenticity and local difference. Who wants to fly to Polynesia or the Caribbean to stay in a hotel similar to the ones available in London, Sydney or New York, to stay inside their air conditioned rooms and see the sea only from the top of a concrete wall? Environmental change has serious consequences indeed for coastal tourism.

There are many challenges in managing climate change, mainly arising from the fact that it is the result of many small actions taken by a myriad of agents. For this reason this chapter has reviewed a range of actions that public and private sector agents operating in marine coastal areas can take to manage their immediate environment that in turn will contribute to the short-term improvement of the local environmental quality, and that, in turn, will contribute to lessening tourism's pressure on the global environment. To facilitate these individual actions, a range of key factors can be identified that are important for an integrated approach to tourism and coastal and marine management (Table 6.11).

Table 6.11 Key factors in an integrated approach to tourism and coastal and marine management

---

1   Operation of good spatial planning mechanisms linked to long-term strategic planning, with inclusion of sustainability considerations as an integral part of decision making. These mechanisms should provide an overall plan of where to allow development, how much and of what kind. They would include integrated planning of tourist development balanced with other uses/developments and with the need to maintain areas free of development, so that environmental quality is maintained to attract tourists and sustain fisheries. Elements include:
  • co-ordination between relevant agencies and public authorities
  • incorporation of Environmental Impact Assessments/Strategic Environmental Assessments combined with an ecosystem-based approach to island management
  • open and transparent planning procedures that include full consultation with stakeholders – including inputs of local knowledge, innovation and practices in planning – for all developments, without exception
  • consideration of cumulative effects of development for both large and small-scale developments when taking development and planning decisions
  • making informed decisions on the basis of a) good baseline information on natural resources/features and any trends in these, b) international obligations, and c) understanding of local/indigenous use of natural resources and their impacts on ecosystem functioning so as to maintain ecosystem services
  • establishment of limits of acceptable change/carrying capacities based on key sustainability indicators such as availability of freshwater sources, infrastructure capacity and local socio-economic considerations
  • designation of land for appropriate uses, including 'zoning' as part of longer-term planning, based on assessment of 'carrying capacities' for tourism and other potential activities, and which take a long-term view
  • clear and unambiguous legislation setting responsibilities for public authorities and enforcement agencies.

2   Monitoring of changes and trends before, during and after developments are introduced, including effects on local/indigenous communities and use of natural resources.

3   Introduction of effective systems to monitor compliance with laws, and to take rapid enforcement action in cases of non-compliance.

4   Building coherence between different policy areas, government departments and public authorities to minimise conflicts between objectives, and to co-ordinate administration, including adoption of integrated pollution prevention and control.

5   Use of mechanisms to resolve resource use conflicts where the same resources are the basis for different socio-economic activities, including involvement of all stakeholders.

6   Promotion of appropriate environmentally sound technologies and management approaches.

7   Education and awareness raising programmes for the public and private sector actors involved in tourism, as well as for tourists and host communities.

8   Strengthening of linkages within the local economy to increase the benefits of tourism without increasing the number of tourists.

References

Agnew, M and Viner, D. (2001) 'Potential impacts of climate change on international tourism', *Tourism and Hospitality Research*, 3(1): 37–60.

Burns, W. (2000) *The possible impact of climate change on Pacific Island ecosystems*. Occasional paper of the Pacific Institute for studies in development, environment and security, Oakland, CA: Pacific Institute.

Doornkamp, J. (1998) 'Coastal flooding, global warming and environmental management', *Journal of Environmental Management*, 52: 327–33.

Feifer, M. (1986) *Tourism in History*. New York: Stern and Day.

Giles, A. and Perry, A. (1998) 'The use of a temporal analogue to investigate the possible impact of projected global warming on the UK tourism industry', *Tourism Management*, 19(1): 75–80.

Hansom, J. (2001) 'Coastal sensitivity to environmental change; a view from the beach', *Catena*, 42: 291–305

Hoagland, P., Schumacher, M. and Gaines, A. (1995) *Toward an Effective Protocol on Land-based Marine Pollution in the Wider Caribbean Region*. WHO 1-95-10 Prepared by Woods Hole Oceanographic Institution for the US. EPA, Office of International Affairs. Woods Hole, MA: Marine Policy Centre, WHOI.

Ker, R. (1999) 'Big El Niños ride the back of slower climate change', *Science*, 283: 1108–9

Kuijper, M. (2003) 'Marine and coastal environmental awareness building within the context of UNESCO's activities in Asia and the Pacific', *Marine Pollution Bulletin*, 47: 265–72.

Timmerman, P. and White, R. (1997) 'Megahydropolis: coastal cities in the context of global environmental change', *Global Environmental Change*, 7(3): 205–34.

Tol, R., Fankhauser, S. and Smith, J. (1998) 'The scope for adaption to climate change; what can we learn from the impact literature?' *Global Environment Change*, 4(2): 109–23.

UNEP (1994) *Ecotourism in the Wider Caribbean Region; An Assessment*. CEP Technical Report no. 31. Kingston, Jamaica: UNEP Caribbean Environment Programme.

UNEP (1997) *Coastal Tourism in the Wider Caribbean Region; Impacts and Best Management Practices*. CEP Technical Report no. 38. Kingston, Jamaica: UNEP Caribbean Environment programme.

Viner, D. and Agnew, M. (2000) 'Climate change and tourism', in A. Lockwood (ed.) *Proceedings of the International Conference on Tourism and Hospitality in the 21st Century*, Guildford: University of Surrey.

Wall, G. (1992) 'Tourism alternatives in an era of global climate change', in V. Smith and W. Eadington (eds) *Tourism Alternatives*. Chichester: John Wiley & Sons, 198–215.

7 Deserts and savannah regions

Robert Preston-Whyte, Shirley Brooks and William Ellery

Introduction

Arid landscapes have been constructed in many forms over the past two centuries. As landscapes of the mind, they take shape through the collection of senses that structure experiential space. As landscapes of ideology, they have been manipulated by power interests from the British and French colonial empires to Saddam Hussein, from the settlement hopes of western pioneers in the USA to the visions of ecotourist planners, from the certainty of scientific interventions to the chaos of social transformations. As landscapes of opportunity, drylands have been invented and reinvented by wave upon wave of interests, from exclusionary conservationists to enthusiastic ecotourist operators. As landscapes of Edenic dreaming, they have been painted as a romantic frontier, a space where mind and heart converge. As landscapes of distress they have been associated with the hurt and disruption of forced removals, from native American communities to Aboriginal Australians, from the San people in Botswana to apartheid victims in South Africa. These are spaces ruptured by multiple constructions, by multiple heterotopias, and by multiple realities, yet they retain a magnetism that is tangible, albeit elusive.

> The greatest part of my satisfaction was animal pleasure: the remoteness of the site, the grandeur of the surrounding mesa-like mountains and rock cliffs, the sunlight and the scrub, the pale camels in the distance, the big sky, the utter emptiness and silence, for round the decay of these colossal wrecks the lone and level sands stretched far away.

writes Theroux (2003: 81), referring to the temple complex at Al Nagger in the Sudan. This feast of sensory stimulation, coupled with the tantalising mystery of lost cultures, unfamiliar landscapes and unusual plants and animals, goes a long way to explaining why arid environments may be considered one of the 'wonders of the world' in their extent and stark beauty. In addition to the ways they have been constructed, manipulated and used in the recent past, these landscapes contain some of the most impressive examples of human occupation of the earth, from ancestral hominid fossils in eastern and southern Africa, to the pyramids of Giza and the lines of Nasca.

Particularly in Africa, drylands provide the habitat for vast herds of animals, predator and prey. As 'safari' landscapes of masculine identity, they have for centuries attracted sport hunters, adventurers and, more recently, wildlife tourists. But all this can change, and indeed *is* changing rapidly in some places, as tourists become the agents of destruction of cultural artefacts (Keenan 2002), as fauna and flora are threatened by competition with land-hungry humans and changing climate, and as global economic interests as well as local authoritarian states and marginalised people undergoing rapid social change become embroiled in violent conflicts over land use and resources.

What then of the future? It is difficult to be optimistic about tourism in arid lands. These are fragile environments, both in terms of their cultural heritage and biodiversity. The nomadic pastoralist communities that have successfully inhabited them for centuries are threatened with sedentarisation and exclusion from their former lands. Contemporary land use, with its razor-sharp divisions between productive and unproductive land, raises questions about sustainability. Issues of population pressure and the possibility of future warmer and drier climates modifying ecosystems in various ways are also troubling, and the concomitant stress on soil, water, ecology and food resources may cause already fragile social and governance systems to reach breaking point. These stresses are exacerbated by the globalisation of markets that tend further to impoverish marginalised communities: the environmental impact of extractive activities in dryland environments can be seen from space (Girard and Isavwa 1990; Tucker *et al.* 1991). Add to this an additional load of uncaring tourists and the burden may prove intolerable.

This chapter is about tourism and change in dry environments. The triangular linkage between natural environments, host communities, and tourist providers and consumers is one of interdependence. Break any one and the triangle collapses. Given these linkages, we discuss in separate sections the nature of arid environments and changing biodiversity, the attraction of these environments for tourists and their impacts on local communities and, finally, tourism as an agent of sustainable development.

Arid landscapes, biodiversity and desertification

The various constructions of arid landscapes cannot avoid the limits imposed on people, plants and animals by lack of water. As early as 1894 Albrecht Penck identified dry areas as those in which evaporation was either equal to or greater than rainfall. In their global map of areas of interior drainage basins, which included non-flowing or arheic regions, de Martonne and Aufrere (1927) recognised the link between endorheism and the fact that rainfall was insufficient to force drainage to the sea. As such, deserts were associated with endorheism. For mid-twentieth century soil scientists, rainfall insufficient to flush out soluble carbonates from surface layers to a depth beyond the reach of plant roots was considered a primary indicator of aridity (Shantz 1956).

Climatologists initially solved the problem of defining arid lands by relying on the work of botanists since maps of world vegetation predated maps of global climate. The remarkably similar vegetation structure of widely separate dry environments – widely spaced, single- or multiple-stemmed trees, low shrubs, grasses and/or short-lived herbaceous plants – is indicative of independently evolved modes of plants using sparse water and surviving erratic droughts. It was found that the distribution of vegetation based upon plant life forms could be used to map climate (Raunkaier 1934). For example, Köppen (1923), using the distribution of xerophytes (plants adapted to dry conditions), devised a classification of global dry climates based on an existing vegetation classification system and map. Thornthwaite (1948) devised a series of indices that showed the relationships between precipitation (water supply) and evapotranspiration (water demand), thus introducing the concept of precipitation effectiveness into the debate about how to define dry climates. In the late 1940s, the United Nations requested that a map of the arid lands be produced and, in response, Meigs (1953) extended Thornthwaite's method to calculate a moisture index that represented the relationship between precipitation and evapotranspiration. Various modifications since then (United Nations 1977a; Rogers 1981; United Nations Environment Programme 1992) show approximately 39.7 per cent of the global land area to be arid or semi-arid (Figure 7.1).

In dry environments the main constraint on biological productivity and human carrying capacity is the absence of water. In these environments plant life varies from true desert, dominated by short-lived annual plants that survive unfavourable periods as seed, to landscapes dominated by shrubs that are often succulent, to a combination of shrubs and grasses, to grasslands with or without scattered trees. Animal life varies along a similar gradient, with body size tending to increase with increasing moisture availability. The most abundant animal life occurs in savannahs, where the combination of plant productivity and forage nutritional value is optimal. Towards the drier end of the spectrum, plant productivity tends to limit

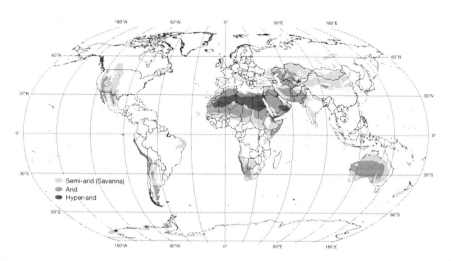

Figure 7.1   Global distribution of hyper-arid, arid and semi-arid (savannah) regions

faunal biomass, whereas towards the wetter end of the continuum nutrient concentration in plant tissue limits animal productivity (Ellery *et al.* 1995). It is worth noting that, along with animals, human populations in arid areas have also evolved patterns of migration in response to variability in rainfall and plant and animal production.

Key developments in the scientific understanding of drylands throw new light on the way these ecosystems respond to spatial and temporal environmental variability and disturbance. It is clear that environmental variability occurs on various spatial and temporal scales measured in seasons, years or decades in the case of rainfall variability. Despite these patterns of temporal variation, certain types of vegetation may continue to persist, suggesting a level of system equilibrium. Skarpe (1991) calls these systems 'resilient'. While vegetation systems may be resilient to rainfall variability, human-induced disturbance such as continuous grazing leads to dramatic changes in vegetation composition and structure, leading to system change from one state to another (Noy-Meir 1982). Such systems are termed 'unstable'. There is general agreement that semi-arid and arid ecosystems are resilient but at the same time tend towards instability (Thomas and Middleton 1994), and that recovery following anthropogenic disturbance is particularly slow because of the rarity of occasions on which sufficient moisture is available (United Nations 1977b).

Irreversible anthropogenic disturbance in arid lands leading to alteration of ecosystems is referred to as desertification. The causes of desertification have been widely attributed to agricultural neglect and misuse of the natural resources in arid environments; to land-use practices which degrade natural vegetation cover and impact negatively on the hydrological cycle and soil fertility and structure (Timberlake 1985). Although the effects of various forms of land use are debated (Tiffen *et al* 1994; Leach and Mearns 1996), the consequences of desertification are clear. They include the expansion of annual grasses and herbs at the expense of perennial grasses, increased incidence of short multi-stemmed trees – 'bush encroachment' – and damage to soils through salinisation, nutrient loss, compaction, crusting and erosion (Thomas and Middleton 1994).

Climate induced change also needs to be considered. There is general consensus that the trend towards warmer conditions evident over the past century will continue (Houghton *et al* 1990; IPCC 1997). Under these conditions, rising temperatures would increase evapotranspiration rates in arid areas, leading to a decrease in potential water availability. The effect on biodiversity will differ from one area to another depending on whether regional warming trends are accompanied by concomitant changes in rainfall. Similarly, the effect that increased aridification consequent upon global warming may have on tourism will also vary by region. For example, in South Africa the predicted warmer and drier conditions (Hulme 1996; Hulme *et al.* 2001) may lead to better representation of the eutrophic or arid savannah at the expense of the dystrophic or moist savannah. This may be 'Bad news for tourism because arid systems often do not favour a high diversity of spectacular plants or large mammals that most tourists visit these areas to see' (Midgley *et al.* 2001: 5). On the other hand, the bush encroachment that can accompany desertification, particularly in enclosed protected areas from which human activity has been excluded, may reduce the attractiveness of savannah environments by making wildlife more difficult to see (Watson 1995; Bond *et al.* 2002).

Tourism attraction and impacts

The very fragility of arid environments and their 'extreme' nature is one of the factors that attracts tourists to visit them. For some tourists such environments may be invested with a spiritual significance and a liminal quality that elicit questions about the nature of existence and mortality. On a more prosaic level, tourists may want to observe or photograph the unique plants and animals that have evolved to cope with such extreme conditions, as well as the wide open landscape vistas and features such as the unusual rock formations often found there (White and Nackoney 2003). In contexts like the central Sahara, cultural heritage in the form of prehistoric rock art is an important attraction (Keenan 2002). Such rock art is also widely spread across dry environments in sub-Saharan Africa and elsewhere. In a globalising world, both savannah and more extreme desert environments are widely sought after by international tourism agencies, which package their products in the form of wildlife tourism, ecotourism or adventure tourism.

The special appeal of the savannah environment, in particular the African savannah landscape, for western tourists is well documented (Graham 1973; Anderson and Grove 1987; Norton 1996; Broch-Due and Schroeder 2000; Bassett and Zueli 2003). The African savannah is widely viewed as the archetypal wild environment, a place of unspoilt nature. The appeal of this 'safari' landscape – scattered thorn bushes, wide vistas, yellow grass, herds of animals – is reinforced by countless wildlife documentaries, tourist brochures, films and web advertising (Norton, 1996). As Adams and McShane (1992: 42) observe:

> We cling to our faith in Africa as a glorious Eden for wildlife. The sights and sounds we instinctively associate with wild Africa – lions, zebra, giraffe, rhinos, and especially elephants – fit into the dream of a refuge from the technological age. We are unwilling to let that dream slip away ... the emotional need for wild places, for vast open spaces like the East African Serengeti Plain, persists.

The power of this landscape image, embodying primeval timelessness, is immense and is used in the marketing of these areas as tourism destinations (Brooks 2000).

Yet such images tend to obscure the fact that these are environments over which bitter struggles have been fought for both material and ideological control (Neumann 1995a, 1995b, 1996, 1998). Whether or not visitors and tour operators recognise their presence, the people that inhabit these environments are often clinging to a precarious livelihood in marginal conditions. In east and southern Africa, for example, the colonial and post-colonial state has had a major role in reshaping the landscape, creating huge national parks in conflict with the interests of both settled local communities and nomadic pastoralists who find themselves in competition with wildlife for the use of savannah plains (Homewood and Rodgers

1991; Neumann 1998; Johnsen 2000). As a result, local people in extreme environments often have contested relationships with the state: they have been moved around at its behest and are unlikely to trust power emanating from a distant capital city. Commercial tourism agencies – often closely linked to the very states that created this crisis of marginalisation – may fail to grasp the highly political nature of questions of environmental sustainability in dry environments. They generally fail to understand how the exploitation of nature or cultural sites for tourism plays into local politics and power struggles (Keenan 2003b).

Furze *et al.* (1996: 146) argue that 'tourism may provide a vehicle or conduit for translating the values that others hold for a natural area into benefits for those who live in or near it'. They caution, however, that tourism itself imposes problems and costs of its own, burdening the natural resources and the host society to an extent which may outweigh potential development benefits. Tourism in remote locations, such as those discussed in this chapter, has often taken the form of 'enclave tourism' (Freitag 1994; Brohman 1996). This occurs when tourism-related facilities are not oriented towards the local community, and most local people cannot afford to participate in or enjoy the services offered. Money generated within these 'enclaves' usually has very little impact on the local economy: tourism revenues tend to go to external agencies and may not even remain in the host country. While countries like Botswana have attempted to combat these trends, this is not always successful and often tourist enterprises in fragile environments exist in isolation from the surrounding communities (Mbaiwa 2003, 2004). If local people are expected to bear the costs of tourism development, and get none of the benefits, then it is obvious that problems will arise.

The devastating social effects of forced removals and reduction of grazing lands through the creation of national parks and other protected areas are widely known (Ghimire and Pimbert 1997; Johnsen 2000). It is increasingly recognised that, in the process of expanding conservation areas and tourist operations, local property rights need to be protected as far as possible – that 'the tradition of government appropriating all resource rights and centrally managing them has not worked to the benefit of either conservation or local development' (Furze *et al.* 1996: 156). In the past two decades, various experiments in community-based tourism and attempts to give communities greater local rights over wildlife and land have been pioneered in dry environments. In sub-Saharan Africa's savannah and semi-arid regions, debates about how to integrate national parks and the livelihoods of pastoralists are ongoing (Lindsay 1987; Thompson and Homewood 2002).

Outside of protected areas, landscapes are being reshaped by new discourses of community-based tourism in which communities are encouraged to link their livelihoods more closely – even to fully integrate them – with wildlife. This approach is widely viewed as the only hope for sustainable wildlife tourism (Murphree 1993; Western and Wright 1994; Hulme and Murphree 2001). If property rights are secure and power relations with partners in ecotourism enterprises fairly equal, this approach can work for communities in dry environments (Thomas and Brooks 2003). Yet there are also serious actual and potential costs associated with the linking of livelihoods so directly to tourism, which involves tolerating the presence of wildlife on grazing lands and reducing pastoralist activity. These two land uses – pastoralism and game farming – often prove mutually exclusive, and in dry environments such matters are literally life and death concerns (Keenan 2003a).

Livelihoods in dry environments tend to be constituted through multiple activities. It is often not sensible for those living in dry environments to rely on a single occupation. For example, as Kawatoko (2004) describes in his abstract to the United Nations University International Conference on Marginal Drylands in Tokyo, for life in the Sinai peninsula:

> There has been not only fruit cultivation in the orchards (date palm and others), agriculture, nomadism (camels, goats and sheep), firewood gathering and charcoal burning and drug plant gathering in the wadis, but also hunting in the mountains (wild goats, desert lizards, birds, etc.), hawk capturing, turquoise mining, and fishing and shell gathering on the seacoast.

The introduction of new tourism activities into fragile environments, even if old forms of land use are not totally abandoned, diverts the energies of host communities into other channels, with unpredictable social and environmental effects (Shackley 1996; 1999).

Even if livelihoods are not disrupted and the problems associated with 'enclave' tourism are overcome, the business of attracting tourists to arid environments is not a straightforward matter. Unfortunately the natural and cultural attractions of dryland areas are vulnerable to degradation and destruction by a range of environmental and human agents. If poorly managed, the appeal of dry environments may diminish, as the tourism experience is itself diminished. Externally driven narratives of desertification aside (Swift 1996), some residents of fragile dryland environments predict ecological catastrophe in the near future. Tourism contributes to this outcome, albeit inadvertently (Keenan 2002). In the case of the Sahara, the situation appears grim:

> [the area's] extraordinarily rich cultural and scientific heritage, so much of which seems to have survived over thousands of years the harsh vicissitudes of climate and other natural forces, has been subjected in the last two generations or so to extensive and largely irreversible damage.
>
> (Keenan 2002: 2).

Photographers moisten paintings so as to bring out the colour, thus damaging them irreparably; collectors 'vacuum' the Sahara for artefacts; tour operators open up new sites in an attempt to offer their clients a 'unique' desert experience. The increasing accessibility of such sites and their opening up to tourism sets in motion forces that need to be carefully controlled.

It is hardly surprising that host communities in dry environments may regard tourism with some scepticism. While states often welcome tourism and encourage their citizens to embrace it as a means of livelihood, there remains deep ambivalence as people also see tourism as posing a serious threat not only to their local environment but also to their way of life. The cultural consequences of tourism for people living in arid environments may be severe. Often these are people who, until recently, have been relatively sheltered from the winds of globalisation. Now, however, their cultures and way of life are ruthlessly exposed, transformed into marketable commodities in the cultural tourism marketplace (Kohl 2002; Dinero 2002). Occupants of these areas are rightly concerned about the effect of tourism on the moral and symbolic order of society (Hobbs 1996; Duffy 2002; Grainger 2003).

Some threats to tourism enterprises in such regions lie well beyond tourism management control. One such threat is the political instability and violent conflict often intertwined with the politics of the extraction of natural resources such as petroleum from semi-arid and desert regions (Watts 2001). Areas like these have become, in Watts' phrase, 'globalized local sites' due to the globalization of resource and biodiversity extraction and mining (Watts 2000: 22; Zerner 2000). Competing global and local interests intersect around questions of conservation and resource use in dry environments and tourism is only one part of the picture. Alliances between authoritarian governments, transnational oil corporations and the emerging world order are disturbing: they contribute to a deteriorating security situation and to broader dynamics of violence in fragile environments that are not conducive to tourism development (Watts 2001; Keenan 2003b). Tourists may want an exciting adventure but they do not want to endanger their lives.

As already mentioned another potential threat to tourism in dry environments is the impact of global climate change. In environments already subject to extreme climatic variability people are generally clinging to livelihoods that are more or less precarious. Predictions are that it is the world's drylands that will suffer the greatest reductions in water availability if significant global warming occurs. Scientists are currently engaged in debate as to whether or not recent decades of drought in regions like the Sahara–Sahel indicate real climatic change or simply variations within a medium-term norm (Durand and Lang 1991; Hulme 2001; Hulme *et al.* 2001). It is striking, however, that residents of dry environments do not perceive this threat in the same way scientists do. These people are accustomed to periods of prolonged drought, and their impressions of environmental change are more likely to be shaped by local understandings of the environment than by climate modelling (Benjaminsen 2000; Mortimore and Adams 2001; Graef and Haigis 2001). Keenan (2003b: 5) found that, in the Sahara, local people's ranking of 'climate change' as one of the agents that may 'turn their world upside down' is low. As host communities battle with pressing issues relating to their immediate survival, the hypothetical possibility of climate change is viewed as remote. It is with the contribution of tourism to these issues of sustainability and development that we end our discussion.

Tourism and sustainable development

The debate around global environmental changes brought about by human activities has brought into focus the present and future plight of billions of people who face an uncertain future (United Nations Development Programme 1998). Many of these people live in arid regions where their sustainability cannot be divorced from issues that contribute to poverty, disempowerment and desertification. Tourism in arid lands, with its relatively wealthy clientele often placed in juxtaposition with communities fractured by drought, unemployment, poverty, poor health and crime, presents a significant challenge to institutions wrestling with the notion of sustainability (Hall and Lew 1998; Honey 1999).

Tourism development in savannah and desert areas often brings with it the opening up of the commons to consumption demands and technological requirements that challenge resilience to change of social and ecological systems (Goldman 1998). Ensuring that development becomes sustainable means that complex questions such as demand- and supply-side constraints, corporate pressures, governance responsibilities, environmental impacts and ethical imperatives are raised. When complicated by global environmental changes that disturb the fragile balance between ecological resilience, social equilibrium and human-induced stress, the equation becomes even more difficult to compute. Host communities in such areas may then find themselves becoming increasingly vulnerable to forces that diminish their quality of life.

Adger and O'Riordan (2000: 165) note that 'vulnerability is a function of powerlessness' brought on by the inability of people to cope in situations where they are kept ignorant of the nature of the threats confronting them and, for a variety of reasons, prevented from intervening in their resolution. This situation usually arises where communities with limited access to power and resources find themselves to be the victims of political and economic ideologies that do little to protect local markets, press freedom and local investments in environments, and even less to control macro forces such as rapid urbanisation, decline in soil productivity and arms expenditure (Blaikie *et al.* 1994). The sustainability of such communities is threatened when the ethical, economic and political conditions that favour vulnerability are coupled with damage to the biophysical functions that sustain life. Sustainability is therefore dependent upon 'the coupling of both human resilience and ecological sensitivity into a single, interactive totality' (Adger and O'Riordan 2000: 165).

If tourism is to militate against rather than contribute to host community vulnerability in savannah and desert regions, two issues need to be borne in mind. The first is that there are many pathways to sustainability (O'Riordan *et al.* 2000) and these will be informed by differences between nations in relation to level of economic development, access to technology and resources, and governance strategies. The second is the need to build the four pillars of sustainability: secure wealth creation, stewardship, empowerment and 'revelation' (O'Riordan 2000). Each of these four pillars is considered in turn.

First, the sustainability of host communities must be grounded in their ability to generate wealth through tourism in a continuous and ecologically tolerable manner. Second, while stewardship means taking care of the environment upon which the community subsists, this may well entail the inclusion of second-order issues such as improving education and health, eliminating crime, corruption and environmental injustice. Third, when wealth creation and stewardship are in place community empowerment should follow. O'Riordan and Voisey (1998) see empowerment as a process that leads towards the development of personal self-respect and self-confidence and the development of a civic consciousness that ensures inclusiveness, legitimacy and accountability. Finally, this chain of events leads to the 'revelation' that a sustainable future is achievable through the visioning of a collective future:

> ... involving programmes of reliable wealth creation, accountable distribution and opportunity provision for jobs and basic needs, social acceptance of the need to take care of network and social capital for self-preservationist reasons, and practically sensible but democratically realistic empowerment arrangements.
>
> (O'Riordan *et al.* 2000: 3).

It is difficult to find examples in dry environments where tourism has led to empowerment and sustainable development. Perhaps they exist and lack visibility. More common are examples where the scorecard of sustainability indicators is less than satisfactory. Honey (1999) for example, provides an analysis of the path towards ecotourism sustainability in Tanzania, Kenya and South Africa. For historical, ideological and economic reasons, each country is at a different stage along the development path. In each case, issues like the building of environmental awareness, minimising environmental impacts, the empowerment of local people, the provision of financial benefits for conservation, and respect for local cultures all need attention before tourism can become a force that can make a positive contribution to global and globalising changes and local problems of corruption, mismanagement and poor governance.

Conclusion

When tourists visit desert and savannah areas in sizeable numbers they become part of an inevitable process of change that impacts in various ways on the natural environment and local communities. The perspective developed here envisages tourism, natural environment and community development at each corner of a triangle. The triangle metaphor is used because it brings into focus the elements that structure the relationship between tourism and global environmental change in arid regions, while also emphasising the powerful linkages between them. It also stresses the importance of exogeneous shocks to the natural environment or stability of local communities which, through breaking the links binding the triangle, would severely impact on tourism demand. It is recognised, however, that for a more detailed and thorough understanding of the dynamic issues that inform the interactions between tourists, physical environment and local communities a network metaphor may be a more appropriate analytic tool.

The links between tourism, natural and social environments in a changing world open up a range of scenarios. At one end, tourism may benefit local communities and the natural environment by encouraging a setting where humans and nature learn to live together in a state of mutual respect and understanding. At the other end, tourism may impact on the natural and social environment in ways that are socially destructive and environmentally irreversible. In between, tourism demand for arid landscapes may contribute to scenarios that produce widely ranging mixes of social impact, economic development and biophysical variability. How, when and where these scenarios play themselves out in response to global environmental changes remains to be seen. The curtain has risen; the play has begun.

References

Adams, J.S. and McShane, T.O. (1992) *The Myth of Wild Africa: Conservation without Illusion*. New York: Norton.

Adger, W.N. and O'Riordan, T. (2000) 'Population, adaptation and resilience', in T. O'Riordan, (ed.) *Environmental Science for Environmental Management*. Harlow: Prentice Hall.

Anderson, D. and Grove, R. (eds) (1987) *Conservation in Africa: People, Policies and Practice*. Cambridge: Cambridge University Press.

Bassett, T.J. and Zueli, K.B. (2000) 'Environmental discourses and the Ivorian savanna', *Annals of the Association of American Geographers*, 90(1): 67–95.

Benjaminsen, T.A. (2000) 'Conservation in the Sahel: policies and people in Mali, 1900–1998', in V. Broch-Due and R.A. Schroeder (eds) *Producing Nature and Poverty in Africa*. Stockholm: Nordiska Afrikainstitutet.

Blaikie, P., Cannon, T., Davis, I. and Wisner, B. (1994) *At Risk: Natural Hazards, People's Vulnerability and Disasters*. New York: Routledge.

Bond, W., Woodward, F.I. and Midgley, G.F. (2002) 'Does elevated CO_2 play a role in bush encroachment?' in A.H.W. Seydach, T. Vorster, W.J. Vermeulen and I.J. van der Merwe (eds) *Multiple Use Management of Natural Forests and Woodlands: Policy Refinements and Scientific Progress*. Pretoria: Department of Water Affairs and Forestry.

Broch-Due, V. and Schroeder, R.A. (eds) (2000) *Producing Nature and Poverty in Africa*. Stockholm: Nordiska Afrikainstitutet.

Brohman, J. (1996) 'New directions in tourism for third world development', *Annals of Tourism Research*, 23(1): 48–70.

Brooks, S. (2000) 'Re-reading the Hluhluwe-Umfolozi Game Reserve: constructions of a "natural" space', Special Issue on Land in Africa, *Transformation*, 44: 63–79.

Dinero, S.C. (2002) 'Image is everything: the development of the Negev Bedouin as a tourist attraction', *Nomadic Peoples*, 6(1): 69–94.

Duffy, R. (2002) *A Trip too Far: Ecotourism, Politics and Exploitation*. London: Earthscan.

Durand, A. and Lang, J. (1991) 'Breaks in the continental environmental equilibrium and intensity changes in aridity over the past 20 000 years in the central Sahel', *Journal of African Earth Sciences*, 12(1–2): 199–208.

Ellery, W.N., Scholes, R.J. and Scholes, M.C. (1995) 'The distribution of sweetveld and sourveld in South Africa's grassland biome in relation to environmental factors', *African Journal of Range and Forage Science*, 12: 38–45.

Freitag, T.G. (1994) 'Enclave tourism development for whom the benefits roll?' *Annals of Tourism Research*, 21(3): 538–54.

Furze, B., de Lacy, T. and Birckhead, J. (1996) *Culture, Conservation and Biodiversity*. Chichester: John Wiley and Sons.

Ghimire, K.B. and Pimbert, M. (eds) (1997) *Social Change and Conservation*. London: Earthscan.

Girard, M.C. and Isavwa, L.A. (1990) 'Remote sensing of arid and semi-arid regions: the state of the art', *Desertification Control Bulletin*, 18: 13–18.

Graef, F. and Haigis, J. (2001) 'Spatial and temporal rainfall variability in the Sahel and its effects on farmers' management strategies', *Journal of Arid Environments*, 48(2): 221–31.

Graham, A. (1973) *The Gardeners of Eden*. London: Allen and Unwin.

Grainger, J. (2003) '"People are living in the park". Linking biodiversity conservation to community development in the middle east region: a case study from the Saint Katherine Protectorate, Southern Sinai', *Journal of Arid Environments*, 54(1): 29–38.

Goldman, M. (ed.) (1998) *Privatizing Nature: Political Struggles for Global Commons*. London: Pluto Press.

Hall, C.M. and Lew, A.A. (1998) *Sustainable Tourism. A Geographical Perspective*. New York: Addison Wesley Longman.

Hobbs, J.J. (1996) 'Speaking with people in Egypt's St. Katherine National Park', *The Geographical Review*, 86(1): 1–21.

Homewood, K.M. and Rodgers, W.A. (1991) *Maasailand Ecology: Pastoralist Development and Wildlife Conservation in Ngorongoro, Tanzania*. Cambridge: Cambridge University Press.

Honey, M. (1999) *Ecotourism and Sustainable Development. Who Owns Paradise?* Washington, DC: Island Press.

Houghton, J.T., Jenkins, G.J. and Ephrams, J.J. (eds) (1990) *Climate Change: IPCC Scientific Assessment*. Cambridge: Cambridge University Press.

Hulme, D. and Murphree, M.W. (eds) (2001) *African Wildlife and Livelihoods: the Promise and Performance of Community Conservation*. Oxford: James Currey.

Hulme, M. (ed.) (1996) *Climatic Change and Southern Africa: An Exploration of Some Potential Impacts and Implications in the SADC Region*. Norwich: University of East Anglia.

Hulme, M. (2001) 'Climatic perspectives on Sahelian desiccation: 1973–1998', *Global Environmental Change*, 11(1): 19–29.

Hulme, M., Doherty, R., Ngara, T., New, M. and Lister, D. (2001) 'African climate change: 1900–2100', *Climate Research*, 17: 145–68.

IPCC (Intergovernmental Panel on Climate Change) (1997) *Climate Change 1995: Summary of Policymakers, and Technical Summary of the Working Groups Report*. Cambridge: University of Cambridge Press.

Johnsen, N. (2000) 'Placemaking, pastoralism, and poverty in the Ngorongoro Conservation Area, Tanzania', in V. Broch-Due and R.A. Schroeder (eds) *Producing Nature and Poverty in Africa*. Stockholm: Nordiska Afrikainstitutet.

Kawatoko, M. (2004) 'Port city, monastery and bedouins', Paper Abstract, United Nations University International Conference on Path to the Sustainable evelopment of Marginal Drylands, Tokyo, 19 May. Online. Available at : www.inweh.unu.edu/inweh/ drylands/AbstractsConference2004.htm (accessed 15 September 2004).

Keenan, J. (2002) 'Tourism, development and conservation: a Saharan perspective', in D.J. Mattingly, S. McLaren, E. Savage, Y. el-Fasatwi and K. Gadgood *Natural Resources and Cultural Heritage of the Libyan Desert: Proceedings of a Conference held in Libya, 14–21 December*. London: Society for Libyan Studies.

Keenan, J. (2003a) 'Indigenous peoples, environmental change and tourism in extreme environments'. Online. Available at: www.psi.org.uk/ehb/projectskeenan.html (accessed 8 September 2004).

Keenan, J. (2003b) 'Indigenous peoples, environmental change and tourism in extreme environments project', Annual Report 2003. Online. Available at: www.psi.org.uk/ehb/ docs/annualreport-Keenan.pdf (accessed 8 September 2004).

Kohl, I. (2002) 'The lure of the Sahara: implications of Libya's desert tourism', *The Journal of Libyan Studies*, 3(2): 56–69.

Köppen, W. (1923) *Die Klimate der Erde*. Berlin: Walter de Gruyter and Company.

Leach, M. and Mearns, R. (eds) (1996) *The Lie of the Land: Challenging Received Wisdom on the African Environment*. London: James Currey and Heinemann.

Lindsay, W.K. (1987) 'Integrating parks and pastoralists: some lessons from Amboseli', in D. Anderson, and R. Grove (eds) *Conservation in Africa: People, Policies and Practice*. Cambridge: Cambridge University Press.

Martonne, E. de and Aufrere, L. (1927) 'Map of interior basin drainage', *Geographical Review*, 17: 414.

Mbaiwa, J.E. (2003) 'The socio-economic and environmental impacts of tourism development on the Okavango delta, north-western Botswana', *Journal of Arid Environments*, 54(2): 447–67.

Mbaiwa, J.E. (2004) 'Enclave tourism and its socio-economic impacts in the Okavango delta, Botswana', *Tourism Management* (in press, corrected proof published online 29 December 2003)

Meigs, P. (1953) 'World distribution of arid and semi-arid homoclimates', in *UNESCO Arid Zone Research Series No. 1, Arid Zone Hydrology*. Paris: UNESCO.

Midgley, G., Rutherford, M.C. and Bond, W. (2001) *The Heat is On: Impacts of Climatic Change on Plant Diversity in South Africa*. Cape Town: National Botanical Research Institute.

Mortimore, M.J. and Adams, W.M. (2001) 'Farmer adaptation, change and "crisis" in the Sahel', *Global Environmental Change*, 11(1): 49–57.

Murphree, M.W. (1993) 'Decentralizing the proprietorship of wildlife resources in Zimbabwe's communal lands', in D. Lewis and N. Carter (eds) *Voices from Africa: Local Perspectives on Conservation*. Washington, DC: World Wildlife Fund.

Neumann, R. (1995a) 'Local challenges to global agendas: conservation, economic liberalization and the pastoralists' rights movement in Tanzania', *Antipode*, 27(4): 363–82.

Neumann, R. (1995b) 'Ways of seeing Africa: colonial recasting of African society and landscape in Serengeti National Park', *Ecumene*, 2(2): 151–69.

Neumann, R. (1996) 'Dukes, earls, and ersatz edens: aristocratic nature preservationists in colonial Africa', *Environment and Planning D: Society and Space*, 14: 79–98.

Neumann, R. (1998) *Imposing Wilderness: Struggles over Livelihood and Nature Preservation in Africa*. Berkeley, CA: University of California Press.

Norton, A. (1996) 'Experiencing nature: the reproduction of environmental discourse through safari tourism in East Africa', *Geoforum*, 27(3): 355–73.

Noy-Meir, I. (1982) 'Stability of plant-herbivore models and possible application to savanna', in B.J. Huntley and B.H. Walker (eds) *Ecology of Tropical Savannas*. Berlin: Springer-Verlag.

O'Riordan, T. (2000) 'The sustainability debate', in T. O'Riordan (ed.) *Environmental Science for Environmental Management*. Harlow: Prentice Hall.

O'Riordan, T. and Voisey, H. (eds) (1998) *The Transition to Sustainability: The Politics of Agenda 21 in Europe*. London: Earthscan.

O'Riordan, T., Preston-Whyte, R.A., Hamann, R. and Manquele, M. (2000) 'The transition to sustainability: a South African perspective', *The South African Geographical Journal*, 82(2): 1–10.

Raunkaier, C. (1934) *The life forms of plants and statistical plant geography*. Oxford: Clarendon Press.

Rogers, J.A. (1981) 'Fools rush in, part 3: selected dryland areas of the world', *Arid Lands Newsletter*, 14: 24–5.

Shackley, M. (1996) 'Community impact of the camel safari industry in Jaisalmar, Rajasthan', *Tourism Management*, 17(3): 213–18.

Shackley, M. (1999) 'Tourism development and environmental protection in southern Sinai', *Tourism Management*, 20(4): 543–8.

Shantz, H.L. (1956) 'History and problems of arid lands development', in G.F. White (ed.) *The future of Arid Lands*. Washington, DC: American Association for the Advancement of Science.

Skarpe, C. (1991) 'Impact of grazing in savanna systems', *Ambio*, 20: 351–56.

Swift, J. (1996) 'Desertification: narratives, winners and losers', in M. Leach and R. Mearns (eds) *The Lie of the Land: Challenging Received Wisdom on the African Environment*. London: James Currey and Heinemann.

Theroux, P. (2003) *Dark Star Safari. Overland from Cairo to Cape Town*. London: Penguin Books.

Thomas, D.S.G. and Middleton, N.J. (1994) *Desertification: Exploding the Myth*. Chichester: John Wiley.

Thomas, N. and Brooks, S. (2003) 'Ecotourism for community development: environmental partnerships and the Il Ngwesi Ecotourism Project, northern Kenya', Special Issue on Tourism and Development in Africa, *Africa Insight*, 33: 9–17.

Thompson, M. and Homewood, K. (2002) 'Entrepreneurs, elites and exclusion in Maasailand: trends in wildlife conservation and pastoralist development', *Human Ecology*, 30(1): 107–38.

Thornthwaite, C.W. (1948) 'An approach towards a rational classification of climate', *Geographical Review*, 38: 55–94.

Tiffen, M, Mortimore, M. and Gichuki, F. (1994) *More People, Less Erosion: Environmental Recovery in Kenya*. Chichester: John Wiley.

Timberlake, L. (1985) *Africa in Crisis*. London: Earthscan.

Tucker, C.J., Dregne, H.E. and Newcomb, W.W. (1991) 'Expansion and contraction of the Sahara Desert from 1980 to 1990', *Science*, 253: 299.

United Nations (1977a) *World Map of Desertification*, UN Conference on Desertification, Nairobi, 29 August–9 September.

United Nations (1977b) *Desertification: Its Causes and Consequences*. Oxford: Pergamon Press.

United Nations Development Programme (1998) *Human Development Report 1998*. Oxford: Oxford University Press.

United Nations Environment Programme (1992) *World Atlas of Desertification*. Sevenoaks: Edward Arnold.

Watson, H.K. (1995) 'Mismanagement implications of vegetation changes in the Hluhluwe-Umfolosi Park', *South African Geographical Journal*, 77(2): 77–83.

Watts, M. (2000) 'Contested communities, malignant markets, and gilded governance: justice, resource extraction, and conservation in the tropics', in Zerner, C. (ed) *People, Plants and Justice: The Politics of Nature Conservation*. New York: Columbia University Press.

Watts, M. (2001) 'Petro-violence: community, extraction, and political ecology of a mythic commodity', in Peluso N. and Watts, M. (eds) *Violent Environments*. Ithaca, NY: Cornell University Press.

Western, D. and Wright, R.M. (eds) (1994) *Natural Connections: Perspectives in Community Based Conservation*. Washington, DC: Island Press.

White, R.P. and Nackoney, J. (2003) 'Drylands, people, and ecosystem goods and services: a web-based geospatial analysis', World Resources Institute. Online. Available at: http://biodiv.wri.org/pubs_description.cfm?PubID=3813 (accessed 12 September 2004).

Zerner, C. (ed.) (2000) *People, Plants and Justice: The Politics of Nature Conservation*. New York: Columbia University Press.

8 Tourism urbanisation and global environmental change

C. Michael Hall

At the start of the twenty-first century the world is more urbanised than ever before and the rate of urbanisation continues to grow. On a global scale, about three billion people, or approximately 48 per cent of the world's population live in urban areas and by the year 2030 this is set to rise to 61 per cent (United Nations (UN) 2004). It is estimated that in 2007 more than half of the world's population will live in urban areas. This will be the first time that the world's urban population has exceeded the rural population. The urban population reached one billion in 1960, two billion in 1985 and three billion in 2002. It is projected to rise to 4 billion in 2017 and 5 billion in 2030. During 2000–2030, the world's urban population is projected to grow at an average annual rate of 1.8 per cent, nearly double the rate expected for the total population of the world (almost 1 per cent per year). Given this expected rate of growth, the world's urban population will double in 38 years or in about half the lifetime of a person. However, almost all the expected growth in urban populations will be in the less developed countries, averaging 2.3 per cent population growth per year. In contrast, the urban population of the developed countries is expected to increase from 0.9 billion in 2003 to one billion in 2030, representing an annual growth rate of 0.5 per cent per annum, in contrast to the 1.5 per cent recorded during the previous 50 years. This slowing of the urban population growth rate is hardly surprising given that 74 per cent of the population in developed countries already lived in urban regions in 2003. Nevertheless, this figure is forecast to increase to 82 per cent by 2030 (UN 2004).

There are also substantial regional differences with respect to the degree of urbanisation. For example, Latin America and the Caribbean are highly urbanised, with 77 per cent of their population living in urban areas in 2003. This proportion is twice as high as for Africa and Asia, which had 39 per cent of their populations living in urban areas in 2003. It is expected that Africa and Asia will experience rapid rates of urbanisation during 2000–2030, so that by 2030 54 per cent and 55 per cent, respectively, of their inhabitants will live in urban settlements. Over the same time period, 85 per cent of the population of Latin America and the Caribbean will have become urbanised. In Europe and Northern America, the percentages of the population living in urban areas are expected to rise from 73 per cent and 80 per cent, respectively, in 2003, to 80 per cent and 87 per cent in 2030. The increase in Oceania is likely to be from 73 per cent to 75 per cent over the same period (UN 2004).

It is not just the proportion of the world's population that live in urban areas which is growing, but also the size of those urban areas. Over the past 50 years more and more cities have populations of one million or more inhabitants. With some rare exceptions (for example, Imperial Rome and Edo in the thirteenth century), the city of several million inhabitants is a phenomenon of industrial modernity. In 1950, only 29 per cent of the world's population of 2.5 billion were urban dwellers and 83 per cent of the developing world's people were still living on the land. By 2030 it is expected that over 60 per cent of the world's 8.13 billion global citizens will be living in towns and cities (UN 2004). In 1990, the average size of the world's 100 largest cities was around 5.1 million inhabitants, compared to 2.1 million in 1950, around 700,000 in 1900 and just under 200,000 in 1800 (Sadowski *et al.* 2000).

In the second half of the nineteenth century London became the first city to have several million inhabitants. By 2003, there were 46 cities with more than five million people, including 20 'megacities' with more than 10 million people (UN 2004). With 35 million inhabitants in 2003, Tokyo is by far the world's most populous urban agglomeration. After Tokyo, the next largest urban agglomerations in the world in 2003 were Mexico City (18.7 million), New York/Newark (18.3 million), São Paulo (17.9 million) and Mumbai (Bombay) (17.4 million). In 2015, it is expected that Tokyo will still be the largest urban agglomeration with 36 million inhabitants, followed by Mumbai (22.6 million), Delhi (20.9 million), Mexico City (20.6 million) and São Paulo (20 million). It is projected that by 2015, 22 cities will have populations of over 10 million, with all but five of these megacities being located in the less developed world. The population of these 22 cities in 2015 will be about 358 million – 75 million more than today, but still only about 5 per cent of the expected global population of over seven billion.

In 2003, 33 of the 46 cities with five million inhabitants or more were in less developed countries, and by 2015, 45 out of the 61 cities are expected to be from the less developed regions (UN 2004). Although considerable attention is given to the growth of these megacities, and the problems of urbanisation that they will face, it should be noted that it is predicted that by 2030 three quarters of the world's anticipated population growth will live in cities with populations between one and five million and 16 per cent in cities of over five million people. Given this scale of change it is therefore hardly surprising that the UN (2004: 3) reported:

> [three quarters] of all governments report that they are dissatisfied with the spatial distribution of their populations ... The speed and scale of this [urban population] growth, especially concentrated in the less developed regions, continue to pose formidable challenges to the individual communities as well as the world community.

Given the growth of the world's population in cities it is therefore unsurprising that urbanisation is a significant factor in global environmental change (GEC), constituting not only a major factor in land use and land cover in the immediate area of urbanisation but having relational effects on other regions because of the demands the cities generate for energy, materials, natural resources and food, as well as disposal of urban wastes, particularly in peri-urban areas (Tacoli 1998; Allen *et al.* 1999; Pickett *et al.*, 2001; Gurjar and Lelieveld, 2004). Nevertheless, as Kötter (2004: 8) recognised:

> Urban agglomerations are complex and dynamic systems that reproduce the interactions between socio-economic and environmental processes at a local and global scale. Despite their importance for economic growth, social well-being and sustainability of present and future generations, urban areas have not received the level of attention they require in the study of global environmental change

(see earlier comments by Setchell (1995) and Satterthwaite (1997) in relation to issues of sustainable development).

Urban systems *per se* actually occupy only a small percentage of the world's land surface – just 1 per cent (Grubler 1994). Although clearly there will be bioregions of the world where this figure will be much higher. Indeed, Allen *et al.* (1999) argue that in countries with high population growth rates, urban development can actually reduce land pressures in rural areas that are becoming too densely populated. Nevertheless, the expansion of cities clearly has an environmental impact in terms of land conversion. For example, in São Paulo, Brazil, the urban core grew from an area of 180 km^2 in 1930 to more than 900 km^2 in 1988. The metropolitan region is even larger, covering 8000 km^2. While the extent of land conversion through urbanisation is substantial, it is just as important to recognise that certain types of land are being converted. In the case of São Paulo:

> Prime agricultural land and forest have been converted to urban uses, and development is beginning to move onto steep slopes, which include some of the region's last remaining reserves of natural vegetation. Urban expansion is also threatening the local watershed.
>
> (WRI 1996: 59)

Concentrating economic activity and consumption in cities has both direct and indirect environmental impacts. The direct environmental impacts are the result of producing levels of pollution which environmental resources such as water bodies, acting as waste sinks, cannot sustainably absorb. 'The breakdown of local and global ecosystems and the health consequences of such levels of pollution are manifest' (Forbes and Lindfield 1998: 9). As centres of consumption and production as well as a specific land use cities can have a substantial ecological footprint (Rees 1992). For example, the United Nations Environment Programme – International Environment Technology Centre (UNEP–IETC 2003) states that the ecological footprint of the Greater Tokyo area is 3.5 times the land area of Japan as a whole, while London's is equal to the land area of UK. However, the role of cities in global change is regarded as more than environmental in scope. Processes of contemporary economic and socio-cultural globalisation and associated global change are seen by many commentators as being inextricably linked to the growth of world cities as well as competition to be a world city. 'The result is a "smaller" world, in which our lives are lived and shaped through the global metropolitanism of "larger" cities' (Knox 1995: 232). Significantly for globalisation processes, embedded within larger cities 'are the nodal points of a "fast world" of flexible production systems and sophisticated consumption patterns' that can be contrasted with 'the "slow world" of catatonic rural settings, declining manufacturing regions, and disadvantaged slums, all of which are increasingly disengaged from the culture and lifestyles of world cities' (Knox 1995: 232–3). Indeed, the footprint of the city is clearly more than ecological in scope although it has substantially shifted the dependence of cities on their surrounding bioregion. As Harvey (1996: 412, 413) observed:

> Each bundle of innovations [transportation, communications, etc.] has allowed a radical shift in the way that space is organised and therefore opened up radically new possibilities for the urban process. Breaking with the dependency upon relatively confined bioregions opened up totally new vistas of possibilities for urban growth ... [T]he development of an interrelated and ultimately global network of cities drawing upon a variety of hinterlands permits an aggregate urban growth process radically greater than that achievable for each in isolation.

Tourism urbanisation

Tourism is embedded in processes of urbanisation in two fundamental ways. First, as the main driver behind urbanisation in places that are very specific urban production spaces for tourism and leisure, what Mullins (1994) has described as 'tourism urbanization'. Second, as a routine element of leisure production in urban space in which, although certain parts of urban land use may be substantially geared toward satisfying tourism consumers and leisure mobility, the city's economy is not dominated by tourism and leisure production (Page and Hall 2003). Mullins (1991:, 326) explained the phenomenon of tourism urbanisation as:

> Cities providing a great range of consumption opportunities, with the consumers being resort tourists, people who move into these centres to reside for a short time ... in order to consume some of the great range of goods and services on offer.

The purpose of consumption is pleasure-related. Mullins (1994) described tourism urbanisation as a new urban form, although such an argument is highly debatable given the rapid development of resort towns for holidaymaking and day-tripping in the industrialising countries during the nineteenth century (Towner 1996). Examples would be towns such as Blackpool, Margate and Southend in the United Kingdom and Atlantic City in the United States. However, in the late twentieth century what did become different, and what characterises the 'new urban tourism' (Page and Hall 2003), is the scale, complexity and diversity of consumption experiences which now exist in urban landscapes built specifically for tourism and leisure as a result of processes of space–time distanciation and increased levels of disposable income. Several cities and regions around the world are now economically geared towards such consumption, including the Gold Coast and the Sunshine Coast in Australia, Honolulu, Reno and Las Vegas in the United States, Blackpool in the United Kingdom, Whistler and Niagara Falls in Canada, the Algarve in Portugal, and the Costa Brava in Spain. However, such cities and regions are only the more obvious examples of the many tourism and leisure specialised communities that have developed in the resort regions of the world. Tourism urbanisation 'based on the sale and consumption of pleasure' (Mullins 1991: 331), is identified by a number of characteristics (see also Page and Hall 2003), that makes such places:

- Spatially and functionally different from other urban places. For example, the amenity-driven nature of much tourism means that along many coastal areas tourism urbanisation is highly linear.
- Symbolically different, with various images and symbols as well as a commodified urban environment being used to promote the tourist function.
- Characterised by rapid population and labour force growth in the early stages of development. Even as urban growth slows, the population and labour force tends to have a relatively high degree of transience, underemployment and unemployment as a function of construction cycles and the temporal nature of tourism consumption and production.
- Distinguished by flexible forms of production, particularly in terms of a highly flexible labour force organised to meet daily, weekly and seasonal changes in consumption (usually through high-rates of part-time and casual employment and low-rates of unionisation).
- Dominated by state intervention which has a 'boosterist' tendency, whereby government indirectly invest in the facilities infrastructure with a view to encouraging further inward investment.
- Associated with large-scale pleasure production that simplifies the local economy and requires substantial importation of goods, services, water and energy from outside the resort region in order to meet the demands of a highly mobile population.
- Subject to substantial transformation of amenity landscapes for tourism production and consumption.
- Foci of transport networks because of the need to import and export not only goods and services but also the tourists themselves.

Tourism urbanisation tends to be focused in high value amenity environments and is associated with other forms of amenity-related urbanisation, particularly in coastal areas (Mullins 1990, 1999; Hall 2005). Nearly three quarters of the world's population live within 100km of a sea coast or lake shore (CO-DBP 1999) and urbanisation in coastal areas is increasingly regarded as being a global problem (e.g. German Federal Agency for Nature Conservation 1997; Intergovernmental

Oceanographic Commission 2000; United Nations Centre for Human Settlements (Habitat) 2001). For example, urbanisation and urban sprawl are a major problem along large areas of Europe's temperate and Mediterranean coastlines. More than 70 per cent of the coast from Barcelona to Naples had been developed by 2000 (CO-DBP 1999). Over a third of Europe's total population live within 50km of Europe's coasts and that figure is growing. By 2025, the percentage of the population of Spain, France, Greece, Italy and the former Yugoslavia living in coastal cities is projected to be more than 85 per cent on average, and as high as 96 per cent in Spain (Stanners and Bourdeau 1995; CO-DBP 1999). The environmental impact of urbanisation of the Mediterranean cost is, of course, greatly exacerbated by the scale of visitation by temporary residents who have second homes along the coast, as well as the large numbers of tourists. As the Parliamentary Assembly, Council of Europe (2003: 2.28) state:

There is little doubt that human-induced causes, such as population pressures, urbanisation, over-construction, and ill-planned development (as well as protection) of the Mediterranean coasts have led to much of its deterioration or destruction. Many of these human-induced pressures stem from or are closely linked to tourism.

The World Wide Fund for Nature (WWF 2001) estimate that in 2000 the Mediterranean region received approximately 30 per cent of all international tourist arrivals. In 1999 international tourism totalled 219.6 million arrivals – by 2020 WWF estimate that this figure will have grown to approximately 355 million tourist arrivals, representing about 22 per cent of international tourist arrivals (WWF 2001; De Stefano 2004). These figures do not include the travel behaviour of domestic tourists who also utilise the coastal areas. According to the WWF (2001: 2): 'the projected growth of tourism development in the region will continue to damage landscapes, cause soil erosion, put pressure on endangered species, further strain available water resources, increase waste and pollution discharges into the sea and lead to cultural disruption.'

Tourism and amenity urbanization has therefore contributed to substantial pressures on Mediterranean coastal landscapes. Three quarters of the sand dunes of the Mediterranean coastline from Spain to Sicily have disappeared. According to the WWF (2001) this is mainly a result of urbanisation linked to tourism development. Tourism urbanisation is also held to be primarily responsible for the urbanisation of the Italian coast. In Italy over 43 per cent of the coastline is completely urbanised, 28 per cent is partly urbanised and less than 29 per cent is still free of construction. There are only six stretches of coast over 20km long that are free of construction and only 33 stretches between 10 and 20km long without any construction (WWF 2001).

Tourism urbanisation is also impacting the coast of Tunisia. According to De Stefano (2004), the urbanised coastal area of Tunisia extended for 140km and tourist areas occupied by hotels and second homes occupy approximately a further 80km of the total urbanised linear space. The combined urban coastline of 220km represents approximately 18 per cent of the total Tunisian coastline. De Stefano (2004) argues that when current and planned tourism projects are accounted for, about 150km of the shoreline will eventually be occupied by tourism and leisure facilities and infrastructure. She goes on to note that inappropriate siting of tourist infrastructures on foredunes is accelerating the process of beach erosion and altering the water dynamics of the coastal region. Similarly, in Cyprus, 95 per cent of the tourism industry is located within 2km of the coast (Loizidou 2003) placing the coastal environment under extreme pressure. Under the Cypriot land-use zoning system in 1997–98, 37 per cent of the coastline (in length) was zoned for tourism, 12 per cent for agriculture, 6 per cent for residences and 3 per cent for industry. The remainder (43 per cent) was zoned as open area/protected natural or as archaeological areas. However, it is expected that future revisions of land-use planning zones will see more of the coast zoned for tourism use (Loizidou 2003).

Ecosystem stress

The focus of tourism urbanisation on the coast also means that many of the Mediterranean coastal ecosystems are under severe stress, with over 500 plant species threatened with extinction. Although the most severe environmental pressure is on the land–sea interface, the Mediterranean Sea is also suffering from severe environmental stress with over 10 billion tones of industrial waste per year estimated to be dumped in the Mediterranean with little or no purification. No figures are available specifically on the amount of waste generated by tourists. However, in OECD countries, municipal waste per capita increased from 410kg to 510kg per year from 1980 to 1995, and total waste generated increased from 347 million tonnes to 484 million tonnes within the same period (United Nations Centre for Human Settlements (Habitat) 2001: 69).

The generation of waste and therefore the placing of increased stress on ecosystems is usually exacerbated by the seasonal pressures placed on sewage and water systems by tourism (German Federal Agency for Nature Conservation 1997). Indeed, one of the most substantial impacts of tourism urbanisation in the Mediterranean is regarded to be the overall pressure that is being placed on freshwater supplies, particularly through aquifer overexploitation (CO-DBP 1999). Not only because of direct demands of tourism for the immediate water needs of tourists but also because of the impacts of tourism urbanisation on coastal wetlands and lagoons (De Stefano 2004).

The main causes of impact on freshwater ecosystems because of tourism urbanisation are as follows.

- Higher water consumption due to population increase – this includes both tourists and the flexible workforce required for tourism production. For example, in the Balearic Islands (Spain), water consumption during the peak tourism month in 1999 (July) was equivalent to 20 per cent of that by the entire local population in the entire year (De Stefano 2004).
- Higher consumption of water for tourist facilities.

- Peaks in wastewater volumes and the stresses that causes for wastewater treatment facilities. Scoullos (2003) reports that only 80 per cent of the effluent of residents and tourists in the Mediterranean is collected in sewage systems with the remainder being discharged directly or indirectly into the sea or to septic tanks. However, only half of the sewage networks are actually connected to wastewater treatment facilities with the rest being discharged into the sea. The United Nations Environment Programme Mediterranean Action Plan Priority Actions Programme (UNEP/MAP/PAP 2001) estimated that 48 per cent of the largest coastal cities (over 100,000 inhabitants) have no sewage treatment systems, 10 per cent possess a primary treatment system, 38 per cent a secondary system and only four per cent a tertiary treatment system.
- Inappropriate siting of tourism facilities and infrastructure on foreshores, dune systems and wetlands. UNEP/MAP/PAP (2001: 14) observed that mass tourism exacerbates issues of urbanisation impacts: 'leading to habitat loss for many wildlife species', and estimated that, since Roman times, the wetland area of approximately three million hectares then has been reduced by 93 per cent. Of this, one million hectares has been lost in the last 50 years (Parliamentary Assembly, Council of Europe 2003: 2.12).

Construction

The Mediterranean experience serves to highlight the extent to which tourism urbanisation acts as a particular urban form. Loizidou (2003) refers to a long 'coastal wall' of tourism development on the island of Cyprus. Such urban structures will clearly not only have long-term environmental effects but will also have significant impacts during their construction phase. According to UNEP-IETC (2003), on average the construction industry accounts for over 35 per cent of total global CO_2 emissions (building operation 10.2 per cent, business operation 9.2 per cent, materials production 10.9 per cent, transport 5 per cent, construction work 1.3 per cent), more than any other industrial activity. Significantly, the construction industry also tends to have substantial impacts on the peri-urban area of large cities as bulky and low value materials, such as building materials, are usually drawn from the immediate region of cities. This results in the proliferation of extractive activities such as claypits, quarries, brickworks and gravel pits thereby further extending the ecological footprint of cities. The CO-DBP (1999), in their examination of the impacts of urbanisation in the coastal regions of Europe, estimated that quarrying of sand and mineral aggregates for the construction of urban dwellings represents approximately 20 per cent of the total land lost to urbanisation.

Climate

Urban areas also have significant climatic effects. Urban form interacts with solar and anthropogenic radiation; absorbing and emitting heat, and leading to the urban heat island (UHI) effect (Baker *et al.* 2002; Arnfield 2003; Dixon and Mote 2003;

Samuels 2003). The relationship of urban versus rural climatic data to population change is pronounced over time (Brazel *et al.* 2000). However, the overall contribution of urban effects on twentieth-century globally and hemispherically averaged land-air temperature-time series do not exceed about 0.05°C over the period 1900 to 1991 (IPCC 2001a; see also Jones *et al.* 1990, 1999; Easterling *et al.* 1997) although the IPCC do note that 'greater urbanization influences in the future cannot be discounted' (2001: 105).

The greatest effect of urban areas on climate is their function as heat islands. Yet these are meso- and micro-climatic effects rather than an impact on the global climate. Nevertheless, given the number and distribution of urban areas around the globe, such UHI effects are global in distribution. No literature exists on the specific climatic dimensions of tourism urbanisation, although bioclimatic conditions and the thermal environment have been found to be significant impacts on leisure behaviour in urban parks (Thorsson *et al.* 2004). High-rise, high thermal-mass building canyons like those found in many central business districts also serve to magnify the impact of the built environment on urban climate. In a study of climate variablility of urban sites in Athens, Greece, Santamouris *et al.* (2001) reported that the average daily heat island intensity for the urban sites was approximately 10°C with a maximum value of around 15°C. Increases in temperature also result in increased rates of ozone formation (Duefias *et al.* 2002). However, much of the UHI effect can be mitigated by increasing tree density (Saito *et al.* 1990), which also has benefits in terms of reducing energy emissions and carbon combustion (Akbari 2002).

Differences between urban and rural/natural meso-climates occur on two different scales. First, the urban canopy layer – which is the air contained between the urban roughness elements, usually buildings – and its condition is determined by the nature of the site materials and geometry of the immediate surroundings. Second, the urban boundary layer which is the portion of the planetary boundary layer whose characteristics are determined by the urban region at its lowest boundary, usually regarded as roof level (Oke 1976; Williamson and Erell 2001). There are five main areas of difference between the climate of urban areas and that of the rural/natural surroundings: the radiation budget, sub-surface (storage) heat flux, advection (horizontal convection), anthropogenic heat release, and turbulent heat transfer including the effects of vegetation. There is no evidence to suggest that tourism urbanised area will be any different from other urban areas, with the possible exception of the linear form of tourism urbanised spaces in coastal regions and the deliberate development of some resorts in peripheral or alpine locations. For example, if evidence from other locations with substantial winter snowfall (e.g. Hinkel *et al.* 2003) was to be extrapolated to winter ski resorts then it is highly likely that the urbanised areas of ski resorts experience earlier snow melt and generate a significant heat island effect. Nevertheless, the effects of climate change on human comfort levels arguably need to be better understood in urban areas than they do in other tourism locations because of the complicating effects of urban micro- and meso-climates on comfort. However, climate knowledge has been poorly utilised in urban planning (Eliasson 2000), let alone tourism planning.

Conclusions

Tourism urbanisation and tourism and leisure mobilities in urban areas lead to environmental change on a global scale, with some regions experiencing more substantial change than others by virtue of the concentration of tourism urbanisation in desirable amenity landscapes. These environments, of which the Mediterranean is the most noted example, are undergoing severe environmental stress. Gössling (2002) estimated that an area of more than $514,950km^2$ had been affected by tourism-related land alterations. However, as with nearly all urban areas, the ecological footprint of such tourism land use is clearly far greater. Yet such an assessment does not provide for an understanding of the environmental footprint of tourism development that considers the 'cradle-to-grave' impact of that development upon environmental capacity (Hall 2000; Ravetz 2000). Moreover, tourism urbanisation arguably has certain characteristics that only serve to reinforce other elements of GEC processes because of the extent to which it links social, economic and environmental dimensions of GEC (O'Brien and Leichenko 2000). In addition, tourism urbanisation has spatial and locational characteristics that tend to concentrate land-use impacts in certain environments, particularly coastal and mountain areas, as well as in the peri-urban regions of urban centres as a result of day-trip activity, although the networks that link such areas with generating regions are also clearly significant. It is, therefore, perhaps ironic that it is precisely the areas in which tourism urbanisation is most likely to occur that are arguably the most vulnerable to the environmental and economic impacts of global climate change through sea-level rise and global warming.

Undoubtedly, low lying coastal areas are among the most vulnerable to sea-level increase and storm surges. Coastal populations are continuing to grow rapidly in developed countries as a result of amenity and lifestyle migration and tourism development. However, it should be noted that globally the coastline is a major focal point of human settlement with approximately a quarter of the world's population living in the near-coastal zone (Nicholls 1995; Small and Nicholls 2003). Nicholls (2004) reports on a range of coastal flooding scenarios and highlights that millions of people will be affected, even given assumptions of flood mitigation and protection measures. Obviously, many of these populations will be in developing countries and in major river estuaries such as in Egypt and Bangladesh. Nevertheless, the continuing development of coastal regions as a result of tourism urbanisation, such as in Florida, coastal Mediterranean and south-east Queensland is seemingly likely to be under increased threat as a result of sea-level increases and potential climate variability. Biogeophysical effects include flooding and storm damage, wetland loss and change, coastal erosion, saltwater intrusion and rising water tables (Nicholls and Lowe 2004). Many of these effects will potentially impact future coastal tourism urbanisation. Nevertheless, in many locations the lure of 'a place by the beach' seems to be stronger than the threat of environmental change. In such locations, it may well need a storm event to encourage new behaviours with respect to coastal planning and development rather than predictions.

References

Akbari H. (2002) 'Shade trees reduce building energy use and CO_2 emissions from power plants', *Environmental Pollution*, 116(Supplement–1), S119–26.

Allen, A. with da Silva, N.L.A. and Corubolo, E. (1999) *Environmental Problems and Opportunities of the Peri-urban Interface and Their Impact Upon the Poor*. London: Development Planning Unit.

Arnfield, A.J. (2003) 'Two decades of urban climate research: a review of turbulence, exchanges of energy and water, and the urban heat island', *International Journal of Climatology*, 23: 1–26.

Baker, L.A., Brazel, A.J., Selover, N., Martin, C., McIntyre, N., Steiner, F.R., Nelson, A. and Musacchio, L. (2002) 'Urbanization and warming of Phoenix (Arizona, USA): impacts, feedbacks and mitigation', *Urban Ecosystems*, 6: 183–203.

Brazel, A., Selover, N., Vose, R. and Heisler, G. (2000) 'The tale of two climates – Baltimore and Phoenix urban LTER sites', *Climate Research*, 15: 123–35

CO-DBP (Committee for the Activities of the Council of Europe in the Field of Biological and Landscape Diversity) (1999) *European Code of Conduct for Coastal Zones*. Strasbourg: CO-DBP (99)11, Secretariat General Direction of Environment and Local Authorities.

De Stefano, L. (2004) *Freshwater and Tourism in the Mediterranean*. Rome: WWF Mediterranean Programme.

Dixon, P.G. and Mote, T.L. (2003) 'Patterns and causes of Atlanta's urban heat island-initiated precipitation', *Journal of Applied Meteorology*, 42: 1273–84.

Duefias, C., Femtindez, Cafiete, S., Carrentero, I. and Liger, E. (2002) 'Assessment of ozone variations and meteorological effects in an urban area in the Mediterranean Coast', *The Science of Total Environment*, 299 (1–3): 97–113.

Easterling, D.R., Horton, B., Jones, P.D., Peterson, T.C., Karl, T.R., Parker, D.E., Salinger, M.J., Razuvayev, V., Plummer, N., Jamason, P. and Folland, C.K. (1997) 'Maximum and minimum temperature trends for the globe', *Science*, 277: 364–7.

Eliasson, I. (2000) 'The use of climate knowledge in urban planning', *Landscape and Urban Planning*, 48: 31–44.

Forbes, D. and Lindfield, M. (1998) *Urbanisation in Asia: Lessons Learned and Innovative Responses*. Canberra: AusAID Australian Agency for International Development.

German Federal Agency for Nature Conservation (1997) *Biodiversity and Tourism: Conflicts on the World's Seacoasts and Strategies for Their Solution*. Berlin: Springer Verlag.

Gössling, S. (2002) 'Global environmental consequences of tourism', *Global Environmental Change*, 12: 283–302.

Grubler, A. (1994) 'Technology', in W.B. Meyer and B.L. Turner II (eds) *Changes in Land Use and Land Cover: A Global Perspective*. Cambridge: Cambridge University Press.

Gurjar, B. R. and Lelieveld, J. (2004) 'New directions: Megacities and global change', *Atmospheric Environment*, 39: 391–93.

Hall, C.M. (2000) *Tourism Planning*. Harlow: Prentice-Hall.

Hall, C.M. (2005) *Tourism: Rethinking the Social Science of Mobility*. Harlow: Prentice-Hall.

Harvey, D. (1996) *Justice, Nature and the Geography of Difference*. London: Blackwell.

Hinkel, K.M., Nelson, F.E., Klene, A.F. and Bell, J.H. (2003) 'The urban heat island in winter at Barrow, Alaska', *International Journal of Climatology*, 23: 1889–1905.

IPPC (Intergovernmental Panel on Climate Change) *Climate Change 2001: The Scientific Basis, Contribution of Working Group I to the Third Assessment Report of the Intergovernmental Panel on Climate Change*. Cambridge: Cambridge University Press.

Intergovernmental Oceanographic Commission (2000) *IOC-SOA International Workshop on Coastal Megacities: Challenges of Growing Urbanisation of the World's Coastal Areas Organised in co-operation with the International Ocean Institute (IOI), Malta* Hangzhou, People's Republic of China, 27–30 September 1999. Workshop Report No.166. Paris: Intergovernmental Oceanographic Commission.

Jones, P.D., Groisman, P.Y., Coughlan, M., Plummer, N., Wang, W.C. and Karl, T.R. (1990) 'Assessment of urbanization effects in time series of surface air temperature over land', *Nature*, 347: 169–72.

Jones, P.D., New, M., Parker, D.E., Martin, S. and Rigor, I.G. (1999) 'Surface air temperature and its changes over the past 150 years', *Review of Geophysics*, 37: 173–99.

Knox, P.L. (1995) 'World Cities and the organization of global space', in R.J. Johnston, P.J. Taylor and M.J. Watts (eds) *Geographies of Global Change: Remapping the World in the Late Twentieth Century*. Oxford: Blackwell, pp.232–47.

Kötter, T. (2004) 'Risks and opportunities of urbanization and megacities', PS2 Plenary Session 2 – Risk and Disaster Prevention and Management, PS2.2 Risks and Opportunities of Urbanisation and Megacities, FIG Working Week 2004, Athens, Greece, May 22–27.

Loizidou, X. (2003) 'Land use and coastal management in the Eastern Mediterranean: the Cyprus example', *International Conference on the Sustainable Development of the Mediterranean and Black Sea Environment*, May, Thessaloniki, Greece.

Mullins, P. (1984) 'Hedonism and real estate: Resort tourism and Gold Coast development', in P. Williams (ed.) *Conflict and Development*. Sydney: Allen & Unwin, pp.31–50.

Mullins, P. (1990) 'Tourist cities as new cities: Australia's Gold Coast and Sunshine Coast', *Australian Planner*, 28(3): 37–41.

Mullins, P. (1991) 'Tourism urbanization', *International Journal of Urban and Regional Research*, 15(3): 591–7.

Mullins, P. (1994). 'Class relations and tourism urbanisation: The regeneration of the petite bourgeoisie and the emergence of a new urban form', *International Journal of Urban and Regional Research*, 18(4): 591–607.

Mullins, P. (1999). 'International tourism and the cities of Southeast Asia', in D. Judd and S. Fainstein (eds) *The Tourist City*. New Haven, CT: Yale University Press, pp.245–60.

Nicholls, R.J. (1995) 'Coastal megacities and climate change', *Geojournal*, 37(3): 369–79.

Nicholls, R.J. (2004) 'Coastal flooding and wetland loss in the 21st century: changes under the SRES climate and socio-economic scenarios', *Global Environmental Change*, 14: 69–86.

Nicholls, R.J. and Lowe, J.A. (2004) 'Benefits of mitigation of climate change for coastal areas', *Global Environmental Change*, 14: 229–44.

O'Brien, K. and Leichenko, R. (2000) 'Double exposure: Assessing the impact of climate change within the context of economic globalization', *Global Environmental Change*, 10(3): 221–32.

Oke, T.R. (1976) 'The distinction between canopy and boundary-layer urban heat islands', *Atmosphere*, 14: 268–77.

Parliamentary Assembly, Council of Europe (2003) *Erosion of the Mediterranean coastline: implications for tourism*, Doc.9981 16 October 2003, Report Committee on Economic Affairs and Development, online. Available at http://assembly.coe.int/Documents/WorkingDocs/doc03/EDOC9981.htm (accessed 25 January 2005)

Page, S. and Hall, C.M. (2003) *Managing Urban Tourism*. Harlow: Prentice-Hall.

Pickett, S.T.A., Cadenasso, M.L., Grove, J.M., Nilon, C.H., Pouyat, R.V., Zipperer, W.C. and Costanza, R. (2001) 'Urban ecological systems: Linking terrestrial ecological, physical, and socioeconomic components of metropolitan areas', *Annual Review of Ecology and Systematics*, 32: 127–57.

Ravetz, J. (2000) 'Integrated assessment for sustainability appraisal in cities and regions', *Environmental Impact Assessment Review*, 20: 31–64.

Rees, W.E. (1992) 'Ecological footprints and appropriated carrying capacity: What urban economics leaves out', *Environment and Urbanisation*, 4(2): 121–30.

Sadowski, A., Lau, S. and Mahtab-uz-Zaman, Q.M. (2000) *Megacities: Trends and issues towards sustainable urban development*. Document prepared for Megacities 2000 Conference, MegaCities Research Group, Hong Kong University.

Saito I., Ishimara, O. and Katayama T. (1990) 'Study of the effect of green areas on the thermal environment in an urban area', *Journal of Energy and Buildings*, 15–16: 445–6.

Samuels, R. (2004) Urban heat islands. Submission to the House of Representatives Standing Committee on Environment and Heritage Sustainable Cities 2025 Inquiry, Canberra. Online. Available at www.aph.gov.au/house/committee/ environ/cities/subs/sub34.pdf (accessed 10 May 2005).

Santamouris, M. (2001) 'The canyon effect', in M. Santamouris (ed.) *Energy and Climate in the Urban Built Environment*. London: James & James.

Santamouris, M., Papanikolaou, N., Livada, I., Koronakis, I., Georgakis, C., Argiriou, A. and Assimakopolous, D.N. (2001) 'On the impact of urban climate on the energy consumption of buildings', *Solar Energy*, 70(3): 201–16.

Satterthwaite, D. (1997) 'Sustainable cities or cities that contribute to sustainable development?', *Urban Studies*, 34(10): 1667–91.

Scoullos, M.J. (2003) 'Impact of anthropogenic activities in the Coastal Region of the Mediterranean Sea', *International Conference on the Sustainable Development of the Mediterranean and Black Sea Environment*, May, Thessaloniki, Greece.

Setchell, C.A. (1995) 'The growing environmental crisis in the world's megacities: The case of Bangkok', *Third World Planning Review*, 17(1): 1–18.

Small, C. and Nicholls, R.J. (2003) 'A global analysis of human settlement in coastal zones', *Journal of Coastal Research*, 19(3): 584–99.

Stanners, D. and Bourdeau, P. (eds) (1995) *Europe's Environment: The Dobrís Assessment*. Copenhagen: European Environment Agency.

Tacoli, C. (1998) 'Rural-urban interactions; a guide to the literature', *Environment and Urbanization*, 10(1): 147–66.

Thorsson, S., Lindqvist, M. and Lindqvist, S. (2004) 'Thermal bioclimatic conditions and patterns of behaviour in an urban park in Goteborg, Sweden', *International Journal of Biometeorology*, 48: 149–156.

Towner, J. (1996) *An Historical Geography of Recreation and Tourism in the Western World 1540–1940*. Chichester: John Wiley.

United Nations (2004) *World Urbanization Prospects: The 2003 Revision. Data Tables and Highlights*. New York: Department of Economic and Social Affairs, Population Division.

United Nations Centre for Human Settlements (Habitat) (2001) *The State of the World's Cities 2001*. Nairobi: United Nations Centre for Human Settlements.

UNEP/MAP/PAP (United Nations Environment Programme Mediterranean Action Plan Priority Actions Programme) (2001) *White Paper: Coastal Zone Management in the Mediterranean*. Split: Priority Actions Programme.

UNEP–IETC (United Nations Environment Programme – International Environment Technology Centre) (2003) *Cities Are Not Cities: Need For a Radical Change in Our Attitudes and Approaches to Manage the Environment in Cities*. Osaka: UNEP – DTIE International Environmental Technology Centre.

Williamson, T.J. and Erell, E. (2001) 'Thermal performance simulation and the urban micro-climate: measurements and prediction', *Building Simulation*, Seventh International IBPSA Conference, Rio de Janeiro, Brazil, 13–15 August, pp.159–65.

WRI (World Resources Institute) (1996) *World Resources. A Guide Publication to the Global Environment. The Urban Environment 1996–97*. Oxford: Oxford University Press.

WWF (World Wide Fund for Nature) (2001) *Tourism Threats in the Mediterranean*. Rome: WWF Mediterranean Programme.

Part II

Global issues

9 Tourism, disease and global environmental change

The fourth transition?

C. Michael Hall

For many people the connection with health and tourism is often made with respect to a visit to a spa, a relaxing sea cruise or, given the present rate of obesity in the populations of the developed world, a trip to the 'fat farm'. However, the reality is that tourism is a major contributor to introduction of new diseases to populations as well as contributing to an increased rate of spread of existing disease.

Travel and trade have long been a major transmission vehicle for infectious diseases. Humans are significant vectors for disease, as well as being carriers of pests which may also host disease. Patterns of health and disease are the product of interactions between human biology and mobility and the social and physical environments. However, these patterns are constantly changing. Indeed, the contemporary health situation of the world is somewhat confused because even though death rates in most countries have declined substantially in the past century and people live longer, there is simultaneously the emergence of new disease risks, such as HIV/ AIDS, and the re-emergence of diseases that were at one time thought to have almost been eradicated, such as tuberculosis. Increased human mobility, including tourism, has become a major factor in the current and emerging patterns of disease, although it should be recognised as being just one of the most recent expressions of increased contact between human populations that at one time would have existed in isolation from one another. In addition, it must also be recognised that the spread of pests and disease is not limited to human disease, but that humans can also act as vectors of a range of pathogens that can severely impact other species. Issues surrounding disease spread therefore become a significant factor not only in the health of human populations but also in the maintenance of ecosystem health and biodiversity.

As the number of people in the world who have become mobile has increased, so it has meant that the rate of disease spread has grown as has the potential for populations to have contact with pathogens to which they have hitherto not been exposed. McMichael (2001) recognises three historical transitions in the co-evolutionary relationships between humans and disease (see also Gould 2002). The first stage corresponds to the rise of settled agriculture and the concentration of population in the early civilisations of the Middle East, South and East Asia and the Americas. According to McMichael (2001) this created a new web of relationships among animals, humans and microbes, facilitating the migration of microbes from animal to human populations:

Smallpox arose via a mutant pox virus from cattle. Measles is thought to have come from the virus that causes distemper in dogs, leprosy from water buffalo, the common cold from horses, and so on.

(McMichael 2001: 101)

Most of the infectious 'crowd' diseases appear to have developed during this transition, although it is significant to note that the 'leap' from animal species to humans can still occur today, as evidenced by the recent HIV, SARS and avian bird flu epidemics. Significantly, if a disease is zoonotic – transferred from animal to humans – then it will remain in animal reservoirs, where it may mutate and strike again. Indeed, it is remarkable to note that writing before the 2003 SARS outbreak in southern China, McMichael commented with respect to that region that:

The intimate pig/duck farming culture creates a particularly efficient environment in which multiple strains of avian viruses infect pigs. The pigs act as 'mixing vessels', yielding new recombinant-DNA strains of virus which may then infect the pig-tending humans.

(2001: 88–9)

It was through such co-evolutionary relationships that each human population acquired its own range of locally evolving infectious diseases.

The second transition corresponds to the period of contact among the Eurasian civilisations from around 500BC to AD1500. McMichael builds on McNeill's (1976) thesis that this contact resulted in the transmission and swapping of microbes, leading to episodic epidemics followed by periods of gradual re-equilibration between the infectious agent and the human host population. One of the best examples of this being the complex pattern of outbreaks of the Black Death in mid-fourteenth-century Europe, which appears to be related to an outbreak of bubonic plague in China during the 1330s, as well as to the development of quite intensive trading networks (Flinn 1979; McEvedy 1988; Cohn 2003). As McMichael (2001: 108) notes: 'After many turbulent centuries, this transcontinental pooling resulted in an uneasy Eurasian equilibration of at least some of the major infectious diseases'.

The third great historical transition refers to Europe's exploration and conquest of distant lands, and the export of 'its lethal, empire-winning, germs to the Americas and later to the south Pacific, Australia and Africa' (McMichael 2001: 89) (see also Crosby 1972, 1986; White 1980; Joralemon 1982; Dobyns 1983; Henige 1986; Snow and Lanphear 1988; Snow and Starna 1989; Diamond 1997; Cook 1998, 2004; Potter 2001; Koplow 2003; Kelton 2004). Importantly, in his examination of the ecological imperialism of Europeans in the rest of the world, Crosby (1986) highlighted that, in addition to disease, Europeans also exported a range of other pests that would affect humans, indigenous animal and plant species and the composition of entire ecosystems. The impact of virgin soil epidemics during this period was enormous (Crosby 1976; Wolfe 1982; Cook 1998; see also Jones 2003). The large Aztec and Inca populations had limited numbers of wild animals they could domesticate and remained relatively free of 'crowd' infectious diseases until they were decimated by smallpox, measles and influenza introduced by Europeans (Crosby 1972; McMichael 2001; Fenn 2002; Koplow 2003; Cook 2004). Between 1519 and 1620, it is estimated that Mexico's population declined from 28 million to 1.6 million due to waves of measles, smallpox, typhus and influenza (S. Hunter 2003). *Falciparum malaria* and yellow fever were brought to the Americas in the seventeenth century by the trans-Atlantic slave trade (Curtin 1968; Milner 1980). In a series of outbreaks in the mid-eighteenth century, New Orleans and Memphis lost half their populations to yellow fever. As yellow fever and malaria wiped out the Amerindian population, the areas were subsequently settled by African populations who were adapted to both diseases (McNeill 1976; Kipple and Higgins 1992; S. Hunter 2003).

Similar impacts to those in the Americas were experienced by the indigenous peoples of the Pacific. The Aborigines of Australia suffered severe epidemics following the arrival of Europeans on that continent (Crosby 1986), and within 80 years of Captain Cook's first visit to Hawaii in 1778 the native population declined from around 300,000 to less than 40,000 (McMichael 2001). A smallpox epidemic in Hawaii in 1853 killed thousands of people, an estimated 8 per cent of the population, despite quarantining and vaccinating (Greer 1965; Schmitt 1970). In the case of Siberia, smallpox appeared for the first time in 1630 as Russians travelled there – the death rate among indigenes in a single epidemic could be over 50 per cent, When smallpox first arrived in Kamchatka Peninsula in 1768–69, it is estimated that it killed between 66 per cent and 75 per cent of the indigenous population (Crosby 1986). McMichael (2001) observed that the third transition was more a 'dissemination' of microbes that had co-evolved with the Eurasian population to other parts of the world than an 'exchange' (Crosby 1972). Nevertheless, the third transition represented another major process re-equilibrating the balance between microbes and humans, this time across transoceanic populations and on a global scale.

Are we experiencing a fourth major historical transition today? Around 1970 many experts thought that tuberculosis, cholera and malaria would soon become extinct. The United States Surgeon-General declared that it was time 'to close the book on infectious diseases' (in McMichael 2001: 88). But now these diseases are increasing again, and a host of other diseases or their pathogens have been newly identified – they include Lyme disease, hepatitis C and E, human herpes viruses 6 and 8, hanta virus, cryptosporidiosis, toxic shock syndrome, Ebola virus, Legionnaires' disease as well as various food-borne disease outbreaks. As McMichael (2001: 115) observes, something 'unusual' seems to be happening to patterns of infectious diseases whether they be human or animal (see also Dorolle 1969; Wilson 1994; Garrett 1996). Fidler (1999: 17) describes a 'new pathology of public health in the era of emerging and re-emerging infectious diseases'. Such emerging and resurgent diseases and viruses occur where pathogens have either 'newly appeared in the population or are rapidly expanding their range, with a corresponding increase in cases of disease' (Morse 1993a: 10).

Of great significance to this new historical transition is the role of international trade and travel as a channel for the spread of infectious disease through both human mobility as well as the mobility of vectors, what Morse (1993b) describes as 'viral traffic'. Since 1950 passenger kilometres travelled by planes have increased by a factor of over 50. Air travel increased total mobility per capita 10 per cent in Europe and 30 per cent in the United States between 1950 and 2000 (Ausubel *et al.* 1998). Spatial diffusion via increasingly mobile human vectors (people and their transport) is obviously a significant factor in the appearance of new pathogens in human populations. However, the globalisation of international trade and travel is not singularly responsible for new public health threats. Factors such as cross-species transfer; pathogenic evolution, or changes in the structure and immunogenicity of earlier pathogens (genetic drift and shift); recognition of a pre-existing pathogen; and changes in the environment and, hence, disease ecology are all significant in explaining the emergence of new diseases in human populations. Numerous commentators (e.g. Lederberg *et al.* 1992; McMichael 1993, 2001; Mayer 1996; Fidler 1999) point to the role of contemporary human-induced social-environmental changes which provide new opportunities for pathogens including: urbanisation, intravenous drug use, sexual practices, medical practices – for example, blood transfusion, organ transplants – intensive food production, poverty and inequality, and other changes to the physical environment such as irrigation, deforestation, and eutrophication of rivers. Indeed, McMichael argues that humanity is currently 'depleting or disrupting many of the ecological and geophysical systems that provide lifesupport' (2001: 283). In addition to these factors, Fidler (1999) also notes the roles of deterioating or non-existent national public health capabilities, failure of the internationalisation of public health and a weakening of state control in light of contemporary globalisation (see also Mayer 1996).

Although, in health terms, tourism only constitutes a relatively small fraction of total human movement (Bradley 1989; Wilson 1995; MacPherson 2001; Hall 2005), it is highly significant in terms of its potential to contribute to the spread of pathogens because it is a cross-border phenomena and, unlike migration, it implies a return to the location of origin (Table 9.1). With advances in transport technology and a loosening of economic, temporal, cultural, political and gendered constraints with respect to travel, international tourism has shown steady increase in the post-WWII period, with international tourism trips growing faster than the rate of population increase (Hall 2003a, 2005) (Figure 9.1, Table 9.2). The extent of time-space convergence because

Table 9.1 What is carried by humans when they travel

- Pathogens in or on body, clothes and/or luggage
- Microbiologic fauna and flora in or on body, clothes and/or luggage
- Vectors on body, clothes and/or luggage
- Immunologic sequelae of past infections
- Relative vulnerability to infections
- Genetic composition
- Cultural preferences, customs, behavioural patterns, technology
- Luggage may also contain food, soil, fauna, flora and organic material

---

Source: Wilson 1995; Hall 2004

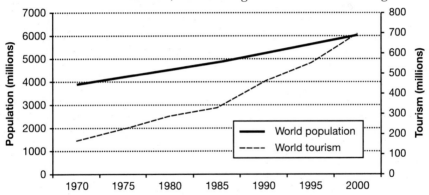

Figure 9.1 Growth in world population versus world tourism

Source: Hall 2004

Table 9.2 World population growth compared to growth in international tourism arrivals

Date	World population (million)	International tourism arrivals (million)	International tourism arrivals as a % of world population
1950	2,520.0	25.3	1.00
1960	3,020.0	69.3	2.30
1970	3,912.1	165.8	4.24
1975	4,205.1	222.3	5.29
1980	4,520.1	287.5	6.36
1985	4,858.8	327.9	6.75
1990	5,222.7	457.3	8.76
1995	5,614.0	552.3	9.84
2000	6,034.5	696.7	11.55
2001	6,122.4	692.7	11.31

Source: United Nations various; WTO various; in Hall 2003a

of changes in transport technology is well illustrated by the observations of Grübler and Nakicenovic (1991) who estimated and plotted the average kilometres travelled daily by the French population over a 200-year period (1800–2000) and found that spatial mobility has increased more than 1000-fold. Similarly, Cliff and Haggett (1995) noted that between the early 1950s and the early 1990s, the size of Australia's resident population doubled while the number of persons moving into and out of Australia increased nearly 100-fold (see also Cliff and Haggett 2004). Such mobility transformations are also significant for international trade. Over 70 per cent of fruit and vegetables consumed in the USA come from less developed countries, while more than 45 per cent of fish and fish products on the international market also come from developing countries, with consequent implications for the migration of pathogenic microbes and food-borne parasites (Fidler 1999).

Epidemics only occur when there is a sufficiently large pool of uninfected people to contract and transmit the disease, that is to say there must be a large enough reservoir of potential victims – or 'susceptibles' – for the disease to spread (Bell and Lewis 2005). For example, S. Hunter (2003) suggests that a minimum population of 250,000 is required to support a measles epidemic. As the disease spreads, it uses up the available pool of 'susceptibles' through either killing them or by conferring immunity on them. At some point, the epidemic will burn itself out but the disease will survive at low levels and re-emerge later as a new epidemic, either in an original or mutant form. For an outbreak to develop into a pandemic (global epidemic), the number of carriers must be large. Urbanisation assists in efficient transmission because of the number of people in close proximity, while geographic mobility enlarges the stock of 'susceptibles', as was the case with influenza in 1918–19 and SARS in 2003 (Bell and Lewis 2005). May's (1958) concept that, for an infectious disease to occur, there must be coincidence in time and space of agent and host is therefore significant for understanding the impact of tourism mobility on disease and environmental health, because an increase in tourist traffic in a given area increases the likelihood of contact occurring with a pool of uninfected people.

The increase in air travel has also increased the potential for disease spread for two reasons: first, the increased speed of travel means that people can travel and pass through airports before disease symptoms become evident; and second, the size of modern aircraft is increasing (Westwood 1980). In the latter case, Bradley (1989) postulated a hypothetical situation in which the chance of one person in the travelling population having a given communicable disease in the infectious stage was 1 in 10,000. With a 200-seat aircraft, the probability of having an infected passenger aboard (x) is 0.02 and the number of potential contacts (y) is 199. If homogeneous mixing is assumed, this means a combined risk factor (xy) of 3.98. If the aircraft size is doubled to 400 passengers, then the corresponding figures are: $x = 0.04$, $y = 399$, and $xy = 15.96$. On the new double-decker Airbus A380 scheduled to come into service in 2006, the passenger configuration ranges from 550 to almost 800 passengers in an all economy-class configuration (Hall 2005). If we assume a 600 seat aircraft then the corresponding figures are: $x = 0.06$, $y = 599$, and $xy = 35.94$. In other words tripling the number of seats available increases the risk factor nine-fold. Therefore, as Cliff and Haggett (1995) observed, new generation jet aircraft are significant for disease spread not only because of their speed but because of their size. Moreover, the larger the aircraft the more likely it will be that it is travelling between transport hubs in large urban areas. The implications of increased rates of mobility with respect to the globalisation of health and disease is discussed below, with the first example being that of an 'old' or 're-emerging' disease – malaria – followed by examples of emerging disease.

Malaria

Malaria affects at least 200–300 million people every year and causes between 1–2 million deaths, mostly children under five and pregnant women in sub-Saharan

Africa (Roll Back Malaria 2004). The economic burden is also extremely high, accounting for a reduction of 1.3 per cent in the annual economic growth rate of countries where malaria is endemic (Sachs and Malaney 2002). It is estimated that malaria costs Africa more than US$12 billion every year in lost GDP, even though it could be controlled for a fraction of that sum (Roll Back Malaria 2004). However, malaria is not limited to Africa – 40 per cent of the world's population is regarded as being at risk of acquiring the disease (Murphy and Oldfield 1996). In addition, each year approximately 30 million people from non-tropical countries visit areas where malaria is endemic (Kain and Keystone 1998), with between 10,000 and 30,000 visitors contracting malaria (Lobel and Kozarsky 1997). Significantly in terms of disease control and assessing the impacts of mobility, around 90 per cent of infected travellers do not become ill until they return home. This 'imported malaria' is easily treated, but only if it is diagnosed promptly (Croft 2000; Kain *et al.* 2001).

Increased tourism mobility to tropical and sub-tropical destinations where malaria is endemic, particularly for nature-based tourism activities, can have substantial consequences for infection rates in the tourist-generating region (Rudkin and Hall 1996; Dos Santos *et al.* 1999; Musa *et al.* 2004). In Canada, for example, Kain *et al.* (2001) reported that between 1994 and 2001 a record number of cases of imported malaria was witnessed, with a peak of 1,036 cases in Canadian travellers reported for 1997. This figure represents a 141 per cent increase since 1994 and a per capita rate about 10 times that reported in the USA (Humar *et al.* 1997). An additional concern in managing malaria is that there has also been a dramatic increase in the number of cases caused by drug-resistant parasites and an approximately 10-fold increase in the number of cases of severe malaria requiring admission to an intensive care unit (Kain and Keystone 1998). Severe malaria is associated with a case fatality rate that often exceeds 20 per cent, even among young, previously healthy adults, and may be complicated by adult respiratory distress syndrome – when this situation arises, case fatality rates often exceed 80 per cent (Kain *et al.* 2001).

AIDS

HIV/AIDS was estimated to claim 3.1 million lives in 2002 (Lee and McKibbin 2004). AIDS has been described as the first great pandemic of the twenty-first century and as the modern equivalent of the 'great plague' (May 2003). The spread of AIDS is interrelated with human mobility. As Garrett (2000: 551) comments:

> It spread swiftly … in a retrovirus form that used human DNA as vehicle and hideaway. Globalized sex and drug trades ensured HIV's ubiquity. And HIV, in turn, facilitated the circumnavigation of new, mutant forms of tuberculosis, the one taking advantage of the weakened human state caused by the other.

Globally, there is a tendency for socially disadvantaged groups to be at greater risk of HIV/AIDS. In the West in the 1980s this included gay men, minorities and injection drug users. The nature of these groups has often meant that AIDS has been portrayed as a disease of morality. However, in Africa the situation is quite different, with long-distance truck drivers and sex workers known as the earliest groups infected (Williams *et al*. 2002). Instead, AIDS has manifested as a primarily heterosexual disease, accounting for 93 per cent of adult infections in sub-Saharan Africa (Webb 1997: 5).

Transmission patterns, distribution, spread and impacts of the virus vary widely among populations, partly as a result of behavioural difference (Lindenbaum 1997). Webb (1997: 11) identified two main modes – Pattern I spread (primarily men having sex with men and injection drug users), that is primarily found in the USA, Western Europe and Australia; and Pattern II spread (heterosexual, with male to female ratio 1:1 and, increasingly, pediatric AIDS through vertical trans-mission), that primarily affects sub-Saharan Africa, Latin America, parts of South America and India. However, Webb (1997: xii) warns: 'The complexity of HIV spread is not to be underestimated,' noting:

> Many of the reasons for the failure of prevention programmes to date lie in this oversimplification, in reducing the epidemic to medical or health terms, to talk of HIV/AIDS as if socioeconomic processes were merely incidental … generalizations negate the reality of the great diversity and variety in the way people react to this unprecedented situation .
>
> (Webb 1997: xii–xiii)

Large-scale movements, including tourism, have increased high-risk sexual behaviour and sexually transmitted disease (STD) infection rates (De Schryver and Meheus 1989; Gössling 2002). This has meant considerable emphasis being placed on controlling the health dimensions of sex tourism as well as under-standing and influencing the sexual behaviours of tourists (Ford and Koetsawang 1991). In Southern Africa, migration plays a major role in the spread of STDs and HIV because of the long history of economic migration from Lesotho, Botswana, Zambia, Swaziland, Mozambique, Malawi, Zambia and Zimbabwe to the gold and diamond mines of South Africa. However, it is not migration alone that is respon-sible for HIV/AIDS spread, rather attention needs to be given to the role of labour return, whether on a permanent basis or, more significantly, trips home to visit family (Williams *et al*. 2002).

The significance of the relationship between mobility – which is often substan-tially gendered – sexual behaviour and HIV infection is indicated in Lydie *et al*.'s (2004) study of an urban population of Cameroon. A representative sample of 896 men and 1,017 women were interviewed and tested for HIV infection and other sexually transmitted infections in Yaounde in 1997. Mobile and non-mobile people were compared with respect to sociodemographic attributes, risk exposure, condom use and prevalence of HIV infection. In terms of mobility, 73 per cent of men and 68 per cent of women reported at least one trip outside of Yaounde in the preceding 12 months. Among men, the prevalence of HIV infection increased with time away from town. Men who declared no absence were five times less likely to be infected than were those away for more than 31 days. Mobile men also reported more risky sexual behaviours – i.e. more partners and more one-off contacts. However, for women, the pattern was less clear with differences in the prevalence of HIV infection being less marked between non-mobile and mobile women – 6.9 per cent versus 9.8 per cent, respectively (P > 0.1).

In the case of HIV/AIDS, its global significance lies not just in its immediate health implications but in its longer-term impacts and ripple effects (Bloom 1999; Bloom and Canning 2003). For example, 'wherever the percentage of HIV-positive adults exceeded ten percent of a given society waves of opportunistic secondary epidemics followed, notably of tuberculosis' (Garrett 2000: 553). In the case of TB it is estimated that about one third of AIDS patients are co-infected with TB, and that TB accounts for approximately one third of AIDS deaths worldwide (May 2003: 7).

According to UNAIDS (the Joint United Nations Programme on HIV/AIDS), at present 42 million people globally live with HIV/AIDS (in Lee and McKibbin 2004). In the southern African regions, the total HIV-related health service costs, based on assumed coverage rate of 10 per cent, is estimated to range from 0.3 to 4.3 per cent of GDP, thus placing enormous strain on longer-term development prospects. As S. Hunter (2003) notes, in economic terms the effect of AIDS on per capita income will make itself felt both through its destruction of existing human capital, a good part of which takes the form of experienced workers, and its weakening of the mechanisms through which new human capital is accumulated.

Arndt and Lewis (2000) estimated that between 1997 and 2010, the AIDS epidemic in South Africa will reduce that country's GDP by 17 per cent and its per capita income by about 8 per cent (or about 0.6 per cent annually). Similarly, MacFarlan and Sgherri (2001) modelled the likely macroeconomic impacts of AIDS in Botswana and conclude that the growth rate of GDP in the non-mining sector will slow by a third or more between 2000 and 2010 due to expected reductions in labour productivity and capital accumulation, stemming in part from a decline in the experience and skills of formal sector workers. In addition, a heavy fiscal impact from higher health spending is also expected (S. Hunter 2003).

SARS

Respiratory infections are the leading cause of human mortality (Raptopoulou-Gigi 2003: 81). In late 2002 the first reports came through in the Western media of an outbreak of a new respiratory disease in southern China which came to be known as SARS (Severe Acute Respiratory Syndrome). The first cases of SARS emerged in mid-November 2002 in Guangdong, China, and the first official report of 305 cases from the area were reported by the World Health Organization on 11 February 2003 – 30 per cent of those affected were health workers. In the six months after the first outbreak in Guangdong province, the SARS disease spread to at least 30 countries/regions including Australia, Brazil, Canada, South Africa, Spain and the USA. By the apparent end of the outbreak on 14 July 2003, the number of probable cases reached 8,437 worldwide – the most affected countries in terms of numbers of cases and deaths were China, Hong Kong, Taiwan, Canada and Russia. The disease killed approximately 10 per cent of those infected, with the global death toll reaching 813 including 348 in China and 298 in Hong Kong (Lee and McKibbin 2004). The disease was noticeable for its rapid global spread which had substantial repercussions for economies and mobility as well as for health. As Raptopoulou-Gigi commented (2003: 81):

> The rapid worldwide spread of the coronavirus that causes SARS and the fact that by May 25th 2003, 28 countries reported cases of this infectious disease, suggested that as for other infectious diseases, evolution and spread is facilitated by the mobility of the society either through air travel or the densely populated urban areas especially in Asia.

SARS is regarded as a serious health threat for several reasons (see Chan-Yeung *et al.* 2003; Donnelly *et al.* 2003; Lipsitch *et al.* 2003; Leung *et al.* 2004; Li *et al.* 2004):

- The disease has no vaccine and no treatment.
- The virus comes from a family that is recognised for its frequent mutations making it difficult to produce effective vaccines.
- The available diagnostic tests have substantial limitations.
- The epidemiology and pathogenesis for the disease are poorly understood.
- The potential impacts on hospital staff presents a major health human resource problem.
- A significant proportion of patients require intensive care.
- The incubation period of 10 days allows spread via air travel between any two cities in the world.

The spread of SARS was checked by a rigorous system of 'old-fashioned public health measures' (Bell and Lewis 2005: 2) on mobility at international borders and health isolation and quarantine procedures. However, preventing the spread of the disease also came at substantial economic cost due to changes in trade and tourism flows as well as associated capital investment. SARS travel bans were put in place for the major affected areas in April 2003, initially China – especially Hong Kong and Guangdong Province – but other cities in Asia and Canada (Toronto) followed. It was estimated that almost half the planned flights to South East Asia were cancelled during the month of April and total visitor arrivals declined by about two thirds over the course of the crisis, with corresponding effects on related parts of the economy, including hotels, restaurants, retail and even shipping (Bell and Lewis 2005). In Hong Kong, tourism declined by 10–50 per cent over the period, with ripple affects throughout the region's economy, including a 50 per cent drop in retail sales.

However, tourism was affected across the entire region, in part because of travel and business connectivities, but also because of inaccurate perceptions in tourism-generating regions of which places were dangerous to travel to. At the height of the epidemic, visitors and tourism declined 80 per cent in Taiwan and almost as much in Singapore (Bell and Lewis 2005: 20). Nevertheless, as Lee and McKibbin

(2004) have stressed in their analysis of the economic impacts of SARS, just calculating the number of cancelled tourist trips and declines in retail trade is not sufficient to get a full picture of the impact of SARS because there are domestic and international trade and capital linkages across sectors and across economies. Bell and Lewis (2005) estimated that SARS cost the region nearly US$15 billion, or 0.5 per cent of GDP, while a 'less readily measurable, but arguably more serious, impact was caused by faltering business confidence' (2005: 21). Nevertheless, Bell and Lewis' estimate is still only relatively short term. Because of patterns of contemporary trade, the economic costs from a disease such as SARS go beyond the direct impacts on the affected sectors in the disease-inflicted countries. As the world becomes more integrated, the global cost of a communicable disease like SARS is expected to rise, and should also include forgone income as a result of disease-related morbidity and mortality (Lee and McKibbin 2004).

Humans as vectors for plant and animal disease

Humans not only act as vectors for pathogens that attack humans but also for a range of plant and animal diseases, for example, foot-and-mouth disease and phylloxera. The February 2002 outbreak of foot-and-mouth disease in the United Kingdom was the first outbreak of the disease in Britain since 1967. However, its effect on the British countryside was devastating. Over 6 per cent of the national livestock herd was slaughtered, with dramatic effects on those farms that lost their stock. Just as seriously, biosecurity measures limited not only the movement of livestock but also the movement of people in the countryside, including tourists. Once a case was confirmed, under EU regulations, a protection zone based on a minimum radius of 3km and a surveillance zone based on a minimum radius of 10km should be established. In addition, a number of controls were imposed to minimise the risk of the public, including visitors to the countryside, spreading the disease. Not only were the majority of footpaths and other rights of way – i.e. other routes over which the public has a legal right to pass, such as bridleways – closed across the country, but also many rural tourist attractions were closed and land owned by the Forestry Commission and the National Trust, along with some national parks and forest areas, even some parks in London, were closed to the public (Sharpley and Craven 2001).

For both agriculture and tourism, the outbreak will have long-term impacts, in agriculture because of restocking costs and customer confidence, for tourism because of consumer perception of rurality and the safety of the countryside (Hall 2005). According to Sharpley and Craven (2001) the overall potential loss to the tourism industry in England alone in 2001 was estimated at £5 billion, while the lost overseas tourism expenditure in Britain was estimated at between £1 billion and £3.5 billion – realistically the final figures will never be known.

Although biosecurity measures are put in place at many national and, in some cases, domestic borders (for example Australia, USA), little attention is paid to biosecurity issues in the tourism industry (Hall 2005). Hall (2003b) examined the New Zealand wine industry with respect to biosecurity measures in place for the growing number of wine tourists in that country and noted that only 17 per cent of wineries which responded to a national survey had any biosecurity measures in place at all. An exploratory assessment of biosecurity risks at wineries looking at how humans act as vectors for grape diseases indicated substantial issues with respect to how tourists perceived wineries and vineyards – particularly with regard to the questions asked about tourist movement in border control procedures, as well as tourist mobility and what they had worn on previous visits to vineyards.

Responding to global health and disease risks

In order to combat the introduction of pests and diseases many countries and regions have introduced biosecurity strategies. Biosecurity refers to the protection of a country, region or location's economic, environmental and/or human health from harmful organisms and involves preventing the introduction of harmful new organisms, and eradicating or controlling those unwanted organisms that are already present (Biosecurity Strategy Development Team 2001). Although biosecurity and tourism are closely entwined, the tourism industry has little overt interest in biosecurity issues, whether on a national or global scale, unless some biosecurity risk, such as foot-and-mouth disease or SARS occurs (Gössling 2002; Hall 2005).

Central to appropriate biosecurity practice by travellers and the tourist industry is an improved understanding of biosecurity and quarantine. Improving awareness of biosecurity may lead to a decrease in the number of prohibited items which cross a border or boundary (Hall 2003b). Biosecurity measures occur on a number of different scales, from the international level – such as agreements on the movement of agricultural produce through border controls – through to biosecurity practices at individual locations, such as farms. Biosecurity strategies can also be categorised in terms of their utility at the pre-border, border and post-border stages (Hall 2005). From a tourism mobility perspective biosecurity strategies occur at different stages of the trip cycle: decision making and anticipation, travel to a tourism destination or attraction, the on-site experience, return travel and recollection of the experience. Each of these five stages will have different implications for how tourism and biosecurity and quarantine organisations establish a relationship with the traveller and assist them in practising good biosecurity (Table 9.3).

However, one of the greatest difficulties faced in managing health and disease risks is that it is not just the passenger that potentially constitutes a risk. Their means of transport may also serve to harbour pathogens or the carriers of pathogens. According to Fidler (1999: 14):

> Even before the advent of air travel, experts recognized that the scope of international travel had rendered national quarantine strategies ineffective. The explosion in global travel facilitated by air technology now threatens national public health strategies in a similar fashion.

Indeed, MacPherson (2001) argues that the inability to detect and contain imported disease threats at national borders requires a shift in immigration, quarantine and

Table 9.3 Pre-border, border and post-border biosecurity strategies

Pre-border
- Identifying threats to ecosystems
- Profiling and modelling the characteristics of damaging or potentially damaging organisms and vectors
- Identifying controls (in the country of origin) for selected organisms that pose a threat to destinations
- Analysing and predicting risk pathways for unwanted organisms
- Identifying and collating databases and expertise on unwanted organisms
- Developing systems for rapid access to appropriate data
- Developing import standards and compliance validation methodologies
- Auditing exporting countries' compliance with destination biosecurity standards
- Identifying and locating biosecurity-related risks to animal, plant and human health
- Analysis of public attitudes and perceptions of biosecurity risks and barriers to biosecurity responses in visitor-generating areas
- Development of educational programmes in exporting regions so as to reduce likelihood of introduction of unwanted organisms in imported goods
- Development of educational programmes for tourists in both generating and destination regions so as to reduce likelihood of introduction of unwanted organisms

Border
- Developing improved systems, including clearance systems and sampling methodologies, and technologies for intercepting unwanted organisms according to import standards
- Developing border containment and eradication methodologies according to import standards
- Developing profiles of non-compliance behaviour to biosecurity requirements

Post-border (includes pest management)
- Developing rapid identification techniques for unwanted organisms
- Designing and developing methodologies for undertaking delimiting surveys for new incursions
- Developing rapid response options for potential incursions of unwanted organisms
- Analysis of public attitudes and perceptions of biosecurity risks and barriers to biosecurity responses in destination areas
- Developing long-term containment, control and eradication strategies

General
- Analysis of economic and political models for the management of biosecurity threats
- Development of rapid-access information systems, collections and environmental databases on unwanted organisms
- Improve export opportunities for 'clean' products
- Development of industry and public biosecurity education programmes

Source: after Hall 2004

public health approaches to health and mobile populations, with a new paradigm being needed to facilitate the development of policies and programmes to address the health consequences of population mobility.

An example of the need for a new way of dealing with global health and disease risk lies in the epidemiological characteristics of AIDS. Unlike cases of SARS or influenza, where infected individuals can be kept from coming in close contact with the healthy population through temporary isolation and quarantine, individuals infected with the HIV virus are currently infectious for life. As Bell and Lewis (2005: 26) note:

> Effective quarantine therefore amounts to lifelong house arrest. This may be a feasible policy at the early stages of the epidemic, but it raises thorny ethical issues. At later stages, the numbers of individuals will be so large as to make it impractical. With this measure ruled out, others assume greater importance.

Therefore, increasing emphasis is being placed on the development of global strategies to deal with global health problems that focus on transfer of health management skills and funds, as well as the availability of pharmaceuticals at affordable prices in the developed world (Fidler 1999). Because of the globalisation of mobility what was once just a health problem in 'another country' now assumes significance as a health risk here.

However, the reality is that the development of systems of international governance of health issues is poorly integrated with governance of other aspects of global environmental change, while the financial contributions from the rich to the poor nations are not sufficient to deal with the health issues that are emerging. Moreover, ongoing urbanisation, changes in land use, land degradation and water pollution all contribute to the potent mix of environmental changes that may serve as the basis for a new wave of epidemics such as the world has not seen for almost a hundred years (McMichael 2001, 2002). Mobility is clearly a major factor in such global environmental change but, to further complicate changes in health and disease risk on a global scale, the potential effects of climate change also need to be considered.

Climate change can have an impact on the overall health of populations so that epidemics may have a greater effect than would have been the case if the population was healthy. An example of this was the plague outbreaks in Western Europe which followed the cooling phase of the climatic cycle that set in at the end of the thirteenth century. The first two decades of the fourteenth century were marked by cold, wet weather and poor harvests. In particular, the famine of 1315–17 left the population of many European centres particularly susceptible to the plague (Genicot 1966). However, contemporary climate change is regarded as having significant implications for disease and health because of its potential to contribute to new distributions of pathogens as a result of a changed environment (Parry 2001; P.R. Hunter 2003; McMichael *et al.* 2003, 2004). For example, with respect to such diseases as dengue fever (Hales *et al.* 2002) and malaria (Rogers and Randolph 2000; Tanser *et al.* 2003; Thomas *et al.* 2004; van Lieshout 2004). Indeed, Epstein (2002: 374) observes that: 'Volatility of infectious diseases may be one of the earliest biological expressions of climate instability'. In addition, human health will be affected because human comfort levels will also be impacted by climate change, along with issues of ozone depletion and skin cancer and effects on food production and water availability (McMichael 2002; Hall and Higham 2005) (Table 9.4).

McMichael (2001, 2002) argues that new patterns of disease emerging today may reflect the fact that humans are stressing ecological life support systems beyond the limits of their tolerance. As Morse (1993b: 23) observes: 'The lesson of AIDS demonstrates that infectious diseases are not a vestige of our premodern past; instead, like disease in general, they are the price we pay for living in the organic world.' There is arguably more focus now from the general public in the developed world on disease than there has been for many years. However, such attention has tended to be generated by concerns about bioterrorist attack (Glass and Schoch-Spana 2002) or the morality of disease (Webb 1997) rather than a considered view of health and disease. Such considerations are not new (e.g. Fenn 2000; Markel 2000). However, the emergence of infectious disease in parallel with other elements of global environmental change clearly has substantial implications for immediate and long-term personal and global security (Bloom and Mahal 1997; Fidler 1999; Hall *et al.* 2003; Eizenstat *et al.* 2004; Haacker 2004).

> Most generations alive today in developed countries are unprepared for the enormous threat infectious diseases now pose to them and their offspring. Even the calamity of AIDS has not dented the complacent attitudes of millions that we have seen the 'end of history' for infectious diseases.
>
> (Fidler 1999: 6)

We are therefore entering what McMichael (2001, 2002) referred to as a fourth transition, marked by disease emergence on a global scale. Not coincidentally, such health risks are also occurring at a time of greater human mobility and travel

Table 9.4 Possible direct and indirect health effects arising from global climate change

- Hyperthermia due to summertime heat-related mortality
- Changes to rate of infection from various diseases as a result of changes in the geographic ranges of pathogens, vectors and reservoirs
- Changes to allergenic reactions as a result of changed geographic ranges of allergic stimulants and environmental changes
- Increases in the rate of respiratory disease due to air pollutants
- Increases in the rate of skin cancer, melanomas, cataracts and immune suppression linked to increases in ambient ultraviolet light
- Impacts of extreme weather events
- Malnutrition and starvation due to changes in location and type of agricultural production
- Population movement as a result of enforced migration from areas affected by major ecological change, e.g. coastlines

Sources: Cliff and Haggett 1995; Epstein 2002; Hall and Higham 2005

to destinations increasingly remote and distant from the traveller's origin. Tourism is an extremely important contributor to such a transition. Unfortunately, while tourism will be dramatically affected by any pandemic, there is very little to suggest that the tourism industry is concerned or even aware of many such global health and disease issues.

References

Arndt, C. and Lewis, J. (2000) 'The macro implications of HIV/AIDS in South Africa: A preliminary assessment', *The South African Journal of Economics*, 69(5): 856–87.

Ausubel, J.H., Marchetti, C. and Meyer, P. (1998) 'Toward green mobility: the evolution of transport', *European Review*, 6(2): 137–56.

Bell, C. and Lewis, M. (2005) 'The economic implications of epidemics old and new', Centre for Global Development Working Paper No.54.

Biosecurity Strategy Development Team (2001) *A Biosecurity Strategy for New Zealand, Strategy Vision Framework Background Paper for Stakeholder Working Groups*. Wellington: Biosecurity Strategy Development Team.

Bloom, D. (1999) 'Economic perspectives on the global AIDS epidemic', *AIDS Patient Care and STDs*, 13(4): 229–34.

Bloom, D. and Canning, D. (2003) 'The health and poverty of nations: From theory to practice', *Journal of Human Development*, 4(1): 47–71.

Bloom, D. and Mahal, A. (1997) 'AIDS, flu, and the Black Death: Impacts on economic growth and well-being', in D. Bloom and P. Godwin (eds) *The Economics of HIV and AIDS: The Case of South and South East Asia*. New Delhi: Oxford University Press, pp.22–52.

Bradley, D.J. (1989) 'The scope of travel medicine: An introduction to the conference on international travel medicine', in R. Steffen, H.O. Lobel and J. Bradley (eds) *Travel Medicine: Proceedings of the First Conference on International Travel Medicine, Zurich, Switzerland, April 1988*. Berlin: Springer-Verlag, pp.1–9.

Chan-Yeung, M., Ooi, G.C., Hui, D.S., Ho, P.L. and Tsang, K.W. (2003) 'Severe acute respiratory syndrome', *International Journal of Tuberculosis and Lung Disease*, 7(12): 1117–30.

Cliff, A. and Haggett, P. (1995) 'Disease implications of global change', in R.J. Johnston, P.J. Taylor and M.J. Watts (eds) *Geographies of Global Change: Remapping the World in the Late Twentieth Century*. Oxford: Blackwell, pp.206–23.

Cliff, A. and Haggett, P. (2004) 'Time, travel and infection', *British Medical Bulletin*, 69: 87–99.

Cohn, S.K. (2003) *The Black Death Transformed: Disease and Culture in Early Renaissance Europe*. New York: Oxford University Press.

Cook, N.D. (1998) *Born to Die: Disease and New World Conquest, 1492–1650*. Cambridge: Cambridge University Press.

Cook, N.D. (2004) *Demographic Collapse: Indian Peru, 1520–1620*. Cambridge: Cambridge University Press.

Croft, A. (2000) 'Malaria: prevention in travellers', *British Medical Journal*, 321: 154–60.

Crosby, A.W. (1972) *The Columbian Exchange: Biological and Cultural Consequences of 1792*. Westport, CT: Greenwood Press.

Crosby, A.W. (1976) 'Virgin soil epidemics as a factor in the aboriginal depopulation in America', *William and Mary Quarterly*, 3rd Series 33: 289–99.

Crosby, A.W. (1986) *Ecological Imperialism: The Biological Expansion of Europe, 900–1900*. Cambridge: Cambridge University Press.

Curtin, P.D. (1968) 'Epidemiology and the slave trade', *Political Science Quarterly*, 83: 190–216.

De Schryver, A. and Meheus, A. (1989) 'International travel and sexually transmitted diseases', *World Health Statistics Quarterly*, 42: 90–9.

Diamond, J. (1997) *Guns, Germs and Steel: The Fates of Human Societies*. New York: Norton.

Dobyns, H.F. (1983) *Their Number Became Thinned: Native American Population Dynamics in Eastern North America*. Knoxville, TN: University of Tennessee Press.

Donnelly, C.A., Ghani, A.C., Leung, G.M., Hedley, A.J., Fraser, C., Riley, S., Abu-Raddad, L.J., Ho, L.M., Thach, T.Q., Chau, P., Chan, K.P., Lam, T.H., Tse, L.Y., Tsang, T., Liu, S.H., Kong, J.H., Lau, E.M., Ferguson, N.M. and Anderson, R.M. (2003) 'Epidemiological determinants of spread of causal agent of severe acute respiratory syndrome in Hong Kong', *Lancet*, 361(9371): 1761–6.

Dorolle, P. (1969) 'Old plagues in the jet age: International aspects of present and future control of communicable diseases', *WHO Chronicle*, 23: 104–7.

Dos Santos, C.C., Anvar, A., Keystone, J.S. and Kain, K.C. (1999) 'Survey of use of malaria prevention measures by Canadians visiting India', *Canadian Medical Association Journal*, 160(2): 195–200.

Eizenstat, S., Porter, J.E. and Weinstein, J. (2004) *On the Brink: Weak States and US National Security*. Report of the Commission on Weak States and U.S. National Security, Washington, DC: Center for Global Development.

Epstein, P.R. (2002) 'Climate change and infectious disease: stormy weather ahead', *Epidemiology*, 13(4): 373–5.

Fenn, E.A. (2000) 'Biological warfare in eighteenth-century North America: Beyond Jeffery Amherst', *Journal of American History*, 86: 1552–80.

Fenn, E.A. (2002) *Pox Americana: The Great Smallpox Epidemic*. New York: Hill and Wang.

Fidler, D.P. (1999) *International Law and Infectious Diseases*. Oxford: Oxford University Press.

Flinn, M.W. (1979) 'Plague in Europe and the Mediterranean countries', *Journal of European Economic History*, 8: 131–48.

Ford, N. and Koetsawang, S. (1991) 'The socio-cultural context of the transmission of HIV in Thailand', *Social Science of Medicine*, 33(4): 405–14.

Garrett, L. (1996) 'The return of infectious diseases', *Foreign Affairs*, 75: 66–79.

Garrett, L. (2000) *Betrayal of Trust: The Collapse of Global Public Health*. New York: Hyperion.

Genicot, L. (1966) 'Crisis: From the Middle Ages to modern times', in M.M. Postan (ed.) *The Agrarian Life of the Middle Ages, The Cambridge Economic History of Europe*, Vol.1. London: Cambridge University Press, pp.660–742.

Glass, T.A. and Schoch-Spana, M. (2002) 'Bioterrorism and the people: How to vaccinate a city against panic', *Clinical Infectious Diseases*, 34: 217–23.

Gössling, S. (2002) 'Global environmental consequences of tourism', *Global Environmental Change*, 12(4), 283–302.

Gould, S.J. (2002) *The Structure of Evolutionary Theory*. Cambridge, MA: Belknap.

Greer, R.A. (1965) 'Oahu's ordeal – the smallpox epidemic of 1853', *Hawaii Historical Review*, 1: 221–42.

Grübler A. and Nakicenovic, N. (1991) *Evolution of Transport Systems, RR–91–8*. Laxenburg, Vienna: International Institute for Applied Systems Analysis.

Haacker, M. (ed.) (2004) *The Macroeconomics of HIV/AIDS*. Washington, DC: International Monetary Fund.

Hales, S., deWet, N., Maindonald, J. and Woodward A. (2002) 'Potential effect of population and climate changes on global distribution of dengue fever: an empirical model', *Lancet*, 360: 830–4.

Hall, C.M. (2003) 'Tourism and Temporary Mobility: Circulation, Diaspora, Migration, Nomadism, Sojourning, Travel, Transport and Home', International Academy for the Study of Tourism (IAST) Conference, 30 June–5 July 2003, Savonlinna, Finland.

Hall, C.M. (2003) 'Biosecurity and wine tourism: Is a vineyard a farm?' *Journal of Wine Research*, 14(2–3): 121–6.

Hall, C.M. (2004) 'Tourism and biosecurity', in C. Cooper, C. Arcodia, D. Soinet and M. Whitford (eds) *Creating Tourism Knowledge, 14th International Research Conference of the Council for Australian University Tourism and Hospitality Education, 10–13 February*, School of Tourism and Leisure Management, University of Queensland.

Hall, C.M. (2005) *Tourism: Rethinking the Social Science of Mobility*. Harlow: Prentice-Hall.

Hall, C.M. and Higham, J. (eds) (2005) *Tourism, Recreation and Climate Change*. Clevedon: Channelview Press.

Hall, C.M., Timothy, D. and Duval, D. (2003) 'Security and tourism: towards a new understanding?' *Journal of Travel and Tourism Marketing*, 15(2–3): 1–18.

Henige D. (1986) 'Primary source by primary source? On the role of epidemics in New World depopulation', *Ethnohistory*, 33: 293–313.

Humar, A., Sharma, S., Zoutman, D. and Kain, K.C. (1997) 'Fatal *falciparum malaria* in Canadian travellers', *Canadian Medical Association Journal*, 156: 1165–7.

Hunter, P.R. (2003) 'Climate change and waterborne and vector-borne disease', *Journal of Applied Microbiology*, 94: 37S–46S.

Hunter, S. (2003) *Black Death: AIDS in Africa*. New York: Palgrave Macmillan.

Joralemon, D. (1982) 'New World depopulation and the case of disease', *Journal of Anthropological Research*, 38(1): 108–27.

Jones, D.S. (2003) 'Virgin soils revisited,' *William and Mary Quarterly*, 3d ser., 60: 703–42.

Kain, K.C. and Keystone, J.S. (1998) 'Malaria in travelers: Epidemiology, disease and prevention', *Infectious Disease Clinics of North America*, 12: 267–84.

Kain, K.C., MacPherson, D.W., Kelton, T., Keystone, J.S., Mendelson, J. and Maclain, J. (2001) 'Malaria deaths in visitors to Canada and in Canadian travellers: A case series', *Canadian Medical Association Journal*, 164(5) (March 6), online. Available at: www.cmaj.ca/cgi/content/full/164/5/654 (accessed 25 January 2005).

Kelton, P. (2004) 'Avoiding the smallpox spirits: Colonial epidemics and southeastern Indian survival', *Ethnohistory*, 51: 45–71.

Kipple, K.F. and Higgins, B.T. (1992) 'Yellow fever and the Africanization of the Caribbean,' in J.W. Verano and D.H. Ubelaker (eds) *Disease and Demography in the Americas*. Washington, DC: Smithsonian Institution.

Koplow, D.A. (2003) *Smallpox: The Fight to Eradicate a Global Scourge*. Berkeley, CA: University of California Press.

Lederberg, J., Shope, R.E. and Oaks Jr., S.C. (1992) *Emerging Infections: Microbial Threats to Health in the United States*. Washington, DC: National Academy of Sciences.

Lee, J.-W. and McKibbin, W.J. (2004) 'Globalization and Disease: The Case of SARS', Brookings Discussion Papers in International Economics No.156.

Leung, G.M., Hedley, A.J., Ho, L., Chau, P., Wong, I.O.L., Thach, T.Q., Ghani, A.C., Donnelly, C.A., Fraser, C., Riley, S., Ferguson, N.M., Anderson, R.M., Tsang, T., Leung, P., Wong, V., Chan, J.C.K., Tsui, E., Lo, S. and Lamm, T. (2004) 'The epidemiology of severe acute respiratory syndrome in the 2003 Hong Kong epidemic: an analysis of all 1755 patients', *Annals of Internal Medicine*, 141(9): 662–73.

Li, Y., Yu, I.T., Xu P., Lee, J.H., Wong, T.W., Ooi, P.L. and Sleigh, A.C. (2004) 'Predicting super spreading events during the 2003 severe acute respiratory syndrome epidemics in Hong Kong and Singapore', *American Journal of Epidemiology*, 160(8): 719–28.

van Lieshout, M., Kovats, R.S., Livermore, M.T. and Martens, P. (2004) 'Climate change and malaria: analysis of the SRES climate and socio-economic scenarios', *Global Environmental Change*, 14: 87–99.

Lindenbaum, S. (1997) 'AIDS: Body, mind, and history', in G.C. Bond, J. Kreniske, I. Susser and J. Vincent (eds) *AIDS in Africa and the Caribbean*. Boulder, CA: Westview Press, pp.191–4.

Lipsitch, M., Cohen, T., Cooper, B., Robins, J.M., Ma, S., James, L., Gopalakrishna, G., Chew, S.K., Tan, C.C., Samore, M.H., Fisman, D. and Murray, M. (2003) 'Transmission dynamics and control of severe acute respiratory syndrome', *Science*, 300(5627) (June 20):1966–70.

Lobel, H.O. and Kozarsky, P.E. (1997) 'Update on prevention of malaria for travelers', *Journal of the American Medical Association*, 278: 1767–71.

Lydie, N., Robinson, N.J., Ferry, B., Akam, E., De Loenzien, M. and Abega, S. (2004) 'Mobility, sexual behavior, and HIV infection in an urban population in Cameroon', *JAIDS Journal of Acquired Immune Deficiency Syndromes*, 35(1):67–74.

McEvedy, C. (1988) 'The bubonic plague', *Scientific American*, 258(February): 118–23.

MacFarlan, M. and Sgherri. S. (2001) 'The Macroeconomic Impact of HIV/AIDS in Botswana', IMF Working Paper WP/01/80, International Monetary Fund.

McMichael, A.J. (1993) *Planetary Overload: Global Environmental Change and the Health of the Human Species*. Cambridge: Cambridge University Press.

McMichael, A.J. (2001) *Human Frontiers, Environments and Disease: Past Patterns, Uncertain Futures*. Cambridge: Cambridge University Press.

McMichael, A.J. (2002) 'Population, environment, disease, and survival: past patterns, uncertain futures', *Lancet*, 359: 1145–8.

McMichael, A.J., Woodruff, R.E., Whetton, P., Hennessy, K., Nicholls, N., Hales, S., Woodward, A. and Kjellstrom, T. (2003) *Human Health and Climate Change in Oceania: Risk Assessment 2002*. Canberra: Commonwealth of Australia, Department of Health and Ageing.

McMichael, A.J., Campbell-Lendrum, D., Kovats, R.S., Edwards, S., Wilkinson, P., Edmonds, N., Nicholls, N., Hales, S., Tanser, F.C., Le Sueur, D., Schlesinger, M. and Andronova, N. (2004) 'Climate change', in M. Ezzati, A.D. Lopez, A. Rogers, and C.J. Murray (eds) *Comparative Quantification of Health Risks: Global and Regional Burden of Disease Due to Selected Major Risk Factors*, Vol.2. Geneva: World Health Organization.

McNeill, W.H. (1976) *Plagues and People*. London: Penguin

MacPherson, D.W. (2001) 'Human mobility and population health: New approaches in a globalizing world', *Perspectives in Biology and Medicine*, 44(3): 390–401.

Markel, H. (2000) '"The eyes have it": Trachoma, the perception of disease, the United States Public Health Service, and the American Jewish immigration experience, 1897–1924', *Bulletin of the History of Medicine*, 74: 525–60.

May, A. (2003) *Social and Economic Impacts of HIV/AIDS in Sub-Saharan Africa, with Specific Reference to Aging*, Institute of Behavioral Science, Population Aging Center, Working Paper PAC2003–0005. Boulder, CO: University of Colorado.

May, J.M. (1958) *The Ecology of Human Disease*. New York: M.D. Publications.

Mayer, J.D. (1996) 'The political ecology of diseases as one new focus for medical geography', *Progress in Human Geography*, 20: 445–56.

Milner, G.R. (1980) 'Epidemic disease in the postcontact southeast: a reappraisal', *Mid-Continent Journal of Archeology*, 5(1): 39–56.

Morse, S.S. (1993a) 'Examining the origins of emerging viruses', in S.S. Morse (ed.) *Emerging Viruses*. New York: Oxford University Press, pp.10–28.

Morse, S.S. (1993b) 'AIDS and beyond: Defining the rules for viral traffic', in E. Fee and D.M. Fox (eds) *AIDS: The Making of a Chronic Disease*. Berkeley, CA: University of California Press, pp.23–48.

Murphy, G.S. and Oldfield, E.C. (1996) 'Falciparum malaria', *Infectious Disease Clinics of North America*, 10: 747–55.

Musa, G., Hall, C.M. and Higham, J. (2004) 'Tourism sustainability and health impact in high altitude ACE destinations: a case study of Nepal's Sagarmatha National Park', *Journal of Sustainable Tourism*, 12(4): 306–31.

Parry, M., Arnell, N., McMichael, T., Nicholls, R., Martens, P., Kovats, S., Livermore, M., Rosenzweig, C., Iglesias, A. and Fischer, G. (2001) 'Millions at risk: defining critical climate change threats and targets', *Global Environmental Change*, 11: 181–3.

Potter, C.W. (2001) 'A history of influenza', *Journal of Microbiology*, 91: 572–9.

Raptopoulou-Gigi, M. (2003) 'Severe acute respiratory syndrome (SARS): A new emerging disease in the 21st century', *Hippokratia*, 7(2): 81–3.

Rogers, D.J. and Randolph, S.E. (2000) 'The global spread of malaria in a future, warmer world', *Science*, 289: 1763–5.

Roll Back Malaria (2004) Malaria in Africa, online. Available at: www.rbm.who.int/cmc_upload/0/000/015/370/RBMInfosheet_3.htm. (accessed 25 January 2005).

Rudkin, B. and Hall, C.M. (1996) 'Off the beaten track: the health implications of the development of special-interest tourism services in South-East Asia and the South Pacific', in S. Clift and S. Page (eds) *Health and the International Tourist*. London: Routledge, pp.89–107.

Sachs, J. and Malaney, P. (2002) 'The economic and social burden of malaria', *Nature*, 415: 680–5.

Schmitt, R.C. (1970) 'The Okuu – Hawaii's epidemic', *Hawaii Medical Journal*, 29: 359–64.

Sharpley, R. and Craven, B. (2001) 'The 2001 foot and mouth crisis – rural economy and tourism policy implications: a comment', *Current Issues in Tourism*, 4(6): 527–37.

Snow, D.R. and Lanphear, K.M. (1988) 'European contact and Indian depopulation in the Northeast: The timing of the first epidemics,' *Ethnohistory*, 35: 15–33.

Snow, D.R. and Starna, W.A. (1989) 'Sixteenth-century depopulation: A view from the Mohawk Valley', *American Anthropologist*, 91: 142–9.

Tanser, F.C., Sharp, B. and Le Sueur, D. (2003) 'Potential effect of climate change on malaria transmission in Africa', *Lancet*, 362: 1792–8.

Thomas, C.J., Davies, G. and Dunn, C.E. (2004) 'Mixed picture for changes in stable malaria distribution with future climate in Africa', *Trends in Parasitology*, 20: 216–20.

Webb, D. (1997) *HIV and AIDS in Africa*. Chicago, IL: Pluto Press.

Westwood, J.C.N. (1980) *The Hazard From Dangerous Exotic Diseases*. London: Macmillan.

White, R. (1980) *Land Use, Environment and Social Change: The Shaping of Island County, Washington*. Seattle, WA: University of Washington Press.

Williams, B., Gouws, E., Lurie, M. and Crush, J. (2002) *Spaces of Vulnerability: Migration and HIV/AIDS in South Africa*. Cape Town: Southern African Migration Project.

Wilson, M.E. (1994) 'Disease in evolution: introduction,' in M.E. Wilson, R. Levins and A. Spielman (eds) *Disease in Evolution: Global Changes and Emergence of Infectious Diseases*. New York: New York Academy of Sciences, pp.1–12.

Wilson, M.E. (1995) 'Travel and the emergence of infectious diseases', *Emerging Infectious Diseases*, 1(2): 39–46.

Wolfe, R.J. (1982) 'Alaska's Great sickness, 1900: An epidemic of measles and influenza in a virgin soil population', *Proceedings of the American Philosophy Society*, 126: 92–121.

10 Tourism and water

Stefan Gössling

Introduction

Fresh water is one of the most essential resources to humanity, and it is becoming increasingly scarce. In 1995, an estimated 450 million people lived under severe water stress and an additional 1.3 billion people under a high degree of water stress (Vörösmarty *et al.* 2000). In the future, fresh water will become even scarcer, making the use and management of water an important political issue (see Clarke and King 2004). Tourism is highly dependent on the availability of fresh water resources (e.g. Orams 1998, Garrod and Wilson 2003). Recreational activities such as swimming, sailing, kayaking, canoeing, diving, fishing, and so on, are often related to lakes and rivers, which also form important elements of the land-scapes visited by tourists (see also Chapter 4, this volume). Fresh water is also needed for the maintenance of tourist infrastructure such as swimming pools, irri-gated gardens, bathrooms, laundry, etc. (see Gössling 2001). Tourism can exacer-bate fresh water problems, because it is often concentrated in regions with limited water resources, such as islands and coastal zones where there are few fossil water resources, low aquifer renewal rates, and few surface water sources. Besides causing a shift in global water consumption from regions of relative water abun-dance to those that are water scarce, tourism also increases total water demand because people use larger quantities of this resource when they are on vacation (Gössling 2002, 2005). Related to these aspects, water quality may often decrease through tourism, as a result of the discharge of untreated sewage, nutrient loads and toxic substances into adjacent water bodies (UN 1995; WWF 2004). In the light of these findings, this chapter seeks to discuss the interdependence of tourism and water resources, the consequences of global shifts in water demand, and the effects of changing water availability in important destinations for the tourist industry.

Global water use and distribution

Fresh water availability is highly unevenly distributed between countries and within countries. For example, within Europe, renewable per capita water resources range from less than 40m^3 per year in Malta to more than 600,000m^3 in Iceland (FAO

2003). Within countries, water availability is dependent on watersheds, which can divide water scarce and water abundant regions (see e.g. Gössling 2001 for Zanzibar, Tanzania). While many countries have vast water resources, desalination has become of major importance in some large industrialised countries such as the USA, Italy and Spain, as well as a range of small islands and island states. Some countries, particularly islands, have also started to import fresh water in tank ships, including the Bahamas, Antigua and Barbuda, Mallorca, the Greek Islands, South Korea, Japan, Taiwan, Nauru, Fiji and Tonga (Clarke and King 2004).

Fresh water use is divided in agricultural, domestic and industrial consumption. On global average, approximately 70 per cent of water use may be for agriculture, 20 per cent for industrial and 10 per cent for domestic purposes (FAO 2003, own calculations). However, there are large differences between countries. For example, agriculture accounts for 1 per cent of the total water use in Belgium or Austria, as opposed to 98 per cent in Afghanistan or Cambodia. Domestic water use, which includes households but also municipalities, commercial establishments and public services, constitutes only 1 per cent of water consumption in Ethiopia, but 83 per cent in Equatorial Guinea. Similarly, industrial consumption is less than 1 per cent in Somalia or Mali, but 89 per cent in Belize. Similar ranges can be observed in terms of per capita water use. For example, daily domestic water consumption varies between 12 litres per capita in Bhutan and 1,661 litres per capita in Australia (WRI 2003). On global average, domestic water consumption is in the order of 160 litres per capita per day (database 1987–1999, WRI 2003). In industrialised countries, 35 per cent of household consumption – constituting the major share of domestic water use – may be for bathing and showering, 30 per cent for flushing toilets, 20 per cent for laundry, 10 per cent for cooking and drinking, and 5 per cent for cleaning (Clarke and King 2004).

Water consumption patterns in tourism

Tourism-related water consumption is still little investigated, and there are few detailed studies of water use in different geographical settings, for different forms of tourism, or for the many forms of accommodation establishments. From what is known, water use varies widely with a range of 100 to 2,000 litres per tourist per day (Lüthje and Lindstädt 1994; UK-CEED 1994; GFANC 1997; Gössling 2001; WWF 2001, 2004). For example, the World Wide Fund for Nature (WWF 2001) reports that the average tourist in Spain consumes 440 litres per day, a value that increases to 880 litres where swimming pools and golf courses exist. A survey of water consumption in the tropical island of Zanzibar showed that water use was lowest in small, locally-owned guesthouses (100 litres per tourist per day), and highest in luxury resort hotels (up to 2,000 litres per tourist per day). The weighted water consumption was found to be 685 litres per tourist per day (Gössling 2001). The survey also assessed water use patterns. Hotels used most water for continuous irrigation of their gardens – 50 per cent, or a weighted average of 465 litres per day per tourist – a result of the poor storage capacity of the soils, high evaporation and use of plant species not adapted to arid conditions. In guesthouses, watering gardens accounted only for 15 per cent of the total water use – 37 litres per tourist per day. The major proportion of water in guest-houses is spent for direct uses including taking showers, flushing the toilet, and the use of tap water – 55 per cent, 136 litres per tourist per day – with a corresponding consumption of 20 per cent or 186 litres per tourist per day in hotels. The higher demand of hotel guests is a result of additional showers taken at pools, more luxurious or better functioning bathroom facilities, etc. Swimming pools represent another important factor of water use, accounting for about 15 per cent of the water demand of hotels (140 litres per tourist per day). Indirectly, swimming pools add to laundry, for example, when additional towels are handed out to guests. Guesthouses in the study area did not have swimming pools, which can partially explain lower water use rates. Laundry accounts for about 10 per cent – 25 litres per tourist per day – of the water used in guesthouses and 5 per cent – 47 litres per tourist per day – in hotels. Cleaning adds 5 per cent to the water demand in both guesthouses – 12 litres per tourist per day – and hotels – 47 litres per tourist per day. Finally, restaurants in guesthouses account for 15 per cent of the water used in guesthouses – 37 litres per tourist per day – and for 5 per cent – 47 litres per tourist per day – in hotels. Further studies need to be conducted to confirm the results of this case study.

Global water use by tourism

Any calculation of global tourism-related water consumption needs to take into consideration the statistical distinction between international/domestic tourists and business/leisure tourists. Statistics provided by the World Tourism Organization (WTO 2003a) only account for international tourist arrivals, summarising both leisure and business tourists. Little is known about domestic tourism, even though there is evidence that it outweighs international tourism by far in terms of volume (e.g. Ghimire 2001). For international tourism, there is only scattered information on the average length of stay, an important parameter for the calculation of global water use. Data is available for the average length of stay of international tourists (business and leisure) in 97 countries, accounting for about 330 of the 692 million international tourist arrivals in 2001 (WTO 2003a). Based on this data (1997–2001), Gössling (2002) calculated a global average length of stay of 8.1 days. It should be noted that this calculation excludes a large number of important tourist countries such as the USA, Italy, China, United Kingdom, Russian Federation, Germany, Austria, Hungary, Hong Kong and Greece. Finally, data on water consumption per tourist is needed for the calculation. Available data suggests that, on global average, daily per tourist water consumption is in the order of 222 litres (for calculations, see Gössling 2005). This can be compared to the global average per capita water use (domestic), which is 160 litres per day (WRI 2003).

Given an average water consumption of 222 litres per day and assuming an average length of stay of 8.1 days, the 715 million international tourists in 2000 may have, for a rough estimate, used 1.3 km^3 of water. This figure excludes domestic tourism as well as indirect water consumption, for example, for the construction of tourist infrastructure or the production of food. It also excludes water used for energy generation. For example, the Worldwatch Institute (2004) reports that it takes 18 litres of water to produce one litre of gasoline, and that air travel entails energy uses of 50–100 litres of fuel per passenger for every 1,000 km of flight distance. Excluding these water uses, international tourism may account for a share of 0.04 per cent of the aggregate global water withdrawal of 3,100 km³ per year (data for 1985, Vörösmarty *et al.* 2000).

Global shifts in water consumption

The major proportion of tourist flows occurs between six regions, North America, the Caribbean, Northern and Southern Europe, North East Asia and South East Asia (WTO 2003b). Of the 715 million international tourist arrivals in 2002, 58 per cent took place within Europe, 16 per cent in North and South East Asia and 12 per cent in North America. Together they represent 86 per cent of all international tourist arrivals. Within sub-regions, about 87 per cent of all international arrivals in Europe are from Europe itself (some 350 million arrivals), while 71 per cent of international arrivals are regionally in the Americas (92 million), and 77 per cent in the Asia Pacific region (88 million). Six major tourist flows characterise international travel: Northern Europe to the Mediterranean (116 million); North America to Europe (23 million); Europe to North America (15 million); North East Asia to South East Asia (10 million); North East Asia to North America (8 million); and North America to the Caribbean (8 million).

Based on the global average length of stay of 8.1 days, each tourist travelling to another region may statistically increase the water use in this region by 1,800 litres, with a concomitant reduction in water use at home. However, there might be great differences between the regions. For example, tourists to the Caribbean might use far more water as a result of the resort character of many hotels in this region, which usually have irrigated gardens and large swimming pools, the main water consuming factors (see Gössling 2001). Travelling abroad also means that water consumption in the source regions is reduced, which needs to be considered when calculating net water increases/decreases. As consumption patterns vary widely, even within industrialised countries, this is a difficult task. The calculation is further complicated by the fact that domestic water use includes municipalities, commercial establishments and public services. The amount of water used for personal purposes may thus be substantially lower. For example, in the USA, per capita use for personal purposes is in the order of 380 litres per day (Solley *et al.* 1998) as compared to 653 litres for domestic purposes (WRI 2003). People on holiday might also reduce only part of their overall household consumption. For example, irrigation of lawns might continue even when the owner of a house is on vacation.

Global shifts in water use can only be calculated by considering the proportion of leisure versus business tourists, their average length of stay, the composition and character of different accommodation establishments, as well as water use patterns at home and in destination countries (Gössling 2005). Currently, no database is available that would allow for such sophisticated calculations. Figures presented in the following are thus based on a number of simplifying assumptions. Table 10.1 shows tourist flows and their respective water use at home –

hypothetical consumption if tourists had stayed at home – and in the destination. It follows that international tourism leads to shifts of fresh water consumption in the order of 470 million m³ (aggregated increase/decrease), and to an increase of global water use in the order of an estimated 70 million m³ per year (Gössling 2005).

As shown in Table 10.1 and Figure 10.1, North America and North East Asia are the regions experiencing a net decrease in water consumption, while Europe, the Caribbean and South East Asia experience a net increase in water consumption. The major shift in water use occurs within Europe, with 116 million tourists travelling to southern Europe, particularly the Mediterranean. These movements account for a net transfer of about 70 million m³.

It should be noted that there is a tendency for tourism to shift water demand from water-rich to water-poor areas both on large regional or continental scales – shifts from Northern Europe to Southern Europe, shifts from Europe and North America to the Caribbean – and on regional or local scales – for example, shifts to coastal zones. Furthermore, tourists may often arrive during the dry season, when rainfall drops to a minimum and water availability is restricted (see Gössling 2001; WWF 2004). Strong seasonality in combination with arrival peaks during dry season might thus put considerable strain on available water resources, particularly in generally dry regions.

Tourism also decreases the quality of water in many regions through the release of untreated sewage, which might also contain toxic components. For example, in

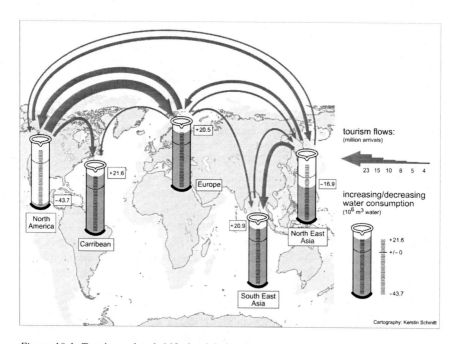

Figure 10.1 Tourism-related shifts in global water use

Source: Gössling 2005

Table 10.1 Global flows of tourists between regions and corresponding water use (2000)

Travel flows between regions	Internat. tourists (million)	Water use home (l/cap/day)	Total home (million l)	Water use destination (l/cap/day)	Total destination (million l)	Increase/decrease by region (million l)	
N. America–Europe	23	300	55,890	222	41,359	N. America	−43,691
N. America–Caribbean	8	300	19,440	222	14,386	Europe	20,510
N. America–N. E. Asia	4	300	9,720	222	7,193	Caribbean	21,579
Europe–N. America	15	150	18,225	222	26,973	N. E. Asia	−16,945
Europe–Caribbean	4	150	4,860	222	7,193	S. E. Asia	20,898
Europe–N. E. Asia	5	150	6,075	222	8,991		
Europe–S. E. Asia	5	150	6,075	222	8,991		
N. E. Asia–N. America	8	200	12,960	222	14,386		
N. E. Asia–Europe	8	200	12,960	222	14,386		
N. E. Asia–S. E. Asia	10	200	16,200	222	17,982		
S. E. Asia–N. E. Asia	5	150	6,075	222	8,991		
Total	95	—	168,480	—	170,831	—	2,351
Europe N. to Europe S.	116	150	140,940	222	208,591	N. Europe	−140,940
						S. Europe	208,591

Source: Gössling 2005

the Mediterranean, it is still a common practice to discharge sewage from hotels directly into the sea (WWF 2004). Similarly, Smith (1997, cited in Kent *et al*. 2002) reported that in the European Mediterranean, only 30 per cent of municipal wastewater from coastal towns received any treatment before discharge. As tourism substantially increases water use in the Mediterranean and other destinations, this sector is also responsible for changes in water quality through the discharge of sewage into adjacent water bodies. However, this problem might be even greater in the tropics, where it is still very common for both municipalities and hotels to release effluents largely untreated into coastal waters with low natural concentrations of nutrients.

Country-specific analysis

In the following, a more detailed database is provided for the world's 50 most important tourism countries (inbound tourism) as well as five small tropical island states for which data on tourism and water use is available. Island states were included because tourism is usually an important pillar of their economies, while they might simultaneously possess limited fresh water resources. The 55 countries contained in the sample account for roughly 584 million (82 per cent) of the 715 million international tourist arrivals in 2002 (Table 10.2).

As Table 10.2 shows, great differences exist between countries in terms of renewable water resources, desalination capacity, use of treated wastewater, and overall water use. For example, in countries such as Bahrain, Barbados, Israel, Malta, Saudi Arabia and the United Arab Emirates, water use greatly exceeds renewable water resources, with up to 15 times the amount of annually available renewable water being consumed (see United Arab Emirates). In these countries, water demand is met by seawater desalination, which can be substantial. Saudi Arabia desalinates some 714 million m^3 per year, which nevertheless represents only a fraction of the total used (17,320 million m^3 per year). Water demand overshooting renewable water resources plus desalinated water capacities thus puts heavy strains on fossil water supplies.

Overall, quite a range of important tourist countries use already a substantial share of their renewable water resources, or even overuse these, as exemplified by Poland (26.3 per cent), Republic of Korea (27 per cent), Ukraine (27 per cent), Mauritius (28 per cent), Germany (31 per cent), South Africa (31 per cent), Cyprus (31 per cent), Spain (32 per cent), India (34 per cent), Morocco (44 per cent), Bulgaria (49 per cent), Tunisia (60 per cent), Barbados (105 per cent), Malta (110 per cent), Israel (122 per cent), Bahrain (258 per cent), Saudi Arabia (722 per cent) and United Arab Emirates (1,538 per cent). Note that these figures represent statistical averages. In many areas water scarcity might already be felt as national water resources are located in remote areas that cannot be tapped. Often, the tourist industry might be able to react flexibly, like, for example, large hotels in tropical islands which are likely to have the financial means to invest in desalination infrastructure to overcome water scarcity. However, small accommodation establishments might not be in the position to adapt to such environmental changes, which raises the question of the

Table 10.2 Country overview statistics

Country	Total natural renewable water resources (million m3/year)	Desalinated water (million m³/year) (note 2)	Reused treated wastewater (million m³/year) (note 2)	Total water use in 2000 (million m³/year)	% of renewable water used	Tourist arrivals 2000 ('000)	Growth rate tourist arrivals	Tourist arrivals 2020 ('000)	Average length of stay, 2000	Water use per tourist per day[3]	Total tourism-related water use, 2000 (million m³)	Tourism-related water use as % of total	Tourism-related water use as % of domestic	Total tourism-related water use 2020 (million m³)[4]
Argentina	814,000	0	0	29,072	3.6	2,909	5.1	8,000[1]	9.9	150	4.3	0.01	0.09	11.9
Australia	492,000	—	—	23,932	4.9	4,931	6.4	17,553[1]	26	300	38.5	0.16	1.09	136.9
Austria	77,700	—	—	2,112	2.8	17,982	1.2	23,100[2]	4.6	150	12.4	0.59	1.68	15.9
Bahrain	116	44.1	8.0	299	258.0	2,420	6.6	7,928[1]	2.4	200	1.2	0.39	0.97	3.8
Barbados	80	0	0	84	104.9	545	4.3	1,265[2]	10.1	400	2.2	2.61	7.34	5.1
Brazil	8,233,000	0	0	59,298	0.7	5,313	5.0	14,100[1]	12.1	300	19.3	0.03	0.16	51.2
Bulgaria	21,300	—	—	10,498	49.3	2,785	4.6	10,600[1]	8.4	150	3.5	0.03	1.10	13.4
Canada	2,902,000	—	—	45,974	1.6	19,627	3.6	40,600[1]	5.2	150	15.3	0.03	0.17	31.7
Cape Verde	300	0	0	28	9.3	83	3.6	168[2]	7	300	0.2	0.63	4.36	0.4
Chile	922,000	0	0	125,39	1.4	1,742	4.7	4,800[1]	10.1	200	3.5	0.03	0.25	9.7
China	2,896,569	—	—	6,302,89	21.8	31,229	7.8	130,000[1]	8.1	200	50.6	0.01	0.12	210.6
Cuba	38,120	0	0	8,204	21.5	1,741	9.2	6,700[1]	10.5	300	5.5	0.07	0.35	21.1
Cyprus	780	0	11.0	244	31.3	2,686	2.5	3,893[1]	11	400	11.8	4.84	16.88	17.1
Czech Rep.	13,150	—	—	2,566	19.5	4,666	4.0	44,000[1]	3.5	200	3.3	0.13	0.31	30.8
Denmark	6,000	—	—	1,267	21.1	2,088	3.8	4,402[2]	8.1	200	3.4	0.27	0.83	7.1
Dominican Republic	20,995	0	0	3,386	16.1	2,972	5.0	6,700[1]	10	400	11.9	0.35	1.09	26.8
Egypt	86,800	25.0	200.0	68,653	79.1	5,116	7.4	17,100[1]	6	400	12.3	0.02	0.23	41.0
Finland	110,000	—	—	2,478	2.3	2,714	3.8	5,722[2]	5.9	150	2.4	0.10	0.71	5.1
France	203,700	—	—	39,959	19.6	77,190	2.3	106,100[1]	7.5	400	231.6	0.58	3.69	318.3
Germany	154,000	—	—	47,052	30.6	18,983	1.2	20,000[1]	8.1	200	30.8	0.07	0.53	32.4
Greece	74,250	—	—	7,759	10.5	13,096	2.1	17,111[1]	8.1	400	42.4	0.55	3.34	55.4
Hungary	104,000	—	—	7,641	7.4	2,992	0.7	24,700[1]	8.1	200	4.9	0.06	0.68	40.0
India	1907,760	0	0	645,837	33.9	2,649	5.9	8,900[1]	31.2	150	12.4	<0.01	0.02	41.7

Table 10.2 Country overview statistics (continued)

Country	Total natural renewable water resources (million m3/year)	Desalinated water (million m³/year) (note 2)	Reused treated wastewater (million m³/year) (note 2)	Total water use in 2000 (million m³/year)	% of renewable water used	Tourist arrivals 2000 ('000)	Growth rate tourist arrivals	Tourist arrivals 2020 ('000)	Average length of stay, 2000	Water use per tourist per day[3]	Total tourism-related water use, 2000 (million m³)	Tourism-related water use as % of total	Tourism-related water use as % of domestic	Total tourism-related water use 2020 (million m³)[4]
Indonesia	2,838,000	0	0	82,773	2.9	5,064	7.7	27,385[1]	12.3	300	18.7	0.02	0.28	101.1
Ireland	52,000	—	—	1,129	2.2	6,737	3.8	14,204[2]	7.4	150	7.5	0.66	2.88	15.8
Israel	1,670	—	—	2,041	122.2	2,417	2.3	3,910[1]	15	300	10.9	0.53	1.73	17.6
Italy	191,300	—	—	44,372	23.2	41,181	2.1	52,451[1]	8.1	400	133.4	0.30	1.65	169.9
Japan	430,000	0	0	88,432	20.6	4,757	4.5	10,055[1]	8	200	7.6	0.01	0.04	16.1
Malaysia	580,000	0	0	9,016	1.6	10,222	5.0	25,046[1]	5.8	200	11.9	0.13	0.78	29.1
Malta	51	31.4	1.6	55	109.6	1,216	2.0	1,831[1]	8.4	400	4.1	7.34	10.21	6.2
Mauritius	2,210	0	0	612	27.7	656	5.3	1,548[1]	10.4	400	2.7	0.45	1.71	6.4
Mexico	457,222	0	0	78,219	17.1	20,641	3.6	48,900[1]	9.9	300	61.3	0.08	0.45	145.2
Morocco	29,000	3.4	0	12,758	44.0	4,113	4.9	8,692[1]	9	200	7.7	0.06	0.72	16.3
Netherlands	91,000	—	—	7,944	8.7	10,003	1.9	14,575[2]	2.7	150	4.1	0.05	0.83	5.9
Norway	382,000	—	—	2,185	0.6	4,348	3.8	9,167[2]	8.1	150	5.3	0.24	1.06	11.1
Philippines	479,000	0	0	28,520	6.0	1,992	7.7	11,293[1]	8.8	300	5.3	0.02	0.11	29.8
Poland	61,600	—	—	16,201	26.3	17,400	4.2	39,619[2]	4.8	150	12.5	0.08	0.60	28.5
Portugal	77,400	—	—	11,263	14.6	12,097	2.1	16,000[1]	6.7	400	32.4	0.29	3.00	42.9
Rep. of Korea	69,700	0	0	18,590	26.7	5,322	4.1	10,272[1]	8.1	200	8.6	0.05	0.13	16.6
Romania	211,930	—	—	23,176	10.9	3,274	2.8	8,500[1]	2.5	150	1.2	0.01	0.06	3.2
Russian Federation	4,507,250	0	0	76,686	1.7	21,169	6.8	48,000[1]	8.1	150	25.7	0.03	0.18	58.3
Saudi Arabia	2,400	714.0	217.0	17,320	721.7	6,296	5.3	12,194[1]	8.1	200	10.2	0.06	0.60	19.8
South Africa	50,000	0	0	15,306	30.6	6,001	8.0	30,523[1]	8.1	300	14.6	0.10	0.57	74.2
Spain	111,500	—	—	35,635	32.0	47,898	2.6	73,867[1]	12.9	400	247.2	0.69	5.16	381.2
Sweden	174,000	—	—	2,965	1.7	2,746	3.8	5,790[2]	8.1	150	3.3	0.11	0.31	7.0

Country	Total natural renewable water resources (million m3/year)	Desalin-ated water (million m³/year) (note 2)	Reused treated waste-water (million m³/year) (note 2)	Total water use in 2000 (million m³/year)	% of renewable water used	Tourist arrivals 2000 ('000)	Growth rate tourist arrivals	Tourist arrivals 2020 ('000)	Average length of stay, 2000	Water use per tourist per day[3]	Total tourism-related water use, 2000 (million m³)	Tourism-related water use as % of total	Tourism-related water use as % of domestic	Total tourism-related water use 2020 (million m³)[4]
Switzerland	53,500	—	—	2,571	4.8	11,000	1.7	17,400[1]	8.1	150	13.4	0.52	2.16	21.1
Thailand	409,944	0	0	87,065	21.2	9,579	6.9	36,959[1]	7.8	300	22.4	0.03	1.03	86.5
Trinidad and Tobago	3,840	0	0	305	8.0	399	4.3	926[2]	8.1	400	1.3	0.42	0.62	3.0
Tunisia	4,560	8.3	20.0	2,726	59.8	5,058	3.1	8,916[1]	6.6	400	13.4	0.49	3.11	23.5
Turkey	231,700	0.5	0	37,519	16.2	9,586	5.5	27,017[1]	10	400	38.3	0.10	0.69	108.1
UK	147,000	0	0	9,541	6.5	25,209	3.4	53,800[1]	8.1	200	40.8	0.43	1.97	87.2
Ukraine	139,550	0	0	37,523	26.9	4,406	4.2	10,032[2]	5.3	150	3.5	0.01	0.08	8.0
UAE	150	385.0	108.0	2,306	1537.5	3,907	7.1	15,404[2]	8.1	200	6.3	0.27	1.19	25.0
USA	3,069,400	—	—	479,293	15.6	50,945	3.5	102,400[1]	8.1	300	123.8	0.03	0.20	249.0
Uruguay	139,000	0	0	3,146	2.3	1,968	5.3	5,528[2]	6.9	150	2.0	0.06	2.55	5.7
Totals	34,076,497	—	—	2,886,370	—	584,066	—	1,357,990	—	—	1,487.3	—	—	3,081.3

[1] WTO 2001.

[2] extrapolation based on growth rate.

[3] weighted average, estimate by author. Categories: countries with i) high share of friends and relative-related tourism, high percentage of small accommodation establishments or city hotels, high share of mountain tourism: 150 litres per tourist per day (t/d), ii) Mediterranean and countries with high percentage of resort hotels: 400 litres t/d, iii) other, individual judgement: 200–300 litres t/d.

[4] extrapolation does not consider increases/decreases in per tourist water use estimates; *global average applied in absence of national data, calculation in Gössling 2002.

Sources: WTO 2001, 2003a; WWF 2001, 2004; www.fao.org/aquastat 2003 (accessed 10 August 2004).

interaction between large-scale technical adaptation and small-scale tourist infra-structure. There are also examples where the water demands of hotels have been favoured over the needs of local populations (see Gössling 2001), reminding us that water access and use are also questions of global environmental justice. In the future, global water demand will depend on population growth, modernisation processes, technology and economic restructuring such as moving from water-intensive to less water consuming economic activities. It is projected that global water demand will substantially increase, putting additional constraints on already water scarce coun-tries (Clark and King 2004).

What role does tourism play in these countries as a water consuming factor? As indicated in Table 10.2, tourism-related water demand depends on water use per tourist, number of international tourist arrivals, and average length of stay of tour-ists. Because country-specific data on these parameters is often not available, some assumptions have to be made for analysis. Water use per tourist per day was esti-mated for three categories of countries: i) countries with a high share of 'friends and relatives'-tourism, a high percentage of small accommodation establishments or city hotels, a large share of mountain-based tourism (for example, Austria, Romania, Switzerland: 150 litres per tourist per day), ii) Mediterranean and coun-tries with a high percentage of resort hotels (for example, Mauritius, Spain; 400 litres per tourist per day), and iii) those not belonging to categories i) and ii). For the latter, water use was estimated based on individual judgements (for example, Philippines, Mexico, Denmark, with water uses of 200–300 litres per tourist per day). Note that most estimates need to be seen as rather conservative, as for example WWF (2004) calculates an average water consumption of 440 litres per tourist per day for Mediterranean countries.

The results show that tourism generally accounts for less than 1 per cent of the total water consumption in the sample countries. Malta (7.3 per cent), Cyprus (4.8 per cent) and Barbados (2.6 per cent) are exceptions, even though they indicate that temperate and tropical islands with high tourist arrival numbers and limited water resources are more likely to face water conflicts. This becomes even more obvious looking at the importance of tourism-related water uses in comparison to domestic water use. While tourism remains a negligible factor of water use in most countries, it might be respon-sible for up to 16.9 per cent of domestic water use in countries like Cyprus. Clearly, tourism has an important influence on water consumption patterns. This also becomes obvious looking at tourism's absolute water consumption, which varied between 0.2 million m³ (Cape Verde) and 247.0 million m³ (Spain) in 2000.

Extrapolating the results of Table 10.2, global water use by international tourism might be in the order of 1.8 km³. This estimate is somewhat higher than the top-down assessment of 1.3 km³ provided above. By 2020, tourism-related water use is likely to increase with international tourist numbers and higher hotel stan-dards (see Gössling 2001). Even though the development of international tourist arrivals has been irregular during the last years, the World Tourism Organization (2004) maintains projections of its *Tourism 2020 Vision* (WTO 2001), which fore-casts over 1.56 billion international arrivals by the year 2020. Of these arrivals, 1.2 billion will be intraregional and 0.4 billion will be long-haul travellers. Distribution by region shows that, by 2020, three regions will receive the majority of tourists: Europe (717 million tourists), East Asia and the Pacific (397 million) and the Americas (282 million), followed by Africa, the Middle East and South Asia. According to the forecast, East Asia and the Pacific, South Asia, the Middle East and Africa will experience growth rates of over 5 per cent per year, compared to the world average of 4.1 per cent per year. Europe and the Americas are anticipated to show lower than average growth rates. Extrapolating projected tourist arrival numbers to water use, and not considering potential changes in water use or average length of stay, international tourism might account for the use of more than 3.8 km^3 of fresh water by 2020.

Vulnerability

In the future, water resources will decrease in many countries; both as a result of the overuse of renewable water supplies and climate change leading to new precipitation patterns. While it is clear that population growth and modernisation processes will lead to increasing water stress, pushing up the number of people living in water scarce countries, it is less obvious how climate change will affect water availability (see Arnell 2004). The results of the models used by Arnell indicate that:

> Climate change increases water resources stresses in some parts of the world where runoff decreases, including around the Mediterranean, in parts of Europe, central and southern America, and southern Africa. In other water-stressed parts of the world – particularly in southern and eastern Asia – climate change increases runoff, but this may not be very beneficial in practice because the increases tend to come during the wet season and the extra water may not be available during the dry season.
>
> (Arnell 2004: 31)

Decreasing water resources will most certainly affect tourism in many areas, particularly in small developing islands and already water scarce areas. However, the consequences for the tourist industry will depend on several factors, including the relative scarcity of fresh water in tourism areas, competition with other economic sectors, the structure of the tourist industry (for example, small guest-houses or large resort hotels), and options to adapt to these changes, for example, through technological change including water saving measures, desalination and wastewater reuse. Technological adaptation might be costly, however, if for example desalination technology is involved. As pointed out earlier, it might also be a solution feasible only for larger accommodation establishments. Water desalination for tourism also contributes to global environmental change, because it entails high energy use. Reverse osmosis, the technical standard for hotels, involves 6 kWh of electricity per m^3 of water – corresponding to emissions of approximately 1.9 kg CO_2 – for pumps forcing water through a membrane to separate fresh water from brine. Distillation, this is heating water to create steam, which distils as fresh water, leads to an even greater energy use of 25–200 kWh per m³ (Clarke and King 2004).

Clarke and King (2004) identify a range of countries that will be chronically short of water by 2050, including the Netherlands, Germany, Tunisia, Malta, Morocco, South Africa, Cyprus, Maldives, Singapore, Antigua and Barbuda, St. Kitts and Nevis, Dominica and Barbados. For these countries, it will be increasingly difficult to provide fresh water for tourism, even though this might certainly prove to be a bigger challenge for the Maldives than for the Netherlands and Germany. Many other small islands not included in the list above are also in jeopardy of overusing their water capacity. In particular, small tropical islands might face comparatively higher costs in adapting to water scarcity, because they compete more directly with other sectors of the national economy for generally scarcer water resources.

It seems difficult to predict in which areas of the world water scarcity will have serious or even severe consequences for tourism, even though it might generally be possible to identify coastlines, small – often coralline – islands, and arid/semi-arid zones as those in greatest danger of water stress. Even though some of these areas have been identified, such as the Mediterranean (WWF 2004), Mallorca (Essex *et al*. 2004) or Zanzibar (Gössling 2001), further country-specific studies are needed in order to better understand the vulnerability of the tourist industry in various locations.

Conclusion

This chapter has outlined the linkages between tourism, water and global environmental change. Tourism is highly dependent on water resources, both in terms of providing the physical framework for tourism (oceans, lakes and rivers), and to maintain the tourism production system (fresh water used for swimming pools, bathrooms, food preparation, etc.). In some areas, tourism is already jeopardised because of increasing water stress, while over-consumption and global environmental change are likely to be felt in many regions in the short-term future.

International tourism seems to play a minor role in global water use, with between 1.3 and 1.8 km³ of fresh water being used annually by international tourists (calculation for 2002), corresponding to 0.04–0.06 per cent of global fresh water consumption. Note that this excludes the water use of domestic tourists as well as indirect water uses, which can be assumed to add significantly to the total. It should also be considered that tourism's water use is highly concentrated in time and space, often occurring in dry and water scarce regions, and usually during the dry season when the recharge of aquifers is limited. In such regions, tourism might often be the most important factor contributing to water consumption. Overall, tourism can thus be said to be highly dependent on fresh water availability, while simultaneously contributing to the depletion of renewable water resources, particularly in already water scarce areas.

Water scarcity can be addressed in various ways. Reducing water use seems the most feasible option in many areas, as this involves lower costs and can often be done with simple measures. For example, irrigation of gardens is usually the most water-consuming factor, accounting for up to 50 per cent of total water use. The use of recycled water can substantially reduce the total amount of water used, as can the use of less water-dependent vegetation. Because swimming-pools seem to be the second most important water-consuming factor, it is clear that the design of hotels should take into consideration fresh-water related issues, moving away from 'pool-landscapes' of large, interconnected swimming pools. A technical option for fresh water production is desalination. However, desalination is costly, and contributes to substantial emissions of greenhouse gases because desalination plants consume considerable amounts of energy. Desalination might also be an option only for larger accommodation establishments.

This chapter has also rendered prominent the complexity of the interaction between tourism and other aspects of global environmental change. Climate change, to which tourism contributes, leads to changing rainfall patterns and might thus put additional stress on water resources in many tourist destinations. Tourism and climate change can thus be seen as self-reinforcing processes, with water scarcity being a case in point: in areas where fresh water resources have been overused, desalination might be the preferred technological option to maintain the tourism infrastructure. However, desalination increases the energy-intensity of tourism, and thus contributes to climate change. Environmental change related to water over-consumption might also affect biodiversity. For example, global freshwater wetlands have diminished in area by about half over the past century, resulting in the loss of important habitats (Clarke and King 2004). Furthermore, some 10,000 freshwater fish species are reported to be at risk of extinction (Worldwatch Institute 2004). Sewage release by hotels also decreases water quality, which in turn can affect human beings and biodiversity. Obviously, the availability of water is also linked to political and social issues, because tourism might use water resources at the expense of local populations. Such cases of environmental injustice have, for example, been reported in Zanzibar, Tanzania (Gössling 2001), where inequalities in water use have been backed by political and economic elites financially involved in the tourist industry.

References

Arnell, N.W. (2004) 'Climate change and global water resources: SRES emissions and socio-economic scenarios', *Global Environmental Change*, 14: 31–52.

Clarke, R. and King, J. (2004) *The Atlas of Water. Mapping the World's Most Critical Resource*. London: Earthscan.

Essex, S., Kent, M. and Newnham, R. (2004) 'Tourism development in Mallorca. Is water supply a constraint?' *Journal of Sustainable Tourism*, 12(1): 4–28.

FAO (Food and Agriculture Organization) 2003. Review of water resource statistics by country. Available at: www.fao.org/ag/agl/aglw/aquastat/water_res/index.stm (accessed 9 August 2004).

Garrod, B. and Wilson, J.C. (2003) *Marine Ecotourism: Issues and Experiences*. Clevedon: Multilingual Matters Limited.

GFANC (German Federal Agency for Nature Conservation) (1997*) Biodiversity and Tourism. Conflicts on the World's Seacoasts and Strategies for Their Solution*. Berlin: Springer.

Ghimire, K.B. (2001) *The Native Tourist: Mass Tourism Within Developing Countries*. London: Earthscan Publications.

Gössling, S. (2001) 'The consequences of tourism for sustainable water use on a tropical island: Zanzibar, Tanzania', *Journal of Environmental Management*, 61(2): 179–91.

Gössling, S. (2002) 'Global environmental consequences of tourism', *Global Environmental Change*, 12(4): 283–302.

Gössling, S. (2005) 'Tourism's contribution to global environmental change: space, energy, disease, and water', in C.M. Hall and J. Higham (eds) *Tourism, Recreation and Climate Change: International Perspectives*. Clevedon: Channel View Publications.

Kent, M., Newnham, R. and Essex, S. (2002) 'Tourism and sustainable water supply in Mallorca: a geographical analysis', *Applied Geography*, 22: 351–74.

Lüthje, K. and Lindstädt, B. (1994) *Freizeit- und Ferienzentren. Umfang und regionale Verteilung. Materialien zur Raumentwicklung*, Heft 66. Bonn: Bundesforschungsanstalt für Landeskunde und Raumordnung.

Orams, M. (1998) *Marine Tourism. Development, Impacts and Management*. London: Routledge.

Smith, B. (1997) 'Water: a critical resource', in R. King, L. Proudfood and B. Smith (eds) *The Mediterranean: Environment and Society*. London: Edward Arnold, pp.227–51.

Solley, W.B., Pierce, R.R. and Perlman, H.A. (1998) *Estimated Use of Water in the United States in 1995*. U.S. Geological Survey Circular 1200, Denver, CO: US Government Printing Office.

UK CEED (UK Centre for Economic and Environmental Development) (1994) *A Life-Cycle Analysis of a Holiday Destination: Seychelles*. British Airways Environment Report No.41/94. Cambridge: UK CEED.

UN (United Nations) (1995). *Guidebook to Water Resources, Use and Management in Asia and the Pacific. Volume One: Water Resources and Water Use*. Water Resources Series No. 74, New York: UN.

Vörösmarty, C.J., Green, P., Salisbury, J. and Lammers, R.B. (2000) 'Global water resources: Vulnerability from climate change and population growth', *Science*, 289: 284–8.

Worldwatch Institute 2004. Rising Impacts of Water Use. Available at: www.worldwatch.org/topics/consumption/sow/trendsfacts/2004/03/03 (accessed 4 March 2004).

WRI (World Resources Institute) (2003) *World Resources 2002–2004, Data Tables*. Available at: www.wri.org (accessed 4 September 2003).

WTO (World Tourism Organization) (2001). *Tourism 2020 Vision*. Madrid: WTO.

WTO (2003a) *Compendium of Tourism Statistics*. Madrid: WTO.

WTO (2003b) *Climate Change and Tourism*. Proceedings of the 1st International Conference on Climate Change and Tourism. Djerba, Tunisia, 9–11 April.

WTO (2004). Available at: www.world-tourism.org/market_research/facts/menu.html (accessed 10 August 2004).

WWF (World Wide Fund for Nature) (2001) *Tourism Threats in the Mediterranean*. WWF Background information. Switzerland: WWF.

WWF (2004) *Freshwater and Tourism in the Mediterranean* Available at: www.panda.org/downloads/europe/medpotourismreportfinal_ofnc.pdf (accessed 12 November 2004).

11 Extreme weather events

Chris R. de Freitas

Introduction

Weather and climate and play a significant role in influencing tourism and recreation behaviour. Many tourist destinations, especially those in the tropics and subtropics, rely heavily on environmental assets such as sun, sea and sand and a generally agreeable climate to attract visitors. Climate change, including changed variability, whether natural or anthropogenic, could modify these assets and how they are perceived by potential visitors. Any change in local weather patterns affecting average and extreme conditions has well-known direct effects on visitors, but may also have indirect effects, such as the influence of storms on the structure of beaches. Both of these categories of impact hold implications for patterns of tourist distribution, including the numbers and types of visitors, and the places they visit.

Given the prospect of changing climate, the past may no longer be an adequate guide for the future, and this needs to be taken into account in tourism planning and investment decisions. Yet the topic of future climate is plagued with uncertainty. At the same time it appears that decision makers who deal with this uncertainty are preoccupied by the prospect of damaging or undesirable changes in climate. It is argued here that this point of view is no more helpful in tourism planning than that which treats climate as being a constant or simply a permanent feature of the physical setting. This chapter puts these views into context. It aims to briefly address the state of scientific knowledge on future climate variability, the sorts of changes that could arise and shifts in risk that might occur.

Evidence, expectations and perceptions of change

One major determinant of risk is the perceived trends in climate extremes. Given that, first, weather hazards are the most significant natural hazards in most places (Downing *et al.* 1996; Dlugolecki *et al.* 1996; Kattenburg *et al.* 1996; Coates 1998; McCarthy *et al.* 2001) and that, second, in the latter part of the twentieth century insured losses have been unprecedented (Dlugolecki *et al.* 1996; Kattenburg *et al.* 1996), the possibility of worsening trends has, understandably, attracted the attention of the insurance industry. There are a number of possible reasons for increasing losses:

- a greater concentration of people and high value property in vulnerable areas, mainly coastal;
- business processes have become more susceptible to damage; or
- that changes have occurred in the frequency and severity of extreme climatic events.

The last of these is in line with expectations of climate change resulting from an enhanced greenhouse effect and this too has attracted the attention of the insurance industry. Dlugolecki *et al.* (1996: 541) comment: 'It is a common perception in the insurance industry that there is trend toward an increasing frequency and severity of extreme climate events.' The important question arises as to the extent to which these expectations are justifiable.

The ongoing debate on climate change and global warming highlights the possibility of changed magnitude and frequency of extreme weather events globally. But it is a widespread misconception that science generally predicts more severe weather events will accompany greenhouse gas-enhanced climate change. The research so far has produced conflicting results. It is notable that the general circulation models of global climate (GCMs) or scenarios referred to in the reports of the United Nations Intergovernmental Panel of Climate Change (IPCC) do not explicitly quantify changes in daily weather extremes (Houghton *et al.* 2001a). Moreover, reflecting on current trends, the IPCC stated: 'Overall, there is no evidence that extreme weather events, or climate variability, has increased, in a global sense, through the twentieth century' (Houghton *et al.* 1996: 173). This contrasts with the very real upward trend in damage due to weather extremes, particularly hurricanes, in recent decades. But the increasing dollar cost of storm and other weather related events is accounted for by a rise in the value of development and number of properties, especially in tropical cyclone prone areas, rather than by an increased frequency of events (Changnon *et al.* 1997; Pielke and Landsea 1998; Kunkel *et al.* 1999). In fact, there is a great deal of research which, taken together, suggests that extreme climate events may become both less frequent and less severe if the planet warms. Some of the elements of this debate are discussed here.

Table 11.1 summarises a number of categories of weather and climate relevant to tourism that might be expected to change in response to increasing concentrations of greenhouse gases in the atmosphere. It outlines the nature of the extreme events, parent climate variables used to identify them, how they may be quantified and the nature of their impact on the tourism sector. The information in Table 11.1 indicates that tourism is, for the most part, sensitive to extremes in air temperatures and windstorms, predominantly the latter. Storms have the greatest impact on coastal areas and, because of this, Perry (1997) points out the tourist industry is particularly vulnerable to these types of extreme events, as 70 per cent of holidays are coast oriented. Tourists also tend to be more vulnerable than

Table 11.1 Definitions and measures of climate and weather extremes, and impact classes

Climate extreme	Parent variable	Measure of extreme (including probability of occurrence)	Impact category
Hurricane/tropical storm	Storm track Storm strength Rainfall Storm surge	Number of events with wind speeds greater than a given threshold	Property damage Loss of life Added insurance cost
Windstorm	Wind speed (maximum gust)	Number of events with wind gusts greater than a given threshold	Property damage Loss of life Added insurance cost
Thunderstorm	Rainfall	Number of events with rainfall greater than a given threshold	Property damage Loss of life Added insurance cost
Heat wave	Air temperature and humidity	Number of days above a given threshold	Thermal discomfort Heat stress Energy use
Cold wave	Air temperature	Number of days below a given threshold	Thermal discomfort Cold stress Energy use
High temperature	Air temperature	Number of degree days above a given threshold	Reduced ski season (re. snowmaking)

locals because they are unfamiliar with the place they are visiting. Aspects of the weather themes outlined in Table 11.1, their relative significance and apparent trends, are discussed below.

Hurricanes

Hurricanes, also known as typhoons and tropical cyclones, are the most intense form of extreme weather and pose the greatest threat to tourists and tourism infrastructure, especially in tropical areas. It is widely held that global warming may increase the odds in favour of more intense and more frequent hurricanes. Hurricanes require warm water to form and global warming means more warm water. But even without global warming, there is ample warm water in tropical oceans. Some of the highest sea surface temperatures in the world are found in the tropical mid-Atlantic and Caribbean, yet in most years there are less 15 hurricanes and frequently less than five per year (Figure 11.1). Given the great expanse of warm water globally in any given year, the important question is why do hurricanes only form some of the time, and why are there so few? At the same time, it is too early to

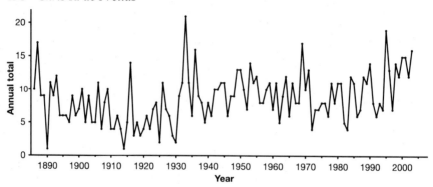

Figure 11.1 The number of hurricanes and tropical storms in the tropical North Atlantic
Basin per year, 1886–2003

Source: United States, National Climate Data Center

dismiss the role of changes in sea surface temperature. For example, Goldenberg *et al.* (2001) showed that hurricanes occur in distinct multi-decadal cycles and are linked to sea surface temperature anomalies in the Atlantic Ocean's main hurricane development region. Warm anomalies are associated with increased major hurricane activity; cold anomalies with suppressed activity. But overall, no significant upward trends have been identified.

Hurricanes need exactly the right conditions to form, and warm water and resulting high water vapour levels are just two of the ingredients. So many factors influence the formation of hurricanes that it is difficult to predict how global warming will influence their frequency and intensity. In the tropical mid-Atlantic and Caribbean, for example, the number and strength of easterly pressure waves coming across the Atlantic and the change in wind with height through the atmosphere (wind shear) are as important as warm sea surface temperatures to hurricane formation. It is not known how global warming will affect these processes.

It could be argued that changing ocean and atmospheric conditions due to global warming might make historical weather patterns less useful for long-range climate forecasts, but recent and past statistics of events are revealing. In the tropical North Atlantic Basin the number of intense hurricanes declined during the 1970s and 1980s, and the period 1991–1994 experienced the smallest number of hurricanes of any four years over the past half century (Idso *et al.* 1990; Murphy and Mitchell 1995; Bengtsson *et al.* 1996; Landsea *et al.* 1996; Zhang and Wang 1997).

Easterling *et al.* (2000) showed that, overall, occurrences of Atlantic hurricanes do not reveal a statistically significant long-term trend over the twentieth century; but Landsea *et al.* (1999) found a statistically significant decrease in the high intensity Atlantic hurricanes that are most destructive. Gray *et al.* (1997) showed that there have been large interdecadal variations of hurricane activity over this period. From 1944 to the mid-1990s, the number of intense and Atlantic hurricanes that made landfall declined (Figure 11.2). As for the future, several researchers have suggested reductions in hurricane frequencies in response to global warming (Wilson 1999; Elsner *et*

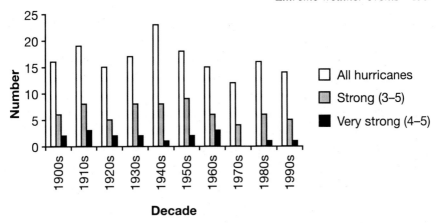

Figure 11.2 USA hurricane strikes by decade, 1900–1999

Source: United States, National Climate Data Center

al. 2000; Liu and Fearn 2000; Parisi and Lund 2000; Singh *et al.* 2000, 2001; Boose *et al.* 2001; Elsner and Bossak 2001; Muller and Stone 2001).

Trends are different in the Pacific. Chu and Clark (1999) analysed the frequency and intensity of intense tropical storms that occurred in the region of the central North Pacific (0–70°N, 140–180°W) over the 32-year period 1966–1997. They found that storm activity has risen, which amounts to an increase of about 3.2 storms over the 1966–1997 period. Accompanying the increase in intense tropical storms is a similar increase in maximum hurricane intensity. In contrast, the results of experiments by Sugi *et al.* (2002) using a GCM indicated that the number of intense tropical storms may be significantly reduced due to the global warming. As for the maximum intensity of intense tropical storms, they state that 'no significant change has been noted'. Nguyen and Walsh (2001) simulated the occurrence of hurricanes in the Australia region using a GCM that assumes a tripling of the atmospheric concentration of CO_2. The results showed that the numbers of hurricanes declined and that the decline is statistically significant.

As a first approximation of hurricane activity over the next two or three decades, one can simply extrapolate past variations in occurrences, assuming there is not some periodicity in the data. Landsea (2000) examined the data for the Australian, the Northwest Pacific and the Atlantic basins in enough detail to allow some suggestions to be made as to what the first decade of the twenty-first century may bring. As regards hurricane frequency, he reported that the Australian basin showed a decline since the late 1960s, the Northwest Pacific showed an increase after experiencing a decrease in frequency from the late 1950s through 1980, while the Atlantic was fairly constant since the mid-1940s. For mean intensity, he reported that there was little or no trend in the Australian basin, the Northwest Pacific showed a downward trend during the 1960s and 1970s and an upward trend in intensity of events since (Landsea 2000). Looking to the future, Landsea concluded that there is no convincing evidence for systematic changes to occur in the frequency, mean intensity, maximum intensity and area of occurrence of hurricanes.

Extratropical storms

Climate change scenarios do not explicitly quantify changes in daily weather extremes. Despite this, it is often implied that an increase in greenhouse gas concentration could create conditions that favoured severe storms over mid-latitude continental areas. This view, however, is not widely supported in the scientific literature. For example, Balling and Cerveny (2003) reviewed the scientific literature on United States weather records on thunderstorms, hail events, intense precipitation, tornadoes, hurricanes and winter storm activity during the modern era of greenhouse gas build-up in the atmosphere and found that, although there has been an increase in heavy precipitation, trends in other severe storm categories are downward.

The most important energy source for extratropical storms is the temperature difference between the tropics and the poles. Most GCMs suggest that the greatest warming would occur over the high latitudes in winter with relatively little warming in the tropics and around equatorial latitudes. This implies reduced temperature variation, since such variations result from air moving from one latitude to another. Thus, according to these predictions, the future contrast between the polar and equatorial latitudes will lessen, producing a weaker gradient and fewer and less intense storms. Consistent with this, the IPCC 2001 *Summary for Policymakers* (Houghton *et al.* 2001c) notes that no significant upward trends have been identified in tropical or extratropical storm intensity and frequency.

There is evidence from Europe that suggests there has been a decline in the number of mid-latitude storms. For the period 1896–1995, Bielec (2001) analysed thunderstorm data obtained at Cracow, Poland, which he states is one of the few continuous records in Europe with an intact, single place of observation and duration of over 100 years. From 1930 onward the trend is negative, revealing a linear decrease of 1.1 storms per year from 1930 to 1996. Bielec also reported that there has been a decrease in the annual number of thunderstorms with hail over the period of record, and there has been a decrease in the frequency of storms producing precipitation greater than 20mm.

Pirazzoli (2000) analysed storm surges, atmospheric pressure and wind change and flooding probability on the Atlantic coast of France over the period 1951–1997. He found that climate variability is decreasing. Specifically, his work showed that the number of atmospheric depressions and strong winds that cause storm surges in this region are becoming less frequent resulting in reduced frequency and severity of coastal flooding.

Air temperature extremes

How air temperatures might change in a greenhouse enhanced world is contentious and research results are conflicting. For example, Frich *et al.* (2002) presented

Table 11.2 Day-to-day air temperature variability for the USA, People's Republic of China and the former Soviet Union, shown as mean linear trend (°C per decade) in daily temperature variability values

Air temperature	USA	China	USSR
Maximum			
January	−0.19	−1.13	0.07
July	−0.13	0.06	−0.02
Minimum			
January	−0.26	−1.32	−0.37
July	−0.19	0.06	−0.08

Source: Michaels *et al.* 1998

results that suggest that trends in the second half of the twentieth century show that the world has become warmer and wetter, heavy rainfalls more frequent, but that cold temperature extremes have become less frequent over the same period. There has been speculation too that global warming will increase climate variability and thus the frequency of heat waves, but Michaels *et al.* (1998) have shown there is no universal support for this, nor for the popular perception that temperatures have become more variable. Michaels *et al.* examined daily maximum and minimum temperatures from the USA, China and the former Soviet Union for day-to-day variability in January and July and most of the trends indicated declining variability (Table 11.2). Other works produce similar results. Overall, however, it is fair to say that relatively little work has been done on changes in high frequency extreme temperature events such as heat waves, cold waves, and number of days exceeding various temperature thresholds (Easterling *et al.* 2000).

Karl *et al.* (1995) point out that an increase in the atmospheric concentration of CO_2 should decrease temperature variability. Balling (1998) examined changes in the spatial variability of mean monthly and daily temperatures that have occurred during the historical climate record. His research showed that, overall, the spatial variability in temperature anomalies has declined, and that the interannual variability in temperature anomalies is negatively correlated to mean hemispheric temperatures. The work of both Balling *et al.* (1998) and Michaels *et al.* (1998) show that as the atmosphere warms, the month-to-month variability also declines (Figure 11.3). The entire June 2003 issue of the scientific journal, *Natural Hazards*, was devoted to assessing whether global warming causes extreme weather. The editors of the special issue concluded that most studies found no such connection.

Shifting risks

The 'normal' characteristics of weather and climate need to be considered as a backdrop to the elements of climatic extremes outlined in Table 11.1. Extreme weather events are a permanent feature of 'normal' climate and constitute an important factor in both financial terms for tourism operators and the personal

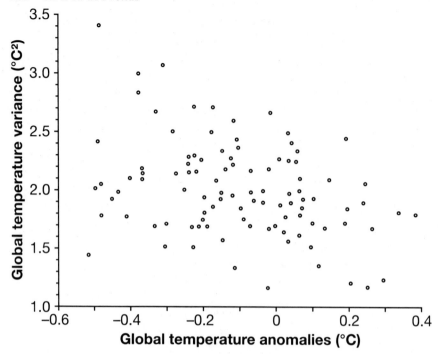

Figure 11.3 Inter-annual surface temperature variability versus global temperature anomalies for the 1897–1997 time series showing the warmer the surface temperature, the less variable climate becomes

Source: Michaels *et al.* 1998

experiences of tourists. Many holiday destinations have 'tourism appeal' that is a
function of the destination's climate. Tourism administrators promote places
based on this appeal and potential tourists make decisions on whether to visit them
based on their perception or expectation of climate conditions. For holiday trip
decisions in which climate of the destination is not a motive for travel, extreme
weather and climate set limits. Financial losses can result from weather and
climate variations and for many tourist activities there are limits or limiting conditions beyond which there is increasing risk of one sort or another (Figure 11.4).

It is possible that climate change due to global warming will shift 'normal' risk
thresholds. It is also recognised that changes in the severity and frequency of
extreme events are likely to be more important than changes in the average climate.
Clearly, tourists respond to events or real conditions rather than to averages – their
response to extremes, such as heat waves and storms, is different from their
response to a change in the mean climate. In the case of weather extremes, the
response time is shorter and the response itself is likely to be greater. By and large,
it is reasonable to expect that the impacts of climate change are likely to be more
severe due to changes in the occurrence of extreme events than due to a change in
the mean climate.

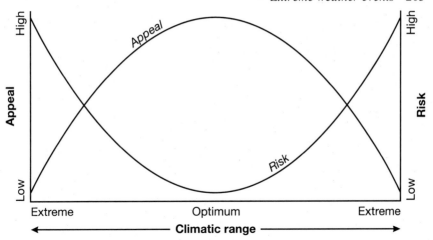

Figure 11.4 A schematic representation of relationships between climatic range and tourism potential. The climate potential of a particular location is a function of its climate and of the risks (e.g. to safety, profit making) weather may impose

Source: Adapted from Perry 1997

In this context, the impact of weather extremes is as much a product of the social, political and economic environment as it is of the natural environment. It follows, therefore, that the risk of these impacts is in part a social construct perceived differently by all of us and must be defined with this in mind (de Freitas 2002). For example, risk is defined by Emergency Management Australia (1995) as the perceived likelihood of given levels of harm. Thus, one major determinant of risk is the perceived trends in weather extremes. In a seminal paper by Hoyt (1981), the theory of extremes as applied to weather records is reviewed and then compared to the actual frequency of record weather events in the USA. He concluded that fewer extremes of temperature are being set in recent years contrary to the popular view. He cautions against using extreme weather events that set new records as evidence of climate change. Hoyt (1981: 248) explains thus:

> Because the probability of establishing a new weather record never drops to zero, then every year some region will establish a new precipitation or temperature record. Even in the warmest years on a global scale there will be some locations where a new monthly mean low temperature record will be set. Even in the coolest years some locations will have periods of record warmth. Individual temperature and precipitation records by themselves tell us nothing about climatic change.

The statistics of climate variability and assumptions used in assessing them are all important in detecting and planning for global climate change. But just as extreme weather cannot be used as evidence of climate change, neither can trends in mean conditions be used as a predictor of trends in extreme occurrences – unless there is no change variability of the predictor variable in question, which is unlikely. Robeson (2002) addresses this issue by questioning the use of near surface air temperature trends in the USA to determine the impacts on extremes. Robeson (2002: 205) points out that, even though extremes are important:

> It is impossible to determine the impacts of changes in mean air temperatures on extremes, however, without simultaneously analysing changes in the variance of air temperature probability distributions ... In addition to its importance for extreme events, the detection of changes in air-temperature variance plays a fundamental role in helping us to understand how the climate system may be changing.

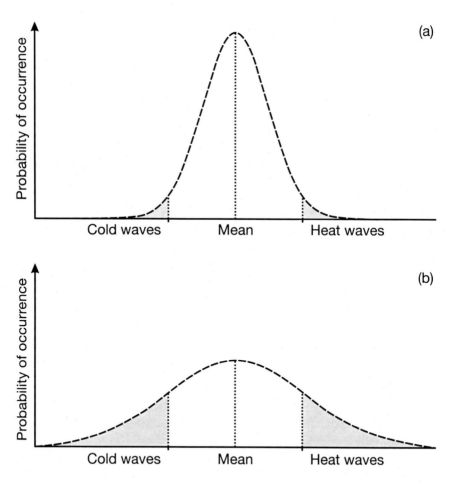

Figure 11.5 These two distributions have the same mean, but the variance is larger for (b). There are more occurrences of extremely high and low temperatures in (b) than in (a).

Aspects of the above are summarised in Figure 11.5 which schematically portrays the statistics of extremes. The distribution of daily temperatures for a given location typically takes the shape of a 'normal' bell-shaped distribution in which temperature on the majority of the days is close to the mean. The number of days with temperatures above or below the mean decreases with movement away from the mean. The normal distribution is characterised by both its mean value and the 'variance', which is a statistic that indicates how close most of the observations are to the mean. A small variance means that most temperatures are quite close to the mean, while a large variance means that a large number of observations are different from the mean. Both Figure 11.5a and Figure 11.5b distributions have the same mean, but the variance is larger in Figure 11.5b, where there are more occurrences of heat waves and cold waves.

If global climate changes, it may do so in a number of ways. The first of these is illustrated in Figure 11.6. Karl *et al.* (1997) point out that GCMs predict temperatures will be confined to a tighter range, which was confirmed by Easterling *et al.* (1997) who found that most of the increase in global temperatures has been occurring during the winter and at night. Summer maximum air temperatures in the northern hemisphere showed no statistically significant trend. If these forecasts are correct, variability in the data will shrink more than the predicted mean warming and give a distribution shown schematically in Figure 11.6.

Figure 11.7 builds on the above by showing different global warming scenarios in which the impact of a rise in mean air temperature is dependent on accompanying changes in the variance, since the probability of extreme temperature occurrence (heat waves and cold waves) can vary greatly depending on the exact nature of the

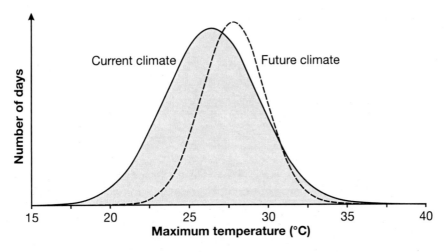

Figure 11.6 A forecast of future maximum air temperature distribution. If interpretations by Easterling *et al.* (1997) and Karl *et al.* (1997) are correct, variability in the data will shrink more than the predicted mean warming and give the distribution shown schematically here

Source: After Michaels and Balling 2000

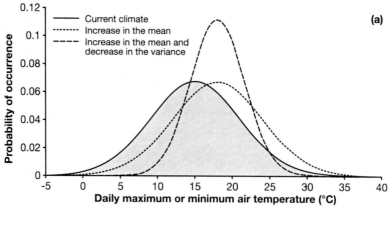

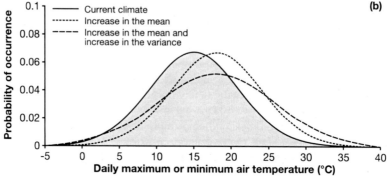

Figure 11.7 Response of probability distributions to changes in mean and variance of daily air temperature. The original climate is warmed by an increase in the mean of 3°C. In the case of (a), a rise in air temperature accompanied by a decrease in variance would lead to a reduction in the number of extreme temperatures. In (b) a rise in temperature accompanied by an increase in variance would lead to a far greater number of extremely hot days. These probability distributions are substantially different from those that include changes in mean temperatures only.

Source: Adapted from Robeson 2002

changes in the distribution. Figure 11.7 shows that a change in the mean can have a disproportionate and non-linear effect on the fraction of extremes beyond critical thresholds. In addition, there may be a non-linear relationship between a change in the mean of a distribution at the extremes because the other aspects of the distribution (variance, kurtosis) have also changed. A rise in mean temperature accompanied by an increase in the variance leads to a big rise in the probability of heat waves. Conversely, a rise in the mean temperature that is accompanied by a decrease in the variance has the opposite effect. While changes in the probability distributions shown here assume a normal distribution, changes in the tails of the curves may not be symmetrical, in which case other parameters such as skewness may need to be evaluated. Robeson (2002) points out that given an increasing mean temperature – 'global warming' – and decreasing variability, it is likely that the lower tail of the curve would be increasing faster than the upper tail (Figure 11.7). Depending on the magnitude of the warming and accompanying change in the variance, upper-tail temperatures that are considered to be extreme in present-day climate conditions may become more probable as temperature variability is reduced.

Clearly, a key to assessing future impacts of climate change is understanding the relationship between mean temperature and daily temperature variance. In the case of the USA, Robeson (2002) concludes his results by suggesting that in most places a warming climate should produce either reduced air temperature variability or no change in air temperature variability. If, however, a negative variance response accompanies global warming, lower-tail temperatures would rise even more than they would be expected to under no change in variance. If this is the case, some areas of the USA could experience reductions in snow cover that could impact negatively on the skiing tourism industry.

Conclusion

Worldwide, intense storms are the most severe form of extreme weather and pose the greatest threat to tourists and tourism infrastructure, especially in tropical areas where hurricanes occur, which are the deadliest and most costly extreme weather events. Because so much coastal tourism is concentrated in tropical regions, hurricanes are the greatest threat to tourism should climate change lead to an increase in their frequency. Although some climate modelling indicates this is a possibility, model performance has not been verified, so the results are speculative. Moreover, current trends are not toward increasing frequency of extreme storm events. The IPCC (Houghton *et al.* 2001b) stated:

> Based on limited data, the observed variations in the intensity and frequency of tropical and extra-tropical cyclones and severe local storms show no clear trends in the last half of the twentieth century, although multi-decadal fluctuations are sometimes apparent.

Commenting on future trends, the same IPPC reports states:

> There is little consistent evidence that shows changes in the projected frequency of tropical cyclones and areas of formation. However, some measures of intensities show projected increases, and some theoretical and modelling studies suggest that the upper limit of these intensities could increase.

There is still much work to be done in determining whether there will be more or fewer extreme weather events in the future and, if so, what the significance of this will be for tourism. What we do know is that climate is naturally variable and always changing. The notion of constant climate is misleading. We can be confident that future climate will be different from the present. What is unknown is the extent to which significant change may take place over the short to medium term. According to Easterling *et al.* (2000: 2071), the lack of established definitions for what constitutes an extreme is one of the biggest problems in analysing extreme weather events and determining whether changes in these are consistent with what should be expected in the future. They say another big problem is the absence of high-quality, long-term climate data for many parts of the world, with the time resolution appropriate for analysing extreme events. This means it will be difficult to determine if extremes have changed, and how they may change in the future.

References

Balling Jr., R.C. (1998) 'Analysis of daily and monthly spatial variance components in historical temperature records', *Physical Geography*, 18: 544–52.

Balling Jr., R.C. and Cerveny, R.S. (2003) 'Compilation and discussion of trends in severe storms in the United States: Popular perception vs. climate reality', *Natural Hazards*, 29: 103–12.

Balling Jr., R.C., Michaels, P.J. and Knappenberger, P.C. (1998) 'Analysis of winter and summer warming rates in gridded temperature time series', *Climate Research*, 9: 175–81.

Bengtsson, L., Botzet, M. and Esch, M., (1996) 'Will greenhouse gas induced warming over the next 50 years lead to a higher frequency and greater intensity of hurricanes?' *Tellus*, 48A: 57–73.

Bielec, Z. (2001) 'Long-term variability of thunderstorms and thunderstorm precipitation occurrence in Cracow, Poland, in the period 1896–1995', *Atmospheric Research*, 56: 161–70.

Boose, E.R., Chamberlin, K.E. and Foster, D.R. (2001) 'Landscape and regional impacts of hurricanes in New England', *Ecological Monographs*, 71: 27–48.

Changnon, S.A., Changnon, D., Fosse, E.R., Hoganson, D.C., Roth, R.J. and Totsch, J.M. (1997) 'Effects of recent weather extremes on the insurance industry: Major implications for the atmospheric sciences', *Bulletin of the American Meteorological Society*, 78: 425.

Chu, P.-S. and Clark, J.D. (1999) 'Decadal variations of tropical cyclone activity over the central North Pacific', *Bulletin of the American Meteorological Society*, 80: 1875–81.

Coates, L. (1998) 'Deaths and ENSO: Fate, Chance or Change?' *Natural Hazards Quarterly*, 4(4): 2–3.

Dlugolecki, A.F., Clark, K.M., Knecht, F., McCauley, D., Palutikof, J.P. and Yambi, W. (1996) 'Financial Services', in R.T. Watson, M.C. Zinyowera and R.H. Moss (eds) *Climate Change 1995 – Impacts, Adaptations and Mitigation of Climate Change.* Contribution of Working Group II to the Second Assessment Report of the Intergovernmental Panel of Climate Change. New York: Cambridge University Press, pp.539–60.

Downing, T.E., Olsthoorn, A.A. and Tol, R.S.L. (eds) (1996) *Climate Change and Extreme Events*. ECU Research Report Number 12. Amsterdam: Vrije Universiteit.

Easterling, D.R., Horton, B., Jones, P.D., Peterson, T.C., Karl, T.R., Parker, D.E., Salinger, M.J., Razuvayev, V., Plummer, N., Jamason, P. and Folland, C.K. (1997) 'Maximum and minimum temperature trends for the globe', *Science*, 277: 364–67.

Easterling, D.R., Evans, J.L., Groisman, P.Ya., Karl, T.R., Kunkel, K.E. and Ambenje, P. (2000) 'Observed variability and trends in extreme climate events: A brief review', *Bulletin of the American Meteorological Society*, 81: 417–25.

Elsner, J.B. and Bossak, B.H. (2001) 'Secular changes to the ENSO-U.S. hurricane relationship', *Geophysical Research Letters*, 28: 4123–6.

Elsner, J.B., Liu, K.-b. and Kocher, B. (2000) 'Spatial variations in major U.S. hurricane activity: Statistics and a physical mechanism', *Journal of Climate*, 13: 2293–305.

Emergency Management Australia (1995) *National Emergency Management Competency Standards*. Canberra: EMA.

Freitas, C.R. de (2002) 'Perceived change in risk of natural disasters caused by global warming', *Australian Journal of Emergency Management*, 17(3): 34–8.

Frich, P., Alexander, L.V., Della-Marta, P., Gleason, B., Haylock, M., Klien Tank, A.M.G. and Peterson, T. (2002) 'Observed changes in climatic extremes during the second half of the twentieth century', *Climate Research*, 19: 193–212.

Goldenberg, S., Landsea, C.W., Mestas-Nuñez, A.M. and Gray, W.M. (2001) 'The recent increase in Atlantic hurricane activity: Causes and implications', *Science*, 293: 474–9.

Gray, W.M., Scheaffer, J.D. and Landsea, C.W. (1997) 'Hurricanes', in H.F. Diaz and R.S. Pulwarty (eds) *Climate and Socioeconomic Impacts*. Berlin: Springer Verlag.

Houghton, J.T., Meira Filho, L.G., Callander, B.A., Harris, N., Katenberg, A. and Maskell, K. (eds) (1996) *Climate Change 1995: the Science of Climate Change*. Contribution to Working Group 1 to the Second Assessment Report of the Intergovernmental Panel on Climate Change. New York: Cambridge University Press.

Houghton, J.T., Ding, Y., Griggs, D.J., Noguer, M., van der Linden, P.J. and Xiaosu, D. (eds) (2001a) *Climate Change 2001: The Scientific Basis*. Contribution to Working Group 1 to the Third Assessment Report of the Intergovernmental Panel on Climate Change (IPCC). New York: Cambridge University Press.

Houghton, J. T., Ding, Y., Griggs, D. J., Noguer, M., van der Linden, P. J. and Xiaosu, D. (eds) (2001b) *Technical Summary, Third Assessment Report. Climate Change 2001: The Scientific Basis*. New York: Cambridge University Press.

Houghton, J. T., Ding, Y., Griggs, D. J., Noguer, M., van der Linden, P. J. and Xiaosu, D. (eds) (2001c) *Summary for Policymakers, Third Assessment Report. Climate Change 2001: The Scientific Basis*. New York: Cambridge University Press.

Hoyt, D.V. (1981) 'Weather "records" and climate change', *Climate Change*, 3: 243–9.

Idso, S.B., Balling, R.C. and Cerveny, R.S. (1990) 'Carbon dioxide and hurricanes: Implications of Northern Hemispheric warming for Atlantic/Caribbean Storms', *Meteorology and Atmospheric Physics*, 42: 259–63.

Karl, T.R., Knight, R.W. and Plummer, N. (1995) 'Trends in high frequency climate variability in the twentieth century', *Nature*, 377: 217–320.

Karl, T.R., Nicholls, N. and Gregory, J. (1997) 'The coming climate', *Scientific American*, 276: 79–83.

Kattenburg, A., Giorgi, F., Grassl, H., Meehl, G.A., Mitchell, J.F.B., Stouffer, R.J., Tokioka, T., Weaver, A.J., Wigley, T.M.L. (1996) 'Climate models – Projections of future climate', in J.T. Houghton, *et al.* (eds) (1996) *Climate Change 1995: the Science of Climate Change*. Contribution to Working Group 1 to the Second Assessment Report of the Intergovernmental Panel on Climate Change. New York: Cambridge University Press, pp.285–357.

Kunkel, K. E., Pielke, R.A. and Changnon, S.A. (1999) 'Temporal fluctuations in weather and climate extremes that cause economic and human health impacts: A review', *Bulletin of the American Meteorological Society*, 80: 1077.

Landsea, C.W., Nicholls, N., Gray, W.M. and Avila, L.A. (1996) 'Downward trends in the frequency of intense Atlantic Hurricanes during the past five decades', *Geophysical Research Letters*, 23: 1697–700.

Landsea, C.W., Pielke Jr., R.A., Mestas-Nuñez, A.M. and Knaff, J.A. (1999) 'Atlantic basin hurricanes: indices of climatic changes', *Climatic Change*, 42: 89–129.

Landsea, C.W. (2000) 'Climate variability of tropical cyclones: past, present and future', in Pielke, R.A. Sr and Pielke, R.A. Jr, (eds) *Storms*. New York: Routledge, pp.220–41.

Liu, K.-b. and Fearn, M.L. (2000) 'Reconstruction of prehistoric landfall frequencies of catastrophic hurricanes in northwestern Florida from lake sediment records', *Quaternary Research*, 54: 238–45.

McCarthy, J.J., Canziani, O.F., Leary, N.A., Dokken, D.J. and White, K.S. (eds) (2001) *Climate Change 2001: Impacts, Adaptation, and Vulnerability*. Contribution of Working Group II to the Third Assessment Report of the Intergovernmental Panel on Climate Change. New York: Cambridge University Press.

Michaels, P.J and Balling, R.C. Jr. (2000) *The Satanic Gases*. Washington, DC: Cato Institute.

Michaels, P.J., Balling, R.C. Jr., Vose, R.S. and Knappenburger, P.C. (1998) 'Analysis of trends in the variability of daily and monthly historical temperature measurements', *Climate Research*, 10: 27–33.

Muller, R.A. and Stone, G.W. (2001) 'A climatology of tropical storm and hurricane strikes to enhance vulnerability prediction for the southeast U.S. Coast', *Journal of Coastal Research*, 17: 949–56.

Murphy, J.M. and Mitchell, J.F.B. (1995) 'Transient response of the Hadley Centre coupled ocean-atmosphere model to increasing Carbon Dioxide. Part II: Spatial and temporal structure of response', *Journal of Climate*, 8: 57–80.

Nguyen, K.C., and Walsh, K.J.E. (2001) 'Interannual, decadal, and transient greenhouse simulation of tropical cyclone–like vortices in a regional climate model of the South Pacific', *Journal of Climate*, 14: 3043–54.

Parisi, F. and Lund, R. (2000) 'Seasonality and return periods of landfalling Atlantic basin hurricanes', *Australian & New Zealand Journal of Statistics*, 42: 271–82.

Perry, A.H. (1997) 'Recreation and tourism', in R.D. Thompson and A. Perry (eds) *Applied Climatology: Principles and Practice*. London: Routledge, pp.240–8.

Pielke, R.A. and Landsea, C.W. (1998) 'Normalized hurricane damages in the United States 1925–1995', *Weather Forecasting*, 13: 621–31.

Pirazzoli, P.A. (2000) 'Surges, atmospheric pressure and wind change and flooding probability on the Atlantic coast of France', *Oceanologica Acta*, 23: 643–61.

Robeson, S.M. (2002) 'Relationships between mean and standard deviation of air temperature: implications for global warming', *Climate Research*, 22: 205–13.

Singh, O.P., Ali Khan, T.M. and Rahman, S. (2000) 'Changes in the frequency of tropical cyclones over the North Indian Ocean', *Meteorology and Atmospheric Physics*, 75: 11–20.

Singh, O.P., Ali Kahn, T.M. and Rahman, S. (2001) 'Has the frequency of intense tropical cyclones increased in the North Indian Ocean?' *Current Science*, 80: 575–80.

Sugi, M., Noda, A. and Sato, N. (2002) 'Influence of the global warming on tropical cyclone climatology: an experiment with the JMA global model', *Journal of the Meteorological Society of Japan*, 80: 249–72.

Wilson, R.M. (1999) 'Statistical aspects of major (intense) hurricanes in the Atlantic basin during the past 49 hurricane seasons (1950–1998): Implications for the current season', *Geophysical Research Letters*, 26: 2957–60.

Zhang Y. and Wang, W.C. (1997) 'Model simulated Northern winter cyclone and anticyclone activity under a Greenhouse warming scenario', *Journal of Climate*, 10: 1616–34.

12 Tourism, biodiversity and global environmental change

C. Michael Hall

Biodiversity (biological diversity) refers to the total sum of biotic variation, ranging from the genetic level, through the species level and on to the ecosystem level. The concept therefore indicates diversity within and between species as well as the diversity of ecosystems. The extent or quantity of diversity can be expressed in terms of the size of a population, the abundance of different species, as well as the size of an ecosystem (area) and the number of ecosystems in a given area. The integrity or quality of biodiversity can be expressed in terms of the extent of diversity at the genetic level, and the resilience at the species and ecosystem level (Martens *et al.* 2003).

Biodiversity loss is a major policy issue and, as with climate change with which it intersects in both environmental and regime terms, is the subject of an international convention (McNeely 1990; Rosendal 2001; Kim 2004). The extinction of species is a natural process (Lande 1993, 1998). However, species and ecosystem loss has accelerated as a result of human activity. The United Nations Environment Programme (UNEP) estimate that almost 4,000 mammal, bird, reptile, amphibia and fish species are threatened with extinction, while about 600 species of animals are on a critically endangered species list (UNEP 2002; Nielsen 2005). Pitman and Jørgensen (2002) estimate that between 24 and 48 per cent of the world's plant species are presently faced with extinction. Wilson (1992) estimated that one species was being lost every 20 minutes, with approximately 27,000 species being lost per year. In contrast, Pimm *et al.* (1995) estimated that human-induced extinction of species was as high as 140,000 per year. At that rate, half of the existing species will be extinct in 70 years.

Although the exact rate of biodiversity loss is disputed (Purvis and Hector 2000) there is no doubt that human domination of the natural environment has led to a decline in biodiversity at all levels, with an acceleration in the rate of species extinction in recent years. According to Martens *et al.* (2003):

> The current speed of extinction of species through human intervention is approximately 100–1,000 times faster than the natural speed of extinction. In many groups of organisms 5–20% of all species are already extinct.

Also critical is recognition of endemic biodiversity versus the total amount of biodiversity for a given region. Endemic biodiversity refers to the biodiversity that is indigenous (native) or endemic for a given region as opposed to introduced or alien biodiversity, which is that biodiversity which is present in a specific environment because of human interference in natural systems and human mobility. The focus on biodiversity conservation is nearly always on the maintenance of endemic biodiversity, with the exception being when an introduced species or variety is endangered or extinct in its naturally occurring range. There are essentially six main tenets to be found in the desire for conserving biodiversity (see Soulé 1985; Callicot 1990; Wilson 1992):

1   The diversity of organisms and habitats on different scales (for example, genetic, species, ecosystem) is positive.
2   The untimely extinction of organisms and habitats on different scales is negative.
3   Ecological complexity is good.
4   Allowing evolutionary processes to occur is positive.
5   Biodiversity has extrinsic or anthropocentric value in terms of the goods and services it provides humankind.
6   Biodiversity has intrinsic or biocentric value.

There are three main mechanisms by which biodiversity is being lost:

- reduction in the size and fragmentation of natural areas
- changes in ecosystem conditions
- deliberate extinction of species.

Conversion to agriculture, forest clearance and urbanisation are the main causes of the loss of natural areas on the global scale. For example, in 2004 the United Nations Environment Programme (UNEP) warned the governments of the Congo, Rwanda and Uganda that, according to satellite studies, the Virunga National Park was being colonised by farmers at the rate of $2km^2$ a day. The park is home to half of the world's population of mountain gorillas of which there are only an estimated 700 left (Radford 2004) The most important causes of changes to ecosystem conditions are fragmentation, disruption and isolation of natural areas, eutrophication, pollution, climate change, erosion and the introduction of diseases and species (Martens et al. 2003). Species are also being deliberately extinguished. Not only through poaching of high profile species such as tigers, rhinoceros and elephant, but also though the use of herbicides and biocides and hunting, fishing and farming practices. As Andrew Purvis, a conservation biologist, commented:

Other species generally have their numbers limited by competitors, predators, parasites and pathogens ... Any competitors, we get rid of those pretty quickly, even if they are just competing with things like crop plants, or our livestock, or our golf courses. We are also doing things to eliminate parasites and pathogens.

(quoted in Radford 2004)

Historic and current loss of biodiversity is related to growth in human popula-tion and consumption. Pressure on natural resources is occurring not only because of existing consumption in developed countries but also because of increased consumption levels in the less developed countries.

> Each human needs roughly two hectares of land to provide food, water, shelter, fibre, currency, fuel, medicine and a rubbish tip to sustain a lifespan. So the more land humans take, the less that is available for all other mammals, birds, reptiles and amphibians. ... humans and their livestock now consume 40 per cent of the planet's primary production, while the planet's other 7 million species must scramble for the rest. No other single species on the planet – except possibly some termites and the Antarctic krill – can match human numbers. People are having such an impact: we are sharply reducing the numbers of other things and very quickly you can go from large numbers to nothing.
>
> (Radford 2004)

Tourism represents a significant part of the consumption practices that impact biodiversity (e.g. German Federal Agency for Nature Conservation 1997; Gössling 2002; Christ *et al.* 2003). However, the impact of tourism development, for example, through tourism urbanisation, and habitat and species disturbance is not all negative. In many locations, tourism provides an economic justification to establish conservation areas, such as national parks and private reserves, as an alternative to other land uses such as logging, clearance for agriculture, mining or urbanisation. Often such tourism is described as ecotourism, safari, wildlife or nature-based tourism or even sustainable tourism (e.g. Cater and Lowman 1994; Hall and Lew 1998; Fennell 1999; Newsome *et al.* 2002; Hall and Boyd 2005). Regardless of the name that is used, it is apparent that charismatic mega-fauna, for example, such animals as dolphins, elephants, giraffes, gorillas, lions, orang-utan, rhinoceros, tigers and whales, do serve as a significant basis for tourism in a number of parts of the world, while national parks and reserves can also be signifi-cant tourist attractions in their own right. Indeed, tourism is seen as a mechanism to directly benefit biodiversity and the maintenance of natural capital through several means (Brandon 1996; Christ *et al.* 2003; Hall and Boyd 2005), including:

- an economic justification for biodiversity conservation practices, including the establishment of national parks and reserves (public and private)
- a source of financial support for biodiversity maintenance and conservation
- an economic alternative to other forms of development that may negatively impact biodiversity and to inappropriate exploitation or harvesting of wildlife, such as poaching
- a mechanism for educating people about the benefits of biodiversity conservation
- potentially involving local people in the maintenance of biodiversity and incor-porating local ecological knowledge in biodiversity management practices.

In 2003 Conservation International, in collaboration with the United Nations Environment Programme (UNEP), produced a report on the relationships between tourism and biodiversity that focused on the potential role of tourism in biodiversity 'hotspots' – 'priority areas for urgent conservation on a global scale' (Christ *et al.* 2003: vi). Hotspots are areas that both support a high diversity of endemic species and have been significantly impacted by human activities. Plant diversity is the biological basis for designation as a biodiversity hotspot, according to Christ *et al.* (2003: 3): 'a hotspot must have lost 70 percent or more of its original habitat. Overall, the hotspots have lost nearly 90 percent of their original natural vegetation'. The biodiversity hotspots identified by Conservation International 'contain 44 percent of all known endemic plant species and 35 percent of all known endemic species of birds, mammals, reptiles, and amphibians in only 1.4 percent of the planet's land area' (Christ *et al.* 2003: 3). The report highlighted several key issues:

- Although most biodiversity is concentrated in less developed countries, five tourism destination regions in the developed world were also identified as biodiversity hotspots – the Mediterranean Basin, the California floristic province, the Florida Keys, south-west Australia and New Zealand.
- An increasing number of biodiversity hotspot countries in the less developed world are experiencing rapid tourism growth: 23 of them record over 100 per cent growth in the last 10 years, and more than 50 per cent of these receive over one million international tourists per year; 13 per cent of biodiversity hotspot countries receive over five million international tourists per year.
- Although receiving fewer tourists overall than the developed countries, many biodiversity-rich countries in the less developed world receive substantial numbers of international tourists. Thirteen of them – Argentina, Brazil, Cyprus, the Dominican Republic, India, Indonesia, Macao, Malaysia, Mexico, Morocco, South Africa, Thailand and Vietnam – receive over two million foreign visitors per year, while domestic tourism is also of growing significance in some of these countries.
- More than half of the world's poorest 15 countries fall within the biodiversity hotspots and, in all of these, tourism has some economic significance or is forecast to increase according to the World Tourism Organization and the World Travel and Tourism Council.
- In several biodiversity hotspots in less developed countries – for example, Madagascar, Costa Rica, Belize, Rwanda, South Africa – biodiversity or elements of biodiversity, such as specific wildlife, is the major international tourism attraction.
- Forecast increases in international and domestic tourism suggest that pressures from tourism development will become increasingly important in other biodiversity hotspot countries, for example, in South and South East Asia.

The Conservation International report highlighted some of the key relationships between tourism and biodiversity and stated that:

Biodiversity is essential for the continued development of the tourism industry, yet this study indicates an apparent lack of awareness of the links – positive and negative – between tourism development and biodiversity conservation.

(Christ *et al*. 2003: 41)

Indeed, it went on to note that while many ecosystems serve to attract tourists, for example, coral reefs, rainforest and alpine areas, many of the factors linked to the loss of biodiversity, such as land clearance, pollution and climate change, are also linked to tourism development. Unfortunately, the report failed to adequately emphasise what some of the strategies by which tourism could both contribute to diversity and economic development might be or to state the broader ramifications of those strategies. For example, while the concept of scarcity rent that underlay much of earlier thinking with respect to the value of ecotourism – reduce access to desirable wildlife in the face of high demand and charge more for the experience while reducing the stress on animals and the environment – sounds sensible, it has often foundered on cultural and political values that have historically favoured access. Indeed, for most of their history, national parks agencies have often sought to encourage visitation so as to meet the recreational component of their mandate and to create a political environment supportive of national parks. Unfortunately, in the face of growing populations and increasing personal mobility the access issue is becoming increasingly problematic for many conservation authorities who seek to conserve biodiversity (Budowski 1976; Runte 1987; Hall 1992; Cater and Lowman 1994; Butler and Boyd 2000; Hall and Boyd 2005).

Despite the growth of research and publications on tourism in natural areas, our understanding of the role and effects of tourism in natural areas is also surprisingly limited. Arguably, the majority of studies have examined the impacts of tourism and recreation on a particular environment or component of the environment rather than over a range of environments (Holden 2000; Weaver 2001; Hall and Boyd 2005). There is substantial research undertaken on tourism with respect to rainforest, reefs and dolphins and whales, for example, but very limited research undertaken on what are arguably less attractive environments, such as wetlands, or animal species that are not the charismatic mega-fauna that are a key component of wildlife viewing tourism but which are just as important a part of the ecosystem (Newsome *et al*. 2002; Hall and Boyd 2005). Moreover, the scale on which interactions between tourism and biodiversity are examined is also critical. The Conservation International report on tourism and biodiversity (Christ *et al*. 2003) can only serve to highlight relationships at the macroscopic level of biological provinces, it does not serve as a useful management tool at the level of ecosystems, let alone individual species. Such a comment is not to denigrate the report because it serves an extremely useful function in terms of policy debate, but the harsh reality is that knowledge of the structure and dynamics of the geographic range of species in terms of abundance, size and limits is extremely limited even before the implications of human impact, including tourism, on range and abundance is considered (Gaston 2003).

As noted above, species extinction is a natural process. It has long been recognised that extinction and colonisation of habitats is an ongoing process. But just as importantly it has also been recognised that without human interference such processes lead to equilibrium between extinction and immigration (e.g. see MacArthur and Wilson 1967; Whitehead and Jones 1969). Human impact is changing this natural balance with respect to both extinction and immigration of new species at a rate that is making it extremely difficult, if not impossible, for new equilibria to be established. Tourism's contribution to the present human-induced mass extinction of species (May *et al.* 1995; Pimm *et al.* 1995; Hilton-Taylor 2000) is several-fold and will be examined in the following sections. Although tourism rarely directly kills off species, tourism-related developments and land use contributes to species range contraction and extinctions through habitat loss and fragmentation. Tourism and other forms of human mobility also introduce alien organisms into areas beyond the natural limits of their geographical range, thereby creating new competition among species. Tourism also affects biodiversity through its contribution to climate change. Finally, we can raise issues over the extent to which national parks and other protected areas can be used to conserve biodiversity.

Habitat loss and fragmentation

Tourism directly affects habitat through processes of tourism urbanisation. As Chapter 8 notes, such processes are spatially and geographically distinct and are often related to high natural amenity areas such as the coast, where coastal ecosystems are subject to urbanisation, land clearance and the draining and clearance of wetlands. Tourism also contributes to habitat loss and fragmentation via its ecological footprint in terms of resource requirements and pollution and waste.

The loss of endemic biodiversity through species extinction can be expressed in relation to changes in the size of the geographic range of a species and its total population size as the total number of individuals in a species declines. Four different idealised forms of these relationships can be presented (Wilcove and Terborgh 1984; Schonewald-Cox and Buechner 1991; Lawton 1993; Gaston 1994, 2003: 168–74), although, as Gaston notes:

> Declines in extinction are often likely to walk much more varied paths through abundance-range space than these simple models might imply, given the complexities of the abundance structure of species' geographic ranges and of the processes causing reductions in overall population size.
>
> (2003: 174)

- The geographic range size remains approximately constant as the number of individuals declines and overall density declines with time. Activities such as hunting, pollution or climatic change can all lead to declines in the number of individuals of a species without affecting the overall geographic range in which they are found.

- The number of individuals and range size decline simultaneously so that the density remains constant. Two circumstances can be identified in which such a situation may occur. First, reduction of losses of individual members of a species may be balanced by losses of lower density areas. Gaston (2003: 168), for example, notes: 'declines in the local abundances of persistent populations accompanied by the loss of small, and often peripheral, populations appear to be a widespread phenomena'. Second, the total habitat area may be eroded without loss in the quality of the remaining habitat. One of the best examples of this situation that is often related to coastal tourism development is the loss of individual wetlands which are drained, leaving other wetlands as yet undeveloped.
- The number of individuals of a species and the size of their range decline simultaneously such that the density of individuals declines with time. Gaston (2003) recognises three cases in which such a situation might exist. First, with respect to broad-scale environmental change, including the effects of an increase in the proportion of 'edge' as a habitat area is fragmented therefore leading to micro-climatic changes as well as other changes to patterns of predation and species invasion (Laurance 2000; Laurance *et al.* 1997, 2000). Such a situation is consistent with the development of edge effects when recreational access is unmanaged in habitat fragments and people do not keep to trails, thereby creating further edges as new walking paths are created. Second, in areas that undergo differential exploitation through, for example, timber extraction. Third, if there is a causal link between abundance of a species and occupancy of a given habitat (see also Gates and Donald 2000; Lawton 1993, 2000).
- The number of individuals of a species and range size decline simultaneously such that the density of individuals increases with time. Such a pattern is most likely to occur when habitat is lost with no compensating increase in density of individuals in other available habitat. This particular model of extinction has substantial implications for conserving biodiversity 'hot spots' which tends to assume (see Christ *et al.* 2003) that protecting such areas in the face of the loss of other areas where the species are present should increase the density of individuals of the species. Yet such density increases may be only short-term if populations in the 'hot spots' were dependent on their relationship with other populations of the species that have been made extinct. As Gaston highlights: 'This emphasizes the need for a regional rather than a site-by-site approach to conservation planning and action, albeit this is at odds with the methodology embodied in some international agreements … for example Ramsar Convention on Wetlands of International Importance … and espoused by some conservation agencies' (2003: 174).

Introduction of foreign organisms

It is estimated that approximately 400,000 species have been accidentally or deliberately introduced to locations that lie beyond the natural limits of their geographic range (Pimentel 2001). The introduction of alien species into an environment is a major influence on biodiversity that is associated with tourism because of the capacity of tourists and the infrastructure of tourism to act as carriers of exotic species. Many introductions have no apparent adverse effects (Williamson 1996), although some introductions, such as deer, rabbits and possum in New Zealand, and cane toads and rabbits in Australia, have caused massive ecological damage and harm to native species (Fox and Adamson 1979).

In the nineteenth and early twentieth centuries many species were deliberately introduced from one part of the world to another by the European colonial powers as a means of economic development and the Europeanisation of the natural environment (Crosby 1986). Although agricultural development and the creation of an ideal environment were the primary motives for such introductions, leisure and tourism were also significant. For example, in New Zealand a number of Australian and European animal species were introduced for hunting purposes but it was not until the 1920s that substantial opposition emerged to the introduction of new species, including widespread indignation and opposition to Lady Liverpool's efforts to introduce grouse into Tongariro Park, New Zealand's oldest national park. Prof. H.B. Kirk, one of New Zealand's leading natural historians, sent an angry letter to the *Evening Post*, which had earlier applauded Lady Liverpool's efforts as likely to 'give added attractions to sportsmen coming to New Zealand from the Old Country': 'No other country would do so ludicrous a thing as to convert the most distinctive of its national parks into a game preserve ... this thing is an insult to the Maori donors and to all lovers of New Zealand as New Zealand'. Kirk's letter appeared to find a supportive response among a wide range of individuals and authorities. By the end of 1924 the New Zealand Legislative Council had 'pushed through a resolution condemning all introduction and proclaiming that the park should be held inviolate' (in Harris 1974: 109–110).

International trade has also served to introduce alien species through accidental carriage in shipping containers and on ships and aircraft (Drake and Mooney 1988; Carlton and Geller 1993). Nevertheless, travellers remain a major source of accidental and intentional species introductions to the point where they are a focus of biosecurity concerns at both international and regional levels (Timmins and Williams 1991; Hodgkinson and Thompson 1997; Hall 2003; Jay *et al.* 2003).

Much of the concern over the introduction of alien species lies in their potential economic damage. Pimentel *et al.* (2000) estimated that the approximately 50,000 exotic species in the USA have an economic impact of US$137 billion per annum in terms of their economic damage and costs of control. For example, since 1998 the State of California has provided US$65.2 million for a statewide management programme and research to combat the glassy-winged sharpshooter and the deadly Pierce's disease (a bacterium, *Xylella fastidiosa*) that it carries. Accidentally introduced in 1989, 15 counties in California have been identified as being infested (Wine Institute of California 2002).

For many grape diseases, humans are a significant vector (Pearson and Goheen 1998), the most notable of which is grape phylloxera, an aphid *Daktulosphaira vitifoliae*, which wreaked havoc on the world's vineyards in the late nineteenth century (Ordish 1987). The economic impact of a phylloxera outbreak on the modern wine industry would be substantial. In Western Australia, it is estimated that phylloxera could cost affected growers A\$20,000/ha in the first five years in lost production and replanting costs (Agriculture Western Australia 2000). Increased personal mobility, particularly through wine and food tourism, is a potential threat to the wine industry because of the potential for the relocation and introduction of pests. Yet, despite recognition of the potential role of humans in conveying grape pests, there is only limited awareness of the biosecurity risks of wine tourism (Hall 2003). However, it is likely that in the future concerns over the risks associated with the introduction of exotic species for endemic biodiversity can only increase as rates of international travel continue to grow and the climate change leads to the creation of environmental conditions conducive to the establishment of alien species.

Climate change

Climate change sets particular challenges for conservation. One of the most significant long-term issues for the global network of protected areas that serve to help maintain biodiversity is how can they 'be established and developed in such a way that it can accommodate the changes in species distributions that will follow from climate change' (Gaston 2003: 181). Substantial research has been undertaken on the implications of climate change for species' geographic ranges which has typically sought to model the relationships between climate and distribution in relation to such issues as habitat fragmentation (e.g. Nakano *et al.* 1996) and loss (e.g. Keleher and Rahel 1996; Travis 2003), species distribution (e.g. Jeffree and Jeffree 1994, 1996), pests (e.g. Baker *et al.* 1996) and disease (e.g. Rogers and Randolf 2000; Lieshout *et al.* 2004). Undoubtedly, the distribution of species is affected by temperature. However, studies of the relationship between species distribution and future climate change scenarios tend to make a number of critical assumptions (Gaston 2003):

- Correlations between climate and species occurrence reflect causal relationships.
- Any influence of other factors on observed relationships between climate and the occurrence of a species, such as competitors, diseases, predators, parasites and resources will remain constant.
- Temporally generalised climatic conditions – for example, seasonal means, annual means, medians – are more important influences on the distribution of species than rates of climatic change and extreme events.
- Spatially generalised climatic conditions derived or interpolated from the nearest climate stations sufficiently characterise the conditions that individuals of a species actually experience.
- Climate change will be relatively simple, in that its influence on species distributions can be summarised in terms of the projected changes in one or a few variables.

- There is no physiological capacity to withstand environmental conditions which are not components of those existing conditions in areas in which a species is presently distributed.
- Range shifts, expansions, or contractions are not accompanied by physiological changes, other than local non-genetic acclimatisation.
- Dispersal limit is unimportant in the determination of the present distribution of species and in their ability to respond to changes in climate.

As Gaston (2003: 185) points out, the reality is that a number of these assumptions will be, and already have been, 'severely violated' (see Lawton 1995, 2000; Spicer and Gaston 1999; Bradshaw and Holzapfel 2001). For example:

> Human activities impose a marked influence on the distribution of species, and how these alter with changes in climate is alone likely to be extremely complicated, and dependent on social pressures and technological developments.
>
> (Gaston 2003: 183)

The above observations are not to deny that climate change will affect the geographic range of species, it clearly will in the future just has it has in the past (e.g. Boer and de Groot 1990; Hengeveld 1990; Huntley 1991, 1994; Huntley and Birks 1983; Huntley *et al.* 1989, 1995). However, the use of relatively simple models based on climate matching approaches is likely to prove misleading in terms of planning conservation regimes that can accommodate future climate change.

Conserving biodiversity

At the start of the twenty-first century the world's biodiversity is threatened as never before – as noted above many species become extinct each year and the number is growing. This section looks at the role of national parks and reserves as present and future refugia. Tourism, and ecotourism in particular, has become a major economic rationale for the establishment of national parks and reserves that serve to conserve and present charismatic mega-fauna and habitats.

The global conservation estate has grown enormously since the first UN List of Protected Areas was published in 1962 with just over 1,000 protected areas. In 1997 there were over 12,754 sites listed. The 2003 edition listed 102,102 sites covering 18.8 million km². 'This figure is equivalent to 12.65 per cent of the Earth's land surface, or an area greater than the combined land area of China, South Asia and South East Asia' (Chape *et al.* 2003: 21). Of the total area protected, it is estimated that 17.1 million km² constitute terrestrial protected areas, or 11.5 per cent of the global land surface, although some biomes, including Lake Systems and Temperate Grasslands, remain poorly represented. Marine areas are significantly under-represented in the global protected area system. Approximately 1.64 million km² comprise marine protected areas – an estimated 0.5 per cent of the world's oceans and less than one tenth of the overall extent of protected areas worldwide (Chape *et al.* 2003).

The size of the global conservation estate raises the question of just how large the global network of protected areas needs to be (Rodrigues and Gaston 2001). The present size of global conservation estate exceeds the IUCN's earlier target of at least 10 per cent of the total land area being set aside for conservation purposes, although there is clearly substantial variation between both countries and biomes in terms of the actual area set aside (Chape *et al.* 2003). Yet commentators such as Soulé and Sanjayan (1998) have noted that the IUCN's target has been dictated more by political considerations than biological science. Rodrigues and Gaston (2001, 2002) observed that the minimum area needed to represent all species within a region increases with the number of targeted species, the level of endemism and the size of the selection units. They concluded that:

- No global target for the size of a network is appropriate because those regions with higher levels of endemism and/or higher diversity will correspondingly require larger areas to protect such characteristics.
- A minimum size conservation network sufficient for capturing the diversity of vertebrates will not be sufficient for biodiversity in general, because other groups are known to have higher levels of endemism (Gaston 2003).
- The 10 per cent target is likely to be grossly inadequate to meet biodiversity conservation needs. Instead, Rodrigues and Gaston (2001) estimated that for a selection unit of $1° \times 1°$ (approximately 12,000 km^2) 74.3 per cent of the global land area and 92.7 per cent of the global rain forest would be required to represent every plant species and 7.7 per cent and 17.8 per cent respectively to represent the higher vertebrates. However, Gaston (2003) also notes that even reserves of 12,000 km^2 may not be large enough for maintaining populations of many species, citing examples from the national parks of Africa (Newmark 1996; Nicholls *et al.* 1996) and the USA (Newmark 1984; Mattson and Reid 1991). Indeed, it must be noted that while there has been a well-considered literature on the size and shape of conservation reserves since the 1970s (e.g. Main and Yadav 1971; Diamond 1975; Slatyer 1975; MacMahon 1976) there has been inadequate utilisation of such knowledge with respect to park and reserve establishment and design and their dual role conservation and tourism roles.

A further concern in terms of biodiversity conservation is the capacity of a national park and reserve system to cope with the impact of global environmental change (GEC) including climate change (Dockerty et al. 2003), surrounding land-use change and anthropogenic pressure (Cardillo *et al.* 2004). Given the migration of species as a result of climate change, present reserves may not be suitable for conservation of target species and ecosystems (Huntley 1994, 1999). Given the potential scale of GEC it is therefore important that sites are identified that can act as refugia from future change. A refuge is a region in which certain species are able to persist during a period in which most of the original geographic range becomes uninhabitable because of environmental change. Historically, such changes have been climatic although in terms of contemporary biodiversity conservation anthropogenic environmental change is just as significant. Although some present national parks and reserves are likely to fill this role, it is also important that sites that are available which have attributes that may potentially fulfil the role of refugia for endangered species in the future also be identified, conserved and managed so as to reduce the impacts of GEC on populations. Such 'future refugia' may then become locations from which future species migration can occur should climate become stabilised.

Conclusion

The loss of biodiversity is one of the most significant aspects of GEC given the extent to which it underpins the global economy and human welfare (Martens *et al.* 2003). Biodiversity, or at least the existence of certain charismatic species (usually mega-fauna) and ecosystems is also significant as an attraction for 'ecotourism' and 'nature-based tourism'. Nevertheless, the interrelationships between tourism and biodiversity are poorly understood in terms of empirical data, although the potential impacts of the loss of some charismatic species such as the polar bear (The Age 2005), or African wildlife, or even entire ecosystems, such as the Great Barrier Reef (Fyfe 2005), on tourism would be dramatic.

The extent to which tourism contributes towards biodiversity loss through tourism urbanisation, habitat loss and fragmentation and contribution to climate change is also dramatic and, arguably, makes a lie out of attempts to paint a picture of tourism as a benign industry. Undoubtedly, tourism can make a contribution to the conservation and maintenance of biodiversity. However, in reality the success stories are few and far between and are generally isolated to individual species and relatively small areas of habitat (e.g. see Newsome *et al.* 2002) rather than a comprehensive contribution to conservation. Such a comment is not to belittle the efforts that have been made with respect to developing a positive contribution from tourism toward biological conservation. Instead, it is to highlight the fact that while tourism has led to biodiversity maintenance at a local level in some instances, the global picture is one in which tourism, like many other industries that have a large ecological footprint and lead to clearance of natural areas, is not a net contributor to biodiversity.

References

Agriculture Western Australia (2000) *Grape Phylloxera: Exotic Threat to Western Australia, Factsheet no.0002–2000.* Perth: Department of Agriculture.
Baker, R.H.A., Cannon, R.J.C. and Walters, K.F.A. (1996) 'An assessment of the risks posed by selected non-indigenous pests to UK crops under climate change', *Aspects of Applied Biology*, 45: 323–30.
Boer, M.M. and de Groot, R.S. (eds) (1990) *Landscape-Ecological Impact of Climate Changes.* Amsterdam: IOS Press.
Bradshaw, W.E. and Holzapfel, C.M. (2001) 'Genetic shift in photoperiodic response correlated with global warming', *Proceedings of the National Academy of Sciences of the USA*, 98: 14509–11.
Brandon, K. (1996) *Ecotourism and Conservation: A Review of Key Issues.* Environment Department Paper no.33. Washington, DC: World Bank.

Budowski, G. (1976) 'Tourism and conservation: conflict, coexistence or symbiosis', *Environmental Conservation*, 3(1): 27–31.

Butler, R. and Boyd, S. (eds) (2000) *Tourism and National Parks*. Chichester: John Wiley.

Callicott, J.B. (1990) 'Whither conservation ethics?' *Conservation Biology*, 4: 15–20.

Cardillo, M., Purvis, A., Sechrest, W., Gittleman, J.L., Bielby, J. and Mace, G.M. (2004) 'Human population density and extinction risk in the world's carnivores', *PloS Biology*, 2(7): 0909–0914

Carlton, J.T. and Geller, J.B. (1993) 'Ecological roulette: the global transport of non-indigenous marine organisms', *Science*, 261: 78–82.

Cater, E. and Lowman, G. (eds) (1994) *Ecotourism: A Sustainable Option?* Chichester: John Wiley.

Chape, S., Blyth, S., Fish, L., Fox, P. and Spalding, M. (compilers) (2003) *2003 United Nations List of Protected Areas*. IUCN, Gland and Cambridge: IUCN and UNEP-WCMC.

Christ, C., Hilel, O., Matus, S. and Sweeting, J. (2003) *Tourism and Biodiversity: Mapping Tourism's Global Footprint*. Washington, DC: Conservation International.

Crosby, A.W. (1986) *Ecological Imperialism: The Biological Expansion of Europe, 900–1900*. Cambridge: Cambridge University Press.

Diamond, J.M. (1975) 'The island dilemma: lessons of modern biogeographic studies for the design of natural reserves', *Biological Conservation*, 7: 129–46.

Dockerty, T., Lovett, A. and Watkinson, A. (2003) 'Climate change and nature reserves: examining the potential impacts, with examples from Great Britain', *Global Environmental Change*, 13: 125–35.

Drake, J.A. and Mooney, H.A. (eds) (1988) *Biological Invasions: A Global Perspective*. New York: John Wiley.

Fennell, D. (1999) *Ecotourism: An Introduction*. London: Routledge.

Fox, M.D. and Adamson, D. (1979) 'The ecology of invasions', in H. Recher, D. Lunney and I. Dunn (eds) *A Natural Legacy: Ecology in Australia*. Rushcutter's Bay: Pergamon Press.

Fyfe, M. (2005) 'Too late to save the reef', *The Age*, February 12.

Gaston, K.J. (1994) 'Geographic range sizes and trajectories to extinction', *Biodiversity Letters*, 2: 163–70.

Gaston, K.J. (2003) *The Structure and Dynamics of Geographic Ranges*. Oxford: Oxford University Press.

Gates, S. and Donald, P.F. (2000) 'Local extinction of British farmland birds and the prediction of further loss', *Journal of Applied Ecology*, 37: 806–20.

German Federal Agency for Nature Conservation (1997) *Biodiversity and Tourism: Conflicts on the World's Seacoasts and Strategies for Their Solution*. Berlin: Springer Verlag.

Gössling, S. (2002) 'Global environmental consequences of tourism', *Global Environmental Change*, 12: 283–302.

Hall, C.M. (1992) *Wasteland to World Heritage: Preserving Australia's Wilderness*. Carlton: Melbourne University Press.

Hall, C.M. (2003) 'Biosecurity and wine tourism: Is a vineyard a farm?' *Journal of Wine Research*, 14(2–3): 121–6.

Hall, C.M. and Boyd, S. (2005) 'Nature-based tourism and regional development in peripheral areas: Introduction', in C.M. Hall and S. Boyd (eds) *Tourism and Nature-based Tourism in Peripheral Areas: Development or Disaster*, Clevedon: Channelview Publications, pp.3–17.

Hall, C.M. and Lew, A. (1998) 'The geography of sustainable tourism development: an introduction', In C.M. Hall and A. Lew (eds) *Sustainable Tourism: A Geographical Perspective*. London: Addison-Wesley Longman, pp.1–12.

Harris, W.W. (1974) 'Three parks: An analysis of the origins and evolution of the national parks movement', unpublished MA Thesis, Christchurch: Department of Geography, University of Canterbury.

Hengeveld, R. (1990) *Dynamic Biogeography*. Cambridge: Cambridge University Press.

Hilton-Taylor, C. (2000) *The 2000 IUCN Red List of Threatened Species*. Gland: IUCN.

Hodgkinson, D.J. and Thompson, K. (1997) 'Plant dispersal; the role of man', *Journal of Applied Ecology*, 34: 1484–96.

Holden, A. (2000) *Environment and Tourism*. London: Routledge.

Huntley, B. (1991) 'How plants respond to climate change: migration rates, individualism and the consequences for plant communities', *Annals of Botany*, 67 (Supplement 1): 15–22.

Huntley, B. (1994) 'Plant species' responses to climate change: implications for the conservation of European birds', *Ibis*, 137: S127-S138.

Huntley, B. (1999) 'Species distribution and environmental change: considerations from the site to the landscape scale', in E. Maltby, M. Holdgate, M. Acreman and A. Weir (eds) *Ecosystem Management: Questions for Science and Society*. Virginia Water: Royal Holloway Institute for Environmental Research. pp.115–29.

Huntley, B. and Birks, H.J.B. (1983) *An Atlas of Past and Present Pollen Maps for Europe: 0–13,000 Years Ago*, Cambridge: Cambridge University Press.

Huntley, B., Bartlein, P.J. and Prentice, I.C. (1989) 'Climatic control of the distribution and abundance of beech (*Fagus L.*) in Europe and North America', *Journal of Biogeography*, 16: 551–60.

Huntley, B., Berry, P.M., Cramer, W. and McDonald, A.P. (1995) 'Modelling present and potential future ranges of some European higher plants using climate response surfaces', *Journal of Biogeography*, 16: 551–60.

Jay, M., Morad, M. and Bell, A. (2003) 'Biosecurity – A policy dilemma for New Zealand', *Land Use Policy*, 20(2): 121–29.

Jeffree, C.E. and Jeffree, E.P. (1996) 'Redistribution of the potential geographical ranges of Mistletoe and Colorado Beetle in Europe in response to the temperature component of climate change', *Functional Ecology*, 10: 562–77.

Jeffree, E.P. and Jeffree, C.E. (1994) 'Temperature and the biogeographical distributions of species', *Functional Ecology*, 8: 640–50.

Keleher, C.J. and Rahel, F.J. (1996) 'Thermal limits to salmonoid distributions in the Rocky Mountain Region and potential habitat loss due to global warming: a geographic information system (GIS) approach', *Transactions of the American Fisheries Society*, 125: 1–13.

Kim, J.A. (2004) 'Regime interplay: the case of biodiversity and climate change', *Global Environmental Change*, 14: 314–24.

Lande, R. (1993) 'Risks of population extinction from demographic and environmental stochasticity and random catastrophes', *American Naturalist*, 142: 911–27.

Lande, R. (1998) 'Anthropogenic, ecological and genetic factors in extinction', in G.M. Mace, A. Balmford and J.R. Ginsberg (eds) *Conservation in a Changing World*. Cambridge: Cambridge University Press, pp.29–51.

Laurance, W.F. (2000) 'Do edge effects occur over large spatial scales', *Trends in Ecology and Evolution*, 15: 134–5.

Laurance, W.F., Laurance, S.G., Ferreira, L.V., Rankin-de Merona, J.M., Gascon, C. and Lovejoy, T.E. (1997) 'Biomass collapse in Amazonian forest fragments', *Science*, 278: 1117–8.

Laurance, W.F., Delamonica, P., Laurance, S.G., Vasconcelos, H.L. and Lovejoy, T.E. (2000) 'Rainforest fragmentation kills big trees', *Nature*, 404: 836.

Lawton, J.H. (1993) 'Range, population abundance and conservation', *Trends in Ecology and Evolution*, 8: 409–13.

Lawton, J.H. (1995) 'The response of insects to environmental change', in R. Harrington and N.E. Stork (eds) *Insects in a Changing Environment*. London: Academic Press, pp.3–26.

Lawton, J.H. (2000) *Community Ecology in a Changing World*. Oldendorf: Ecology Institute.

Lieshout, M. van, Kovats, R.S., Livermore, M.T.J. and Martens, P. (2004) 'Climate change and malaria: Analysis of the SRES climate and socio-economic scenarios', *Global Environmental Change*, 14: 87–99.

MacArthur, R.H. and Wilson, E.O. (1967) *The Theory of Island Biogeography*. Princeton, NJ: Princeton University Press.

MacMahon, J.A. (1976) 'Thoughts on the optimum size of natural reserves based on ecological principles', in J. Franklin and S. Krugman (eds) *Selection, Management and Utilization of Biosphere Reserves*. Proceedings of the USA-USSR Symposium on Biosphere Reserves, Moscow, May 1976, Corvallis: Pacific Northwest Forest and Range Experiment Station, United States Department of Agriculture, Forest Service, pp.128–34.

McNeely, J.A., (1990) 'Climate change and biological diversity: policy implications', in M. Boer and R.S. de Groot (eds) *Landscape-Ecological Impact of Climate Change*. Amsterdam: IOS Press, pp.406–29.

Main, A.R. and Yadav, M. (1971) 'Conservation of macropods in reserves in Western Australia', *Biological Conservation*, 3: 123–33.

Martens, P., Rotmans, J. and Groot, D. de (2003) 'Biodiversity: luxury or necessity?', *Global Environmental Change*, 13: 75–81.

Mattson, D.J. and Reid, M.M. (1991) 'Conservation of the Yellowstone grizzly bear', *Conservation Biology*, 5: 364–72.

May, R.M., Lawton, J.H. and Stork, N.E. (1995) 'Assessing extinction rates', in J.H. Lawton and R.M. May (eds) *Extinction Rates*. Oxford: Oxford University Press, pp.1–24.

Nakano, S., Kitano, F. and Maekawa, K. (1996) 'Potential fragmentation and loss of thermal habitats for charrs in the Japanese archipelago due to climatic warming', *Freshwater Biology*, 36: 711–22.

Newmark, W.D. (1987) 'A land-bridge perspective on mammalian extinctions in western North American parks', *Nature*, 325: 430–2.

Newmark, W.D. (1996) 'Insularization of Tanzanian national parks and the local extinction of large mammals', *Conservation Biology*, 10; 1549–56.

Newsome, D., Moore, S.A. and Dowling, R.K. (2002) *Natural Area Tourism: Ecology, Impacts and Management*. Clevedon: Channelview Publications.

Nicholls, A.O., Viljoen, P.C., Knight, M.H. and van Jaarsveld, A.S. (1996) 'Evaluating population persistence of censused and unmanaged herbivore populations from the Kruger National Park, South Africa', *Biological Conservation*, 76: 57–67.

Nielsen, R. (2005) *The Little Green Handbook: A Guide to Critical Global Trends*. Carlton North: Scribe Publications.

Ordish, G. (1987) *The Great Wine Blight*, 2nd ed. London: Sedgwick and Jackson.

Pearson, R.C. and Goheen, A.C. (eds) (1998) *Compendium of Grape Diseases*. Saint Paul: The American Phytopathological Society.

Pimentel, D. (2001) 'Agricultural invasions', in S.A. Levin (ed.) *Encyclopedia of Biodiversity*. San Diego: Academic Press, pp.71–85.

Pimentel, D., Lach, L., Zuniga, R. and Morrison, D. (2000) 'Environmental and economic cost of nonindigenous species in the United States', *BioScience*, 50: 53–65.

Pimm, S.L., Russell, G.J., Gittleman, J.L. and Brooks, T.M. (1995) 'The future of biodiversity', *Science*, 269: 347–50.

Pitman, N.A. and Jørgensen, P.M. (2002), 'Estimating the size of the world's threatened flora', *Science*, 298: 989.

Purvis, A. and Hector, A. (2000) 'Getting the measure of biodiversity', *Nature*, 405: 212–19.

Radford, T. (2004) 'Life on the brink', *The Age*, 11 August.

Rodrigues, A.S.L. and Gaston, K.J. (2001) 'How large do reserve networks need to be?' *Ecological Letters*, 4: 602–9.

Rodrigues, A.S.L. and Gaston, K.J. (2002) 'Rarity and conservation planning across geopolitical units', *Conservation Biology*, 16: 674–82.

Rogers, D.J. and Randolf, S.E. (2000) 'The global spread of malaria in a future, warmer world', *Science*, 289: 1763–6.

Rosendal, G.K., (2001) 'Overlapping international regimes: the case of the inter-governmental forum on forests (IFF) between climate change and biodiversity'. *International Environmental Agreements: Politics, Law and Economics*, 1(4): 447–68.

Runte, A., (1987) *National Parks: The American Experience*, 2nd ed. Lincoln, NE: University of Nebraska Press.

Schonewald-Cox, C. and Buechner, M. (1991) 'Housing viable populations in protected habitats: the value of a coarse-grained geographic analysis of density patterns and available habitat', in A. Seitz and Loeschke (eds) *Species Conservation: A Population-Biological Approach*. Basel: Birkhauser Verlag, pp.213–26.

Slatyer, R.O. (1975) 'Ecological reserves: size, structure and management', in F. Fenner (ed.) *A National System of Ecological Reserves in Australia*. Canberra, Australian Academy of Science, pp.22–38.

Soulé, M. (1985) 'What is conservation biology?' *Bioscience*, 35: 727–34.

Soulé, M. and Sanjayan, M.A. (1988) 'Ecology-conservation targets: do they help?', *Science*, 279: 2060–1.

Spicer, J.J. and Gaston, K.J. (1999) *Physiological Diversity and its Ecological Implications*. Oxford: Blackwell Science.

The Age (2005) 'Polar bears' days may be numbered', February 3.

Timmins, S.M. and Williams, P.A. (1991) 'Factors affecting weed numbers in New Zealand's forest and scrub reserves', *New Zealand Journal of Ecology*, 15: 153–62.

Travis, J. (2003) 'Climate change and habitat destruction: A deadly anthropogenic cocktail', *Proceedings of the British Royal Society B*, 270: 467–73.

United Nations Environment Programme (UNEP) (2002) *Global Environmental Outlook 3: Past, Present and Future Perspectives*. Nairobi: UNEP.

Weaver, D. (2001) *The Encyclopedia of Ecotourism*. Oxford: CABI Publishing.

Whitehead, D.R. and Jones, C.E. (1969) 'Small islands and the equilibrium theory of insular biogeography', *Evolution*, 23: 171–9.

Wilcove, D.S. and Terborgh, J.W. (1984) 'Patterns of population decline in birds', *American Birds*, 38: 10–13.

Williamson, M. (1996) *Biological Invasions*. London: Chapman and Hall.

Wilson, E.O. (1992) *The Diversity of Life*. New York: Norton.

Wine Institute of California (2002) *Pierce's Disease Update*. San Francisco: Wine Institute of California.

Part III

Stakeholder adaptation and perceptions

13 The role of climate information in tourist destination choice decision making

Jacqueline M. Hamilton and Maren A. Lau

Introduction

The impact of climate change on tourism has been examined quantitatively in several different ways. There are economic theory-based studies that involve estimating the demand for destinations using, among other things, climate variables (see Maddison 2001; Lise and Tol 2002; Hamilton, 2003). Related to these studies are global models of tourism flows that include temperature as a determinant of the flows of tourists between countries (Berritella *et al.* 2004; Hamilton *et al.* 2003). There are also studies that use tourism climate indices to predict the effect of a changed climate on tourism demand (Scott and McBoyle 2001; Amelung and Viner, in press). The latter group of studies combine climate variables in a more complex way to reflect the thermal, physical and aesthetic properties of climate. The former two groups take a more simplistic approach: they include temperature, and up to two other variables. How far does the reduction of climate to one or two variables limit these studies? Moreover, de Freitas (2003) argues that climate data expressed as an average, which is used in the economic studies mentioned above, have no psychological meaning. Nevertheless, the economic theory-based studies and the global models base their analysis on the actual behaviour of tourists, in other words, actual destination choices.

A tourist's choice of destination will be based on what they expect from the chosen destination. What they expect will be driven by the image that they have of the destination. Of course, weather is not experienced as a set of separable and independent attributes but as a complex impression. In terms of climate, this leads us to ask: do tourists have an image of the climate and, if so, how was this image formed? It is unclear whether tourists form a complex picture of climate or if information on a few key attributes tells them enough about climate to construct an image. Lohmann and Kaim (1999) note that there is a lack of empirical evidence on the importance of climate on destination choice decision making. In contrast to the German travel surveys reported by Lohmann and Kaim, we have focused this study on climate image and climate information. As far as the authors of this chapter know, this is the first study of its kind and there is a considerable gap to be filled.

After considering the issues mentioned above, we formulated the following research questions:

A   How decisive is climate as a factor in decision making?
B   At what point in the holiday decision-making process do tourists gather infor-
    mation about climate and weather?
C   What sources of climate information are most frequently used?
D   What are the most frequently used types of climate information?

In order to gather data to answer these questions, a survey of tourists departing
from Hamburg and its vicinity was carried out during July and August 2004. The
survey produced 394 completed self-administered questionnaires. The questionnaire provided details on the current holiday, destination image, information
sources, type and presentation of information and demographic details of the
respondents.

Literature review and hypothesis formulation

Morley (1992) criticises tourism demand studies, which typically focus purely on
economic factors, because they do not consider utility in the decision-making
process (see Witt and Witt 1995; Lim 1995). He suggests an alternative way to
estimate demand based on the expected utility derived from the characteristics of
the product – in this case the destination country is the product. Lancaster (1966 in
Um and Crompton 1990) originally developed the concept that the characteristics
of a good are more important to the consumer than the actual good itself. How
these characteristics are perceived will determine the expected utility. In the case
of tourism, the product is the holiday at a certain destination and at a certain time
and this product will have certain characteristics. Knowledge of destination characteristics will be limited for a first time tourist. As climate can be temporally as
well as spatially defined, even repeat visitors will not necessarily have experienced
all seasons at the destination. Limits to knowledge lead Um and Crompton (1990:
433) to argue that:

> The image and attitude dimensions of a place as a travel destination are likely
> to be critical elements in the destination choice process, irrespective of
> whether or not they are true representations of what the place has to offer.

The final choice of destination is the result of a decision-making process that
involves the use of information, whether from personal experience or through an
active search, to generate an image of the destination. This section develops the
hypotheses related to destination image, decision making and information search
as well as climate information for tourists.

Destination image

There are many different definitions of what destination image actually is (Gallarza
et al. 2002). There is, however, a consensus that destination image plays an important role in destination choice. What role does climate play in destination image? Not

all studies of destination image include climate as an image defining attribute, as can be seen in the extensive review of destination image studies by Gallarza *et al.* (2002). Of the 25 destination image studies reviewed, climate was included as an attribute in 12 studies. Nevertheless, from their list of 20 attributes, climate is the seventh most frequently used attribute. Studies of destination image that include climate/weather as an attribute find that it is one of the most important attributes. There are, however, differences in the preferences shown by different types of tourists and for tourists from different places (Hu and Ritchie 1993; Shoemaker 1994; Kozak 2002; Beerli and Martin 2004).

Only one of the 142 destination image papers reviewed by Pike (2002) specifically deals with weather. This was a study by Lohmann and Kaim (1999), who assess the importance of certain destination characteristics using a representative survey of German citizens. Landscape was found to be the most important aspect, even before price considerations. Weather and bioclimate were ranked third and eighth respectively for all destinations. They found that, although weather is an important factor, destinations are also chosen in spite of the likely bad weather. In a study by Gössling *et al.* (2005) of tourists surveyed in Zanzibar, tourists were asked to rate climate's importance in their decision to travel to Zanzibar. More than half rated climate important but a small share of the respondents (17 per cent) stated that climate was not important at all. Based on the existing literature, it seems that climate is an important factor for tourists when choosing their holiday destination. We have, therefore, formulated the following hypothesis: *Hypothesis A1: Destination climate is an important consideration for the choice of destination.*

Decision making and information search

Fridgen (1984) expanded the five-phase model of recreation behaviour of Clawson and Knetsch (1966). The five phases are anticipation, travel to the site, on site behaviour, return travel and recollection of the trip. The anticipation phase includes decision making and preparation for the holiday. According to Fridgen (1984), tourism decision making involves environmental preferences and the cognitive image of what they expect from the destination. Other models of decision making in the tourism literature contain a number of stages. Among these stages may be the motivation to go on holiday, information gathering and evaluation of the holiday, which may include feedback loops into the next holiday decision (e.g. see Van Raaij 1986; Ahmed 1991; Mansfeld 1992). The temporal aspect of the holiday decision, in other words when to go on holiday, is absent from these models of decision making. Sirakaya and Woodside (in press) distinguish between behavioural and choice set approaches to decision making. According to them, behavioural approaches seek to identify the different stages in the decision-making process and the factors that influence the process. Choice set approaches involve identifying the various destinations that are in the awareness set and, following an active information search, an evoked set develops (e.g. see Um and Crompton 1990). From the latter set, the final destination will be chosen. In both of these models the tourist assesses the destination options available, using information acquired from their search and gradually eliminates the options that do not meet their needs. In both cases and in the studies discussed above, information is gathered in order to make the decision. Hence, we formulate our hypothesis as: *Hypothesis B1: Tourists gather climate information before they make their concrete holiday decision.*

Information on the current weather at the destination or predictions for the weather in the coming week can only be used to make decisions about destination choice at the very last minute. Therefore, we assume that the tourist gathers weather information in order to prepare for their holiday and make any necessary adjustments to the clothing or equipment that they will take with them. They may also do so to adjust their image according to the current situation and so modify their expectations. This leads to the following hypothesis: *Hypothesis B2: Tourists gather weather information in preparation for their holiday.*

Closely related to the time of information gathering is the question of which information sources are used. The destination image studies that take climate and weather into account do not consider this factor, whereas another group of studies focus on information search strategies but do not specifically look at climate information. Three distinct information search strategies are classified by Fodness and Murray (1998, 1999). First, there is a spatial element – the information search can occur internally, that is information from the individual's own memory, or it can occur externally, through the acquisition of information from sources such as travel agents or friends and family. Second, there is a temporal element to the information search. Tourists may continually be gathering information for their holiday or they may do so only when they are planning to go on holiday. The third aspect of the search is operational, which reflects the type and number of sources used. In a survey of American tourists who travelled to Florida, 68 per cent of the tourists used more than one source in their information search (Fodness and Murray 1998, 1999). The sources most likely to be used on their own were: personal experience, travel agencies, and friends and relatives. For a repeat visit, which involves less complex problem solving than a first time visit, Fodness and Murray (1999) argue that personal experience will be favoured. In their results, however, an external source of information – friends and relatives – was the main source. For those with a longer decision period, possibly reflecting a first time visit, friends and relatives are also the main source followed by auto club and travel agent. This study uses the length of planning period but the actual type of decision, that is whether it was a first time visit or a repeat visit, is not made explicit.

Van Raaij (1986) argues that novel destination possibilities and expensive holidays will necessitate an extensive information search. As the following analysis concerns itself with international tourism trips, the holidays under consideration are likely to be one of the major purchases by a household. Not only this, a holiday abroad is a significant event. Therefore, we can assume that the majority of the tourists will use several different information sources. Four information source categories were examined by Baloglu and McCleary (1999). These were professional advice, word of mouth, advertisements, books/movies/news. Word of mouth was ranked highest in terms of its importance in forming an image of the destination.

The least important category was advertisements. In addition, they find the mean number of sources used in their sample to be 3.75. In a study on the destination image of India, tourists used several different information sources. Friends and relations were the main source for more than half of the tourists (Chaudhary 2000). From the above, we have formulated the following hypotheses. *Hypothesis C1: Tourists rely on more than one information source. Hypothesis C2: 'Friends and family' is the dominant information source category for first time visitors. Hypothesis C3: 'Own experience' is the dominant category for repeat visitors.*

Climate information

Types of climate information can be examined in terms of content as well as presentation. De Freitas (2003) classifies climate according to its aesthetic, physical and thermal aspects. The thermal aspect is argued to be a composite of temperature, wind, humidity and radiation. Since climate is complex, we assume tourists are striving for a detailed picture in their information search and therefore formulate the hypothesis as: *Hypothesis D1: Tourists gather climate information on several different attribute types.*

The studies that analyse the demand for destinations in terms of characteristics include variables for temperature and in some cases precipitation and the number of wet days in the demand function (see Loomis and Crespi 1999; Mendelsohn and Markowski 1999; Maddison, 2001; Lise and Tol 2002; Hamilton 2003; Berritella *et al.* 2004). Moreover, in the studies that use tourism indices, such as Scott and McBoyle (2001) or Amelung and Viner (in press), temperature plays a greater role than any other climate variable. The tourism climate index, developed with regard to the biometeorological literature on human comfort, consists of five sub-indices. The sub-indices contain seven climate variables, three of which are temperature ones (mean, maximum and minimum temperature). The two sub-indices that contain the various temperature variables account for 50 per cent of the weighting in the tourism climate index. Because temperature is an important factor in both behavioural and biometeorological studies of tourism and climate, we have formulated the following hypothesis: *Hypothesis D2: Temperature is the dominant attribute for climate information.*

We found little guidelines in the literature on the way that climate information is portrayed. De Freitas (2003) argues that a climate index would be the most appropriate way to present climate information to tourists. Nevertheless, the authors are not aware of actual studies where the preferences of tourists for different formats are tested. From a survey of the internet and print sources of climate information, we can conclude that there are many different ways of presenting such information. There was, however, no clear tendency towards a particular presentation form. For this reason we randomly chose one of the possibilities for our hypothesis, which we have formulated as: *Hypothesis D3: Tourists prefer a textual format for the presentation of climate information.*

The nine hypotheses and the related research questions are shown in Figure 13.1. This figure depicts the phases of potential image change indicated by the various grey shades. The tourist has an image before planning that may change during the actual planning process and even after the decision for a specific destination has been made. Although not examined in this paper, the tourist's image could also change after the experience of the holiday.

Research Design

The fact that this study includes not only the question of information sources and information types but emphasises the time of information gathering led us to

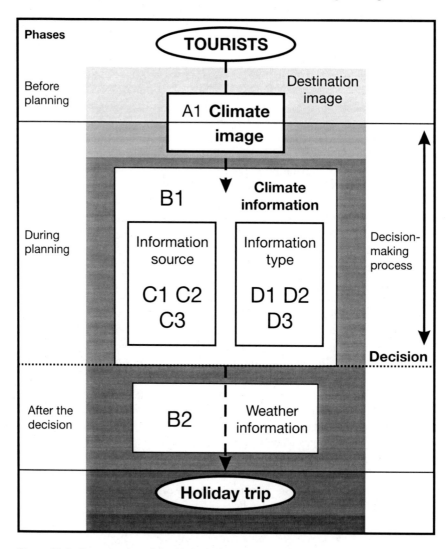

Figure 13.1 Conceptual model with hypotheses of the role of climate information in the tourist decision-making process

choose a specific point in time to survey tourists – shortly before departure. This allowed us to include the phase of preparation for the travel. Our study population was those residents of Germany going on an outbound holiday and departing from Hamburg and its vicinity. Our sampling frame consists of those tourists departing from Hamburg and its vicinity at specific points of departure: the airport, the train station, the international bus terminal and the harbours of Travemünde and Kiel for ferries to Scandinavia. Our convenience sample consists of those tourists travelling on the selected days and on the selected departures. All participants were aged 16 or over and resident in Germany. Additionally, only one member of a travel party was questioned. We purposefully excluded business travellers from the sample.

We paid attention to the following quotas:

1   Destination countries according to the market shares from the Reiseanalyse (FUR 1998, 2004).
2   Transportation mode market shares also from the Reiseanalyse.

The survey was carried out on 20 days spread over the months of July and August 2004. The days and times of the survey were chosen to correspond with departures to the countries with a high quota. The study period covered the main parts of the local school holidays. The schedule and budget of this study did not allow for an inclusion of car travellers according to the market share of about one third of all travellers. The quotas, therefore, corresponded to the relative market shares of the other transport modes.

While creating the questionnaire, we consulted a group of specialists who commented on preliminary versions of the questionnaire. These were tourism experts from academia as well as professionals from the tourism industry and others from the fields of marketing and quantitative research. A two-step pilot study was carried out at the end of June with the target group of tourists leaving from Hamburg Airport and a group of randomly chosen students. This pilot phase yielded valuable insights into intelligibility for the final questionnaire version. The comments of the experts and the results of the pilot phase resulted in the reformulation of individual questions and the questionnaire to improve its intelligibility.

In the following, we give an overview of the relevant questions from the questionnaire that we use in this analysis. The first section of the questionnaire includes general questions on the holiday: the destination country, the length of stay and the organisational form of the trip. We largely oriented this section on the Reiseanalyse (FUR 1998, 2004) in order to guarantee comparability to other studies. As far as possible, these questions are in multiple-choice format. Another section contains two questions that identify the main image attributes and the main information sources. See Table 13.1 for details of the sources used to formulate these questions.

The next section begins with a filter question about whether the tourist had been to the destination country before. Answered positively, the respondents are asked to complete five additional questions. After that another filter question is asked; if the respondent had informed themselves about the climate of their destination. If answered positively, another block of five questions follows. The questionnaire closes with a section containing demographic questions that provide details on the respondent's place of residence, gender, age, and education level.

Hypothesis A1

This hypothesis will be tested by examining if climate is at least the third most important attribute for the choice of destination. In order to assess this we asked respondents to rank the three most important out of 10 attributes. The 10 attributes were chosen according to an analysis of the attributes that were found to be the most important for tourists in studies on destination image (see Table 13.1). (We took the five highest valued attributes from each study and calculated the frequency that each attribute appeared over all the studies. The 10 most frequent were then taken from this list.) We purposefully put this ranking question on the first page of the questionnaire. Respondents were not told in advance the specific focus of the questionnaire. This way the individual's perception of the importance of climate was assessed before the respondent became aware that climate was the main theme of the questionnaire.

Our assumption that climate information is indeed important within the decision-making process purposefully does not implicate a certain quality of climate, for example as Baloglu and Mangaloglu (2001) do when using the attribute of 'good climate'. Although this could mean either a good climate according to the individual's perception or a good climate for certain activities that the tourist prefers to undertake while on holiday, such a formulation may yield a pre-valuation of the climate factor. We assume that some tourists will search for warmer places to go, others may prefer a cooler climate than they experience in their home region at the same time of the year and some may be completely indifferent. Moreover, the individual's perception of the climate at the destination as being 'good' may be influenced by the home weather at the time of booking. In the region of Hamburg, where the survey has been undertaken, the summer 2004 has been

Table 13.1 Sources of attributes for the questionnaire

Tested attributes	Source of attributes
Destination image	Baloglu and Mangaloglu (2001); Baloglu and McCleary (1999); Kozak (2002; Lohmann and Kaim (1999; Gallarza *et al.* (2002); Hu and Ritchie (1993); Yaun and McDonald (1990)
Information sources	Baloglu and McCleary (1999), Chaudhary (2000), Fodness and Murray (1999) and Phelps (1986)
Type and presentation of information	Own research of online weather information providers, online travel guides, information provided online by travel agents, tour operators, foreign offices and tourist boards, and print travel guides
General information on the trip and demographic information	FUR (1998, 2004)

widely perceived as comparatively cold and wet. In order to hold this sort of seasonal deviation at a minimum, we focus on climate and do not value it.

Hypotheses B1 and B2

Motivated by the decision phases formulated in studies of Fridgen (1984), Ahmed (1991) and Mansfeld (1992), we emphasise three distinct phases of information gathering. The first phase is limited to the time before the tourist decides to go on holiday. It is not an active information gathering phase, since an image of the climate of the destination is there already either through previous experience in the country – or comparable climatic regions – or through knowledge gained from a general interest in the area. Phase 2 covers the period after the tourist is motivated to go on holiday but has not made the concrete decision of where and when. In this phase, information will be actively gathered in order to make these decisions. Phase 3 includes information gathering in preparation for the holiday. This is carried out after the decision has been made but before the actual trip.

The hypotheses B1 and B2 are tested using the results of two questions. The first question asks the tourists to state when they informed themselves about climate. There were seven options, which belonged to the following three groups: *before planning, during planning* and *after the decision*, which correspond to the phases 1 to 3 respectively. We gave the tourists the opportunity to choose more than one option. The second question concerns the actual weather at the destination before the trip: we ask the tourists whether they have been following the weather during the week before their holiday.

Pinpointing the time at which information is gathered also contributes to the analysis of the climate as an important factor in decision making (see hypothesis A1). Information gathering *during planning* indicates a decisive character, while *after the decision* indicates, for instance, an adaptation of clothing to the climate and does not play an important role in the decision to go to the destination.

Hypotheses C1, C2 and C3

We included a question on the sources of information about the destination in general. Information sources for general information on a destination may be different from the sources used for climate information. From the review of the studies shown in Table 13.1, we included 12 possible sources of information, including *friends and family* and *own experience*, as well as weather information providers. The latter was included not only because of the purpose of this study but also because such sites contain information about destinations and links to online travel agents, tour operators and airlines.

In order to test the hypotheses, exactly the same sources were included in a question specifically focusing on climate. We asked the tourists to rate, on a five point Likert scale, the actual information sources used according to the importance for the decision. The filter question on previous visits is used to establish the two groups of first time and repeat visitors.

Hypotheses D1, D2 and D3

In these hypotheses, we distinguish between the presentation of the information and the content of the information. An examination of the possible sources of destination information and destination climate information resulted in the inclusion of the following categories: text format, maps, diagrams and numerical data (see Table 13.1). The various information sources provide different types of climate information, which range from several temperature types to precipitation-related information and less frequently mentioned attributes such as humidity or UV-radiation.

Analysis

General results

Not all of the tourists asked to participate in the survey agreed to take part. The response rate differed in two ways, first between the two months and second according to the departure point where the survey was carried out. Generally, July showed a better response rate (of 2:1 and even better) than August. The response rate at the airport was altogether less high than at the bus terminal, train station or ferry terminal. At the airport, the terminals seemed to matter. The survey was easier to carry out in the charter flight terminal, where we had a response rate of 3:1 during August, whereas at the terminal for scheduled flights, on some survey days, we had a response rate of 10:1. In total, we had 413 returned questionnaires. We eliminated 19 questionnaires because core questions were unanswered and so we coded 394 questionnaires in total.

Table 13.2 shows the demographic profile of the tourists surveyed. Compared to the age structure from the Reiseanalyse data from 1998 (FUR 1998), this survey has a more distinctive bi-modal pattern, which can be seen by the larger shares of tourists in the 20–29 and 40–49 age groups and the much smaller percentage in the 30–39 age group. The male/female split corresponds to that of the current population of Germany. Compared to the Reiseanalyse from 1998, this survey has a much larger share of those with a degree or who have a university entrance diploma. A comparison with national statistics is difficult because the statistics cover the age group 25–64.They are also based on completed years of education and include technical qualifications, which are not included in our options.

Questions were also asked about the current holiday, the results of which are presented in Table 13.3. The average length of the holiday is 14.3 days, which corresponds to the average length of holiday (13.7 days) reported for the Reiseanalyse 2004 (FUR 2004). Surprisingly, a large proportion of the holidays were organised independently. The shares for package tours and booking through a travel agent are similar to that of international trips in the Reiseanalyse 2004. As mentioned above, quotas had been used to get a representative spread of holidays to the most popular countries for German tourists. Nevertheless, an important group of tourists, those travelling to their destination by car, could not be included. Countries that are very

Table 13.2 Descriptive profile of respondents (n=394)

	Mean	Frequency
Age (n=377)	40.3	
16–19		9.0%
20–29		28.9%
30–39		18.3%
40–49		21.5%
50–59		10.1%
60–69		11.4%
70–79		0.8%
Gender (n=387)		
Male		48.8%
Female		51.2%
Place of residence (n=362)		
Hamburg		34.5%
Northern Germany		51.7%
Other within Germany		13.8%
Education (n=378)		
Completion of compulsory education		40.2%
University entrance diploma		27.5%
Higher education		31.7%
No qualifications		0.5%

popular but which are typically travelled to by car include: Austria, Switzerland, Poland, Denmark and the Netherlands, and are under-represented in the survey. In addition, the share of long-haul trips is smaller than that of the Reiseanalyse (2004). We must take into consideration however that the Reiseanalyse covers a whole year. This study concentrates on the summer and it is logical that there would be less of a tendency to travel far, when Europe is at its most attractive climatically. Finally, the majority of respondents had visited their destination previously.

Research question A: climate as a factor in decision making

The tourists were asked to pick the three attributes from a list of 10 that were most important in their decision to travel to their destination, and rank them. Ninety-four per cent of the respondents provided a useable ranking of the attributes. From Table 13.4, we can see that only two attributes are chosen more often than they are not chosen, namely climate and access to the sea/lakes. Not only was climate the most frequently chosen attribute, it also achieves the highest ranking of all attributes. The t-test for related samples was used to test if the mean rank value of climate is significantly different from that of sea/lakes, culture/history and nature/landscape, the

Table 13.3 Descriptive profile of holidays (n=394)

	Mean	Frequency
Duration of stay (n=388) in days	14.3	
Less than one week		14.4%
One week		17.5%
One to two weeks		19.1%
Two weeks		27.1%
Two to three weeks		9.8%
Three weeks		4.9%
Three to four weeks		1.0%
Four weeks and more		6.2%
Holiday organisation (n=393)		
Independent		42.5%
Travel agents (but not a package tour)		20.6%
Package tour		32.3%
Other		4.6%
Destination (n=394)		
Spain		25.4%
Greece		8.9%
France		7.1%
Italy		6.3%
Croatia		5.3%
Hungary		5.3%
Turkey		5.3%
Bulgaria		3.8%
Sweden		3.8%
Tunisia		3.8%
Other European		20.1%
Other non-European		4.8%
Previous visit to the destination (n=391)		
No		36.8%
Yes		58.6%
No response but answered the follow up questions		4.6%

three attributes closest in popularity to climate. Table 13.5 presents the results of this test and we can see that the mean of climate is significantly different from the other three attributes. For that reason, we can accept our hypothesis that climate is at least the third most popular attribute. Moreover, we can say that it is the most popular for the tourists in our survey. Almost two thirds of the respondents said that they had informed themselves about climate before their holiday. A further 10 per cent

Table 13.4 Results of the ranking of destination attributes (n=370)

	1st position value=3	2nd position value=2	3rd position value=1	Not chosen value=0	Total chosen	Mean
Access to the sea/lakes	53	79	56	182	188	1.01
Accommodation	14	33	22	301	69	0.35
Climate	91	65	40	174	196	1.20
Cuisine	2	12	10	346	24	0.11
Cultural/historical attractions	60	50	33	227	143	0.85
Ease of access	3	22	23	322	48	0.21
Hospitality	17	38	35	280	90	0.44
Nature/landscape	62	58	36	214	156	0.91
Price	17	61	48	244	126	0.60
Sport and leisure activities	8	22	19	321	49	0.24

Table 13.5 Mean differences between destination attribute rank values (n=370)

	Mean	T-value	2-tail sig.
Climate and nature/landscape	0.28	2.701	0.007
Climate and access to the sea/lakes	0.19	2.228	0.027
Climate and cultural/historical attractions	0.35	3.242	0.001

answered the questions on climate information, even though they said that they did not inform themselves about climate or did not give any answer to the question.

Research question B: Decision-making process and information search

There were seven options, which we converted into three stages: *before planning the holiday*, *during the planning* and *after the decision* has been made to go to the destination. The most common phase for gathering information about climate is during the planning stage (42 per cent). Nevertheless, 'shortly before the holiday' was the most frequently chosen single category (34 per cent) and for those that only chose one category, the split between the three phases *before planning*, *during planning* and *after the decision* is 25 per cent, 35 per cent and 39 per cent respectively. The majority stated only one phase where they gathered climate information. Of the tourists that combined two or more options, 61 per cent combined the phases *during planning* and *after the decision*. We can thus accept our hypothesis B1 that tourists gather climate information before they make their decision but with the caveat that the group of tourists informing themselves after the decision is also considerable.

In addition to the results presented above, we examined whether the respondents had been following the weather at their destination during the week before their departure. The majority of respondents (59 per cent) had been following the weather during the week before their departure. Table 13.6 shows the cross-tabulations of this variable and the groups *before planning*, *during planning* and *after the decision*. The correlations are not significant. It seems that there is no relationship between when the tourists inform themselves about climate and whether they follow the weather. Nevertheless, the relationship between obtaining climate information and following the weather in the week previous to travel is significant. If tourists inform themselves about climate, they also inform themselves about the weather shortly before they travel. We can accept the hypothesis B2 that tourists gather weather information before they travel, because the majority of tourists do this. Nevertheless, we accept this hypothesis with the caveat that a large group of tourists (41 per cent) showed no interest in weather. An examination of different tourists groups and destinations could provide more information on what conditions make weather and climate information important for the tourist.

Research question C: Sources of climate information

The results of the question on information sources are problematic. Tourists were asked to rate 12 different information sources and a thirteenth option of 'other' on

Table 13.6 Cross-tabulations of climate information and the weather in the week before the holiday

		Yes	No	
		Climate information gathered		
	Yes	68%	33%	
	No	33%	67%	
	N	286	91	
Respondent was aware of the weather at their destination during the week before their holiday		*Climate information before planning*		
	Yes	71%	68%	
	No	29%	33%	
	N	68	206	
		Climate information during planning		
	Yes	70%	67%	
	No	30%	33%	
	N	133	141	
		Climate information after the decision		
	Yes	68%	68%	
	No	32%	32%	
	N	132	142	

a scale of one to five only for those sources that they used. The question was answered in two different ways: first, that only the actual sources used were given a rank and second, that all sources were given a rank. For the following analysis we have examined these two groups separately. The first group, those that ranked only the sources that were used, we will call group A. The second group, B, are those that ranked more than 10 sources. Table 13.7 shows the number of climate information sources used. The first column contains the number of sources used by group A. The second column contains the number of sources used for group B, when we exclude those that are ranked lowest. In both cases, we can accept the hypothesis C1 that more than one source is used, given that 21 per cent (A) or 7 per cent (B) of the respondents state only one source. For comparison, the number of sources used as information about the destination is shown. Here there is a greater reliance on only one source (45 per cent).

For the first time visitors of group A, friends and family and travel guides are the most frequently chosen sources with 51 per cent each – more than one response was possible. The second most important sources are travel agent and tour operator. For the group of repeat visitors of group A, own experience was chosen by 69 per cent of the respondents, followed by friends and family (53 per cent) and travel guides (40 per cent). An examination like this is difficult for group B because they rank (almost) all of the sources. From this preliminary analysis, it seems that we can accept our hypothesis C2 that for first time visitors family and friends are the most important source and the hypothesis C3 that for repeat visitors own experience is the most important source. Nonetheless, a more detailed analysis is needed. Table 13.8, shows the cross-tabulations of previous visit (yes/no) with the sources family and friends (yes/no) and with own experience (yes/no), for

Table 13.7 Number of information sources used

	Climate-group A	Climate-group B	Destination
1	21%	7%	45%
2	24%	6%	28%
3	24%	20%	17%
4	17%	19%	8%
5	4%	10%	2%
6	4%	17%	<1%
7	<2%	9%	<1%
8	<2%	14%	
9	<2%	7%	
10	2%	8%	
11		4%	
12		6%	
13		2%	
N	141	138	392

the sources of information about the destination in general – for all tourists – and about the climate for the groups A and B. For destination information and for climate information (group A), there is no statistically significant effect of being a first time visitor on the tourists' likelihood to get information from family and friends. For group B, the effect is significant but counterintuitive. Having visited the destination before has the effect that you are more likely to ask family and friends about the climate. The results are much clearer for own experience. The positive relationship between previous visit and own experience is significant for all groups.

Not only can we examine the most frequently chosen sources, we can also look at the mean of importance value attached to them. There are no statistically significant differences in the means of own experience and family and friends for groups A and B. There are, however, differences in the means if we examine the groups of repeat and first time visitors separately. For group A, there are few first time visitors who used both sources. This makes a comparison of the means difficult, so we will continue with the repeat visitors. For that group, we have a mean difference of

Table 13.8 Cross-tabulations of information sources and the weather and having visited the destination previously

			Previous visit		
			Yes	No	
Sources of information about the destination					
Family and friends	Yes		38.5%	41%	
	No		61.5%	59%	
		N	247		145
Own experience	Yes		53%	2%	
	No		47%	98%	
		N	247		145
Sources of climate information (group A)					
Family and friends	Yes		49%	57%	
	No		51%	43%	
		N	92		47
Own experience	Yes		65%	6%	
	No		35%	94%	
		N	91		47
Sources of climate information (group B)					
Family and friends	Yes		71%	51%	
	No		29%	49%	
		N	83		53
Own experience	Yes		85%	36%	
	No		15%	64%	
		N	85		50

–0.4828 between friends and family and own experience, which is significant at the 5 per cent level. Not only is own experience relied on by more tourists it also is more important. For the first time visitors of group B, friends and family has a higher mean value than own experience and is statistically significant at the 10 per cent level. Again, for the repeat visitors, we see a significant difference in the means and own experience is ranked the more important of the two sources. Other sources that were given a high rank were newspapers and television, travel guides and weather information providers.

Research question D: Types of climate information

An overwhelming majority of the respondents (91 per cent) chose more than one climate attribute. The mean number of attributes chosen is 3.23. We can therefore accept the hypothesis D1 that tourists choose more than one attribute.

In Table 13.9, we can see that temperature is quite clearly the most frequently chosen attribute. Maximum temperature was chosen by two thirds of the respondents, while 32 per cent and 16 per cent of the respondents chose average and minimum air temperature respectively. Other attributes that were chosen by more than half of the respondents were the number of rainy days, duration of sunshine and water temperature. Because respondents were able to chose more than one attribute, we present the frequencies with which the air temperature attributes were chosen both singularly and in combination. As the lower half of Table 13.9 shows, only 12 per cent of the respondents did not choose one of the air temperature attributes. This gives very clear support for hypothesis D2, that temperature is the dominant attribute.

From the five possibilities offered, textual format was the second least preferred option and, if we discount the option 'other', then it is the least preferred. In this case, we can reject the hypothesis D3 that tourists prefer a textual format. Table 13.10 shows the results for all options in two forms: for all respondents and for those only giving one response. In both cases, numerical data is the most popular option.

Discussion and conclusion

This study adds to the evidence that climate is an important factor in destination choice. In addition, it provides clarity over the role of climate and weather information gathering in the various phases of the decision-making process. Our results highlight the importance of information gathering before making a decision. Furthermore, this study shows that information gathering also occurs after the decision. The number of sources used by the tourists is comparable with other studies (Van Raaij 1986; Fodness and Murray 1998, 1999; Baloglu and McCleary 1999; Chaudhary 2000). Moreover, this study gives support for Fodness and Murray's theory (1999) that personal experience will be the main source of information for repeat visitors. The importance of friends and family as an information source for all of the tourists in our sample reflects the results of Chaudhary (2000). The majority of tourists informed themselves about climate from a variety of

Table 13.9 Preferences for information about climate attributes

	Mean	Frequency
Number of attributes chosen	3.23	
Climate attributes chosen		
Maximum temperature		67%
Water temperature		52%
Duration of sunshine		51%
Number of rainy days		50%
Average temperature		32%
Minimum temperature		16%
Amount of precipitation		16%
Humidity		14%
Cloudiness		10%
Wind conditions		7%
UV Radiation		6%
None of these		3%
Air temperature options chosen		
Maximum temperature		27%
Average temperature		19%
Minimum temperature		1%
Maximum and minimum		8%
Maximum and average		25%
Average and minimum		<1%
Maximum, minimum and average		6%
Did not choose any temperature option		12%
N	283	

Table 13.10 Preferences for the presentation of information about climate attributes

	Frequency	
	more than one response	*only one response*
Maps and satellite images	33%	23%
Text	27%	15%
Diagrams	36%	17%
Numerical data	57%	42%
Other	2%	3%
N	283	149

sources. Therefore, the results of this study could also be useful for the providers of tourism information, in that they tailor the information they present to meet the preferences of tourists.

There has been some debate on the effectiveness of using tourism climate indices and demand studies to assess the impact of climate change on tourism. Studies of destination demand have been criticised for simplistically representing climate using single variables, such as temperature and precipitation and not a complex of variables. The results presented in this study support the use of temperature as the main determining variable in destination demand studies. Nevertheless, we cannot claim from these results that temperature alone is enough to represent the considerations of tourists about destination climate. We do not find support for de Freitas' argument (2003) that data presented as averages has no psychological meaning. Travel guides typically present climate data as monthly averages and they were, along with family and friends, the most frequently used source for first time visitors.

The limitations of this study need to be addressed. A major issue is that of the sample used. Time and budget considerations limited the study to easily accessible departure points. Because tourists travelling by car have no common departure point, we had to omit them from our sample. This had the consequence that certain destinations, such as Italy, Denmark, the Netherlands and Austria were under-represented. Nevertheless, climatically comparable destinations were well represented. It is unclear if different information search strategies are related to particular travel mode choices. In addition, a non-random sampling method was used, which limits the generability of the results. The survey period encompassed the school holidays for the states of northern Germany. This peak holiday period can easily be avoided by other groups of tourists who are not tied to institutional holidays. Therefore, the study may be biased towards tourists travelling with children. From other survey sources, it can be seen that older travellers favour the off-peak months (e.g. Oppermann 1995). Despite two pilot studies, certain questions were not formulated clearly enough, which hindered the analysis – see the results for research question C. An interview methodology may be better to examine such complex issues but this would be expensive and time consuming on this scale. Instead of using a self-administered questionnaire, verbally administering the questionnaire could bring more success.

Although they have quite different definitions, the terms weather and climate are used interchangeably by the general public. This can also be seen in some of the images studies that refer to weather, even though what is actually meant is climate. We tried to be clear and to distinguish between weather and climate in our survey. Nevertheless, in some questions it is possible that the respondents misunderstood and gave responses in terms of weather information. This is particularly the case with climate information sources, where some of the sources listed can give information on past weather, the climate, current weather and predicted weather. For example, the weather information providers, which have information on all four, or family and friends, who may also be able to provide information on all four. Again a verbally administered questionnaire could be more effective.

Global climate change is already having an effect on mean temperature and its further course is very likely to have an impact on the tourism industry as well. As the results of this study showed, climate is a defining factor for the destination choice of tourists. When the climate changes, destinations' attractiveness will change and with it – probably with a considerable time lag – also tourists' images of the destinations. An ancillary effect of global warming is that of sea-level rise. Access to the sea will change considerably and the quality of beaches will mostly deteriorate, with intensified erosion and change of slope occurring. As this study shows that access to sea and lakes is the second most important attribute to tourists when choosing a destination, sea-level rise will have a large effect on the tourism industry – tourists will not necessarily adapt to the new situation by changing their preferences, they may prefer to change destinations.

Having carried out this survey, the first of its kind to focus on climate as a specific attribute of destination image and on its role in the decision-making process, we have produced a valuable database that can be used for further research. For instance, the issue of whether the tourists' images of climate are accurate when compared to the climate of their destination can be assessed (Um and Crompton 1990). Some destination image studies found that there were differences in image for different groups of tourists (Shoemaker 1994; Kozak 2002). It would be an interesting extension of this study to examine whether we find different information preferences for different demographic or holiday groups.

Acknowledgements

The CEC DG Research through the DINAS-Coast project (EVK2–2000–22024) provided welcome financial support. We are grateful for the assistance and enthusiasm of Petia Staykova, Christian Röbcke and Aydin Nasseri. All errors and opinions are ours.

References

Ahmed, Z.U. (1991) 'The influence of the components of a state's tourist image on product positioning strategy', *Tourism Management*, 12(4): 331–40.

Amelung, B. and D. Viner (in press) 'The vulnerability to climate change of the Mediterranean as a tourist destination', in: B. Amelung, K. Blazejczyk, A. Matzarakis and D. Viner (eds) *Climate Change and Tourism: Assessment and Coping Strategies*. Dordrecht: Kluwer Academic Publishers.

Baloglu, S. and McCleary, K.W. (1999) 'A model of destination image formation', *Annals of Tourism Research*, 28(4): 868–97.

Baloglu, S. and Mangaloglu, M. (2001) 'Tourism destination images of Turkey, Egypt, Greece, and Italy as perceived by US-based tour operators and travel agents', *Tourism Management*, 22: 1–9.

Beerli, A. and Martin, J.D. (2004) 'Tourists' characteristics and the perceived image of tourist destinations: a quantitive analysis – a case study of Lanzarote, Spain', *Tourism Management*, 25(5): 623–6.

Berritella, M., Bigano, A., Roson, R. and Tol, R.S.J. (2004) *A General Equilibrium Analysis of Climate Change Impacts on Tourism*. Research Unit Sustainability and Global Change Working Paper FNU–49. Hamburg: Hamburg University and Centre for Marine and Atmospheric Science.

Chaudhary, M. (2000) 'India's image as a tourist destination – a perspective of foreign tourists', *Tourism Management*, 21: 293–7.

Clawson, M. and Knetsch, J.L. (1966) *Economics of Outdoor Recreation*. Baltimore, MD: Johns Hopkins Press.

Fodness, D. and Murray, B. (1998) 'A typology of tourist information search strategies', *Journal of Travel Research*, 37: 108–19.

Fodness, D. and Murray, B. (1999) 'A model of tourist information search behaviour', *Journal of Travel Research*, 37: 220–30.

FUR (Forschungsgemeinschaft Urlaub und Reisen e.V.) (1998) *Die Reiseanalyse RA 98*. Köln: Zentralarchiv für empirische Sozialforschung.

FUR (2004) *Reiseanalyse Aktuell*. Online. Available at www.fur.de/downloads/ Reiseanalyse_2004.pdf (accessed 4 November 2004).

Freitas, C. de (2003) 'Tourism climatology: evaluating environmental information for decision-making and business planning in the recreation and tourism sector', *International Journal of Biometeorology*, 48: 45–54.

Fridgen, J.D. (1984) 'Environmental psychology and tourism', *Annals of Tourism Research*, 11(1): 19–39.

Gallarza, M.G., Saura, I.G. and Garcia, H.C. (2002) 'Destination image: towards a conceptual framework', *Annals of Tourism Research*, 29(1): 56–78.

Gössling, S., Bredberg, M., Randow, A., Svensson, P. and Swedlin, E. (2005) 'Tourist perceptions of climate change: a study of international tourists in Zanzibar', *Current Issues in Tourism*, (in press).

Hamilton, J.M. (2003) *Climate and the Destination Choice of German Tourists*. Research Unit Sustainability and Global Change Working Paper FNU–15 (revised). Hamburg: Centre for Marine and Climate Research, Hamburg University.

Hamilton, J.M., Maddison, D.J. and Tol, R.S.J. (2003) *Climate Change and International Tourism: A Simulation Study*. Research Unit Sustainability and Global Change Working Paper FNU–31. Hamburg: Centre for Marine and Climate Research, Hamburg University.

Hu, Y. and Ritchie, J.R.B. (1993) 'Measuring destination attractiveness: a contextual approach', *Journal of Travel Research*, 32(2): 25–34.

Kozak, M. (2002) 'Comparative analysis of tourist motivations by nationality and destinations', *Tourism Management*, 23: 221–32.

Lim, C. (1995) 'Review of international tourism demand models', *Annals of Tourism Research*, 24(4): 835–49.

Lise, W. and Tol, R.S.J. (2002) 'Impact of climate on tourism demand'. *Climatic Change*, 55(4): 429–49.

Lohmann, M. and Kaim, E. (1999) 'Weather and holiday destination preferences, image attitude and experience', *The Tourist Review*, 2: 54–64.

Loomis, J.B. and Crespi, J. (1999) 'Estimated effects of climate change on selected outdoor recreation activities in the United States', in R. Mendelsohn and J.E. Neumann (eds) *The Impact of Climate Change on the United States Economy*. Cambridge: Cambridge University Press, pp.289–314.

Maddison, D. (2001) 'In search of warmer climates? The impact of climate change on flows of British tourists', in D. Msddison (ed.) *The Amenity Value of the Global Climate*. London: Earthscan, pp.53–76.

Mansfeld, N. (1992) 'From motivation to actual travel', *Annals of Tourism Research*, 19: 399–419.

Mendelsohn, R. and Markowski, M. (1999) 'The impact of climate change on outdoor recreation', in R. Mendelsohn and J.E.Neumann (eds) *The Impact of Climate Change on the United States Economy*. Cambridge: Cambridge University Press, pp.267–88.

Morley, C.L. (1992) 'A microeconomic theory of international tourism demand', *Annals of Tourism Research*, 19: 250–67.

Oppermann, M. (1995) 'Travel life cycle', *Annals of Tourism Research*, 22(3): 535–52.

Phelps, A. (1986) 'Holiday destination image – the problem of assessment: an example developed in Menorca', *Tourism Management*, 7(3): 168–80.

Pike, S. (2002) 'Destination image analysis – a review of 142 papers from 1973 to 2000', *Tourism Management*, 23: 541–9.

Scott, D. and McBoyle, G. (2001) 'Using a "Tourism Climate Index" to examine the implications of climate change for climate as a tourism resource', in A. Matzarakis and C. de Freitas (eds) *International Society of Biometeorology Proceedings of the First International Workshop on Climate, Tourism and Recreation*. Retrieved from www.mif.uni-freiburg.de/isb/ws/report.htm

Shoemaker, S. (1994) 'Segmenting the U.S. travel market according to benefits realized', *Journal of Travel Research*, 32(3): 8–21.

Sirakaya, E. and Woodside, A.G. (in press) 'Building and testing theories of decision making by travellers', *Tourism Management*, corrected proof available online 13 August 2004.

Um, S. and Crompton, J.L. (1990) 'Attitude determinants in tourism destination choice', *Annals of Tourism Research*, 17: 432–48.

Van Raaij, W.F. (1986) 'Consumer research on tourism: mental and behavioral constructs', *Annals of Tourism Research*, 13: 1–9.

Witt, S.F. and Witt, C.A. (1995) 'Forecasting tourism demand: a review of empirical research', *International Journal of Forecasting*, 11: 447–75.

Yuan, S. and MacDonald, C. (1990) 'Motivational determinants of international pleasure time', *Journal of Travel Research*, 29(1): 42–4.

14 Restructuring the tourist industry

New marketing perspectives for global environmental change

Szilvia Gyimóthy

Introduction

Depending on the economic role of tourism in their particular area, regional and national governments are anxious to assess the consequences of climate change scenarios in terms of vulnerability or new market opportunities. Alpine skiing regions and seaside resorts prepare for a declining tourism industry, expecting enduring deficiencies in snow and fresh water supply (Parry 2000; Bürki *et al.* 2003). At the same time, cold-water resorts in higher latitudes hope that warmer temperatures could boost tourism in the form of diversified products and a prolonged season. There are a growing number of policy workshops, conferences, expert panels and think tanks established to advise the industry on future development strategies (see Chapter 1). However, in many cases, these are being projected at single cases – destinations or specific market segments – often making the discussion limited in scope. The focus is typically directed at curing already acute or expected direct symptoms of environmental change in order to maintain present levels of tourism activities.

Practitioners in stagnating or declining mass destinations urge immediate adaptation measures: 'We do not have a lack of knowledge; we have a lack of the implementation'(Wolfgang Pfefferkorn, Director of the Future of the Alps initiative). Future tourism agendas in these areas do not only include mitigation and adaptive measures, but also climate-safe projects, such as 'managed retreat', for example, the construction of new beach resorts farther back from the sea on the island of Jersey; or the building of second-generation ski resorts in the Tyrol above the revised alpine snowline. Arguably, both developments are the result of a difficult trade-off between securing local employment structures and the environmental protection of pristine areas. However, giving artificial breath to the regional economy's cash-cow in this form is a myopic, short-term reaction to inevitable environmental change. 'Rescue' developments of this kind attempt to maintain known forms of mass tourism, which are essentially monocultural, unsustainable and sensitive to shifts in demand. The question is not so much about *how*, but rather *whether* present forms of climate-jeopardised tourism activities should be preserved in a global context.

The illustrations above reflect reactive patterns in the process of restructuring the tourism industry, focusing on isolated local responses to changing visitor arrivals. Even if these investments can maintain present levels of tourism or generate additional business in the short term, their long-term consequences may not be in balance with marginal economic gains. Adjustments to the destination product portfolio may influence declining or increasing tourist numbers in some areas, but will hardly address the ecological footprint of another long-haul tourist or a new snow canon. Instead of local measures reacting to a narrowly defined problem, practitioners and decision makers should acknowledge the far reaching implications of climate change for the global tourism system and its components – entailing tourist destinations as well as tourist-generating regions.

Clearly, altered physical destination conditions or product alternatives will affect tourism demand, but the nature of demand is also shaped by climate-induced socio-economic changes (see Aall and Høyer 2005). This complex causality is seldom captured in the academic debate on changing travelling patterns and habits in Western societies. For instance, several sources (Giles and Perry 1998; IPCC 2001; Maddison 2001; Viner and Amelung 2003) name global warming-related opportunities for tourist investment in traditionally marginal – and less favourable – areas and predict that dominant, north-to-south tourist flows may be reversed in Europe (Maddison 2001). However, if traditional tourist economies in the south collapse because of drought and sea-level rises, how could these regions generate affluent demand for the new 'Nordic Costas'? Tourist behaviour may not only be influenced by temperature preferences, as Hamilton suggests (2004), but also by recently experienced weather extremes at home. Summer heat waves followed by a cold and wet season the year after may influence outbound travel flows in often unpredicted ways (see Dewar 2005). Furthermore, travelling patterns are also subject to external market factors, such as increased fuel prices or environmental taxes. If the cost of everyday life becomes more expensive because of higher energy prices, this could greatly influence potential tourists' discretionary income. The present pattern of long-haul short breaks (for example, Europeans on Christmas shopping trips to Shanghai), facilitated by inexpensive flight offers, may also change dramatically if policy makers effectuate planned taxation on air transport. Although it is impossible to account for all factors in drawing future scenarios, the debate would benefit from a more holistic approach, identifying complex, induced effects of climate change reverberating through the global economy and society at large.

The real challenge of climate change, as Higham and Hall (2005) claim, is about making tourism sustainable, by providing macro-level insights and moving beyond short-term planning horizons concerning energy and fossil fuel consumption. Arguably, governments and international tourism organisations play an important role in educating and raising awareness, for example, through 'The Djerba Declaration on Tourism and Climate Change' (TRI 2003). Despite the strategic significance of the declaration – in terms of urging adaptive, constraining and educative measures to cut greenhouse gas emissions – only few practical implementations have seen light since then, and there is little operative co-ordination taking place across national, organisational and sector borders in this field. The lack of specific actions may partly be accounted for by the scope and uncertainty associated with this 'wicked problem' (Stewart *et al.* 2004). At present, there exist several alternative future scenarios, and the expected consequences or prioritised areas are debated by climatologists, politicians and industry experts alike. Furthermore, there is a real challenge associated with the institutionalisation of environmentally responsible practices in a fragmented, neo-liberal tourism industry. Despite their subscription to the Kyoto protocol, several European governments provide generous subsidies or VAT-exemptions to low-cost airlines – like EasyJet, Ryanair and Germanwings – in order to boost regional growth around smaller provincial airports. Furthermore, tourists and tourism suppliers are unlikely to regulate their market behaviour based on the recognition of long-term environmental risks and sustainability principles alone. While there is certainly a growing group of conscious and reflective actors, and numerous policy guidelines and operational monitoring tools exist (for example, Green Globe 21), there is still a long way to go to alter the behaviour of the grand majority from being reactive and economic opportunity driven towards proactive and strategic measures.

If the tourism industry as a whole is to serve as a tool for mass education and a model for global change, it must change its strategic approach. There is first a need to understand the limitations of isolated local rescue projects in a global context and, second, the poor viability of global institutionalised control must be acknowledged. The challenge is to permanently change the attitudes of actors, mobile consumers and stakeholders in the tourism industry, which calls for the application of more subtle methods. This chapter argues for the significance of marketing activities and product development strategies in inducing those attitude changes.

Global responsible marketing

The main objective of any marketing activity – product development, advertising or strategic alliance development – has always been focusing on achieving corporate goals – maximising economic profitability – by inducing changes in market conditions in order to alter consumers' perceptions, motives and eventually behaviour in a favourable way. This implies that marketing activities may play a historical role in changing consumers' environmental consciousness and in catalysing co-ordinating activities among destination actors towards a more sustainable agenda. New marketing perspectives may also energise the creative process of new product development and innovative destination branding, by moving away from traditional representations and commodification practices that dominate the industry today. Tourism marketers should recognise and use their powerful potential in influencing and challenging present constructions of cognitive destination concepts and contemporary practices of tourist activities and geographical flows.

Environmental responsibility is often placed on the individual consumers' shoulders, which in principle corresponds with the consumer trend shift in tourism described by Poon (1993). Poon argues for the rise of a new, diversified generation of tourists, which is essentially autonomous, sophisticated and conscious of its environmental and social impact. Non-price factors, such as quality or fair-trade concepts are central to new tourism and Poon stresses that these consumers are prepared to pay more for good value (for example, ecological products). However, environmental consciousness or leisure sophistication can also be interpreted as a social differentiator to mark cultural superiority. As Hughes (2004) points out, demonstrative ecological consumerism ('ego-tourism') has its roots in a struggle for personal uniqueness and self-actualisation, rather than a deep-seated concern for the environment. Hughes (2004: 503) notes: 'Environmental rhetoric is used by ego-tourists to claim the moral high ground and to obfuscate their part in the exploitation of the destinations they visit.' This implies that 'eco-friendly' individuals may also engage in alternative, ego-enhancing tourism activities during their trips, which border on deliberate negativism – snow scooter safaris, big game hunting, coral reef diving. Because the inherent motive for leisure travel is hedonistic escape from the rationality of the everyday, it is unclear 'how far tourism will be shaped by environmental reflexivity and how far by self-indulgence' (Hughes 2004: 507).

In principle, each tourist can contribute individually to mitigate global climate change by regulating his or her lifestyle and refraining from excessive consumption. However, environmentally conscious 'new' tourism – whether driven by ideological convictions or consumerist trends – is still a niche; the majority of travellers are unintentionally ignorant of their contributional share to environmental problems. If environmental responsibility must be consumer driven, marketers and regulating bodies must define steps towards widening and altering consumers' perspective on travelling as an everyday commodity and a consumer right. Tourist behaviour can primarily be influenced by institutionalised constraints, such as environmental taxation or emissions-based pricing of tourism commodities. These measures would make tourists consider reducing their use of fossil fuels and would also support the development and use of nonpolluting renewable energies – solar, wind, biological. An alternative awareness-raising measure with a consumption implication could include the development of a new generation of tourism packages, configuring various CO_2-neutral travel, accommodation and attraction brands. While tourists' motivations to consume 'green destinations' today are typically driven by moral and ideological arguments, in the future, these places may become more price-competitive, because CO_2-neutral energy consumption will cost less.

Environmental responsibility may also be extended to the entire supply side of the tourist system – tour operators, transport companies, destination and transit suppliers. The Djerba declaration calls for responsible measures aimed at reducing the industry's own contribution to greenhouse gas emission and thus to future climate changes. Arguably, the declaration is an important and visionary statement, however, individual firms may be more easily convinced by economic arguments. Because of increasing energy prices in the future, operation costs of hospitality and attraction products may rise dramatically. Therefore, it is important to consider investing in energy-saving installations now and to educate small entrepreneurial actors to be conscious about their use of fossil fuel resources. Recently, a few pioneering companies, such as the Aspen Skiing Co. have begun marketing 'low-impact' recreation to tourists, in the form of an eco-friendly ski lodge. This includes the building of new resort facilities from recycled materials, as well as operations exclusively powered by renewable energy sources. Other actors in the accommodation sector have also begun to set corporate policies with specific goals concerning the reduction of CO_2 emissions within their operational boundaries. Such developments must be integrated into revenue management strategies that motivate environmentally friendly consumption, for instance charging less for CO_2-neutral products and including additional fees in the price of polluting/energy demanding activities.

National marketing agencies must define strategies addressing the challenges of climate change, even if local scenarios do not offer dramatic environmental transformations. This does not only imply the establishment and educational dissemination of sustainability regulations, but also seizing the consequences of present and future development activities on a global scale. Destination marketers in Scandinavia, for instance, seem to worry little about the global sustainability paradox of attracting new, long-haul segments. Nordic countries are opening tourist representations in the Far East and Pacific Region in order to draw more Chinese and Australian tourists to Swan-labelled hotels or Blue Flag beaches. Tour operators and destination marketers must realise that environmentally-oriented consumerism does not only create new business opportunities or an attractive market positioning argument. Sustainability principles, if rightfully implemented, also impose certain codes of conduct on tourism actors – for instance, giving up some market segments or products – that hamper rather than fulfil short-term economic goals.

Challenges arising from climate changes

Global warming scenarios have implications for almost every tourism activity. With a few exceptions, the implications appear mostly negative. Temperature increases of a few degrees are projected to induce wide-ranging environmental changes: sea-level rises, reduced precipitation, glacier melting, beach erosion, inundation, degradation of ecosystems and saline intrusion. Changes in coastal morphology, hydrological balance, fauna and flora will affect the established mix of attractors of a destination, which in turn reshape economic, political and sociocultural life. It is hard to imagine any region or sector that is not going to be (directly or indirectly) affected by global environmental change. The tourism industry has no choice but to prepare for and adapt to climate-related impacts already in the pipeline, so it is plausible that several countries will follow the Arctic example (ACIA 2004) and integrate adaptation measures into strategic tourism planning.

Arguably, the scope of environmental change in certain areas will reach dimensions at which tourism cannot be regarded as a viable long-term economic option. The implications of rising sea levels, beach erosion and drinking water contamination may be fought by adaptive strategies, but a considerable problem in the immediate future will be responding to the frequency and intensity of extreme weather phenomena. Flooding, storms and hurricanes may boost insurance rates and, in extreme cases, entail uninsurability. Apart from the real costs of insurance and reconstruction, destinations must also calculate the market effects of intense media coverage of tourism disasters (Faulkner 2001), especially significant when an affected area is being described in terms signalling a 'troubled' or 'lost' Paradise.

Many of the genuine, iconic qualities of destinations worldwide are climate-dependent, and destination brand equity is severely affected by environmental change. Fluctuations in accessibility or in perceived quality of established products result in uncertainty about regional brand product portfolios, undermining the consumers' primary 'reason-to-go' perceptions. Unpredictable climatic characteristics and vulnerable tourist assets cannot be considered appropriate and stable unique selling points over time, thus marketers must define the attractiveness of a place in terms of the relative value-in-use for the individual customer. This calls for a more sophisticated or altered marketing rhetoric in destination representations. For instance, advertisements might focus on addressing so called 'push'-motives, such as escaping densely populated agglomerations and urban living style – without the conventional pathos of climatic opposites or stereotyped clichés (such as 'untamed wilderness', 'lost Eden' or 'Paradise on Earth').

Taking a look at any peripheral tour operator, it is striking that many of their products are still promoted by the glamour of 'the last frontier' or 'beyond the modern world': the virtues of stability, familiarity, the supposed authenticity and a 'return to the roots' (Blomgren and Sørensen 1998; Schellhorn and Perkins 2004). Representations of native culture and traditional lifestyle are stereotyped and stress ecological and cultural unchangedness, a construction that resembles Saïd's (1979) depiction of the orientalist discourse. For example, a 15-day trekking tour in Greenland's Disko Bay area includes dog sledge-running, visits to trapper-villages, traditional hunting tours and regional gastronomic experiences with drum-dance performances. But, as the trapper culture slowly disappears in the wake of climatic and societal changes, such packages will become just as inauthentic and superficial as tribal tourism in tropical destinations. Even if Inuit and Saami people take pride in demonstrating their cultural heritage to visitors and, as according to Johnston and Viken (1997), tourism has a positive impact on Greenlandic culture, there are great problems associated with the museification and commodification of traditions, arts and crafts. Although often presented as such in tourism advertising, peripheral regions in a global economy are neither 'primitive' nor 'remote'. Therefore, romantic, ethnicising representations of 'vanishing worlds' must be complemented with images of the modern society, if tourism marketers are to successfully preempt the image consequences of destination changes borne by extrinsic climate factor changes.

Probably the biggest marketing challenge arising from environmental changes apart from the addressing of altered flows of tourism demand is a radical revision of how nature should be redefined and re-presented in a particular destination iconology. According to the constructionist perspective, tourism is grounded in the ideas that people attach to places far away (Saarinen 2004), hence, a 'destination' is a constantly changing product of our perception. This implies that attached meanings can be reconfigured by inventing new brand assets and spatial practices. The image of many tourist places will be radically altered in the next 20–50 years, losing physical characteristics that prompted associations with the 'romantic' – a wide sandy beach bathed in moonlight – 'spiritual' or 'sublime' – a bird's eye view from a snow-covered mountain top. The task for future destination marketers is then to design new representations of a region that use a different language of form to express the 'romantic' or the 'spiritual' or, alternatively, to develop concepts – exploration, recreation, contemplation, leisure hunting – that match a modified shoreline or snowless slopes.

Adaptive strategies: new product development

Long-term changes bear the seeds of new opportunities, product concepts and place identities. A proactive measure on climate change may start with reflecting on what other experiences can be offered to tourists and how a diversified product portfolio can best be developed. In other words, destinations must assess and revise their soft infrastructure and marketing activities. Changes in the environmental profile necessitate adjustments in the tourism product, in order to remain competitive during global change. Changes in the composition of market segments require meeting the expectations of new tourist cultures, in terms of developing new services and products catering for their specific needs. Both consolidated resorts and peripheral destinations on the rise must acknowledge and compare various segments' preferences and potential behaviour in order to evaluate current and future visitor impacts.

Tourism consumption is often described through functionalist arguments, such as breaking free from everyday routines and obligations or looking for otherness and difference. Tourists' motives are often classified along a novelty–familiarity continuum (Cohen 1972; Plog 1987), defining various forms of how the out-of-the-ordinary experience is conceptualised by the individual tourist. The construction of particular tourism geographies was rationalised through the modern individual's quest for the authentic (MacCannell 1976) or a centre 'out there' (Cohen 1979), disillusioned by the urbanisation and social alienation. Despite accelerating globalisation processes in the past decade, *nature* and natural resources are still considered and represented in brochures as offering 'real' and 'authentic' experiences on the peripheries of the modern world (Shaw 2001; Saarinen 2004). However, maintaining the illusion of the 'untouched', 'indigenous' or 'unique' Other in twenty-first-century tourism marketing will become contested and bemusing as sophisticated consumers are presented with opposite images and representations in other, non-tourist media. The question is what alternative representations can replace reassuringly the loss of 'Paradise' and the 'Exotic Other'. How can difference and cognitive opposites be constructed if not through historically and ideologically determined images of cultural, or environmental, otherness?

It is unclear whether post-modern tourists are still looking for preconstructed and referential novelty and difference, as the modernist discourse advocates. Leisure consumption in the post-industrial society is no longer understood through its restorative function and as routinised breaks from everyday life, but through demonstrative individualism and lifestyle management. Societal structures and routines are becoming fragmented, illegibile and fuzzy: spatial and temporal segregation of the classic dichotomy between work and leisure disappears. Thus it will be increasingly difficult to frame and define routinised breaks from work activities. The everyday of the mobile individual in a hypermodern network society is characterised by unstability, uncertainty and unknown variation, which points towards a growing consumer trend towards nostalgia (Goulding 2001), familiarity and a sense of belonging (Prentice 2004). Even if stability, slowness, security and intimate kinships are just as illusory as 'real wild natural areas', these can be marked off and commodified unrelated to the physical environment.

Peripheral areas are becoming metaphors of home and the private domain. The post-modern rural nostalgiascape materialises cosmopolitan yearnings into an idealised fixed point and a refuge in an increasingly fluid and changeable world as well as to the degree of alienation and disorientation in the present (see Allon 2000). The first wave of nostalgic lifestyle products has already hit the market, Slow Food, Città Slow, stress-releasing wellness packages witness the reorganisation of commodities and ideologies into a coherent collection of signs for tradition, continuity and Gemeinschaft. None of these offerings are bounded to traditional natural or climatic resources in a strict sense. The consumption of this new generation of tourism products is little dependent on geographical location and weather conditions: their attractiveness is defined through inherent values for contemporary consumers. This implies that there is a strong future market for innovative product developments, which smartly address the malaise of the hypermodern society rather than focusing on ambient or extreme climatic resources, a major attractor in modernity.

The nostalgic turn in contemporary tourism consumption combined with global environmental changes may entail reversive life cycle development at tourist destinations. Major phase shifts may take place as a result of complex covariating external conditions – altered physical environment, travelling habits, reduced discretionary income, higher energy prices, environmental taxes, perceived higher risks associated with long-haul travel – where tourist activity levels may return to earlier – exploration, involvement or development – life cycle stages. There are some indicators that domestic tourism in the UK and Scandinavia is growing again, and previously popular accommodation forms – inns, youth hostels and campsites – are experiencing a renaissance and new boom in visitor numbers. Such 'retro-tourism' is partly a condition of the larger structural changes in society explained above; but it is also fuelled by demographical factors: the first generation of mass tourists, babyboomers are retiring, looking for easily accessible commodities with emotional bonding – products they can feel nostalgic about. In either way, it is plausible that future consumption trends will be even more characterised by a search for stability, 'back to the roots', bonding with family and friends.

Recommendations for adaptive measures often stress the importance of new, climate-neutral products, achieving a diversification of product portfolios, and thereby spreading the risks related to uncalculated events or deficiencies in natural assets. Monocultural – sea-and-sand or snow – destinations are now following the example of urban destinations, developing year-round, complementary packages of weather independent activities, including hiking, or romantic getaways. In Denmark, provincial destinations urge small-scale thematic developments – like gastronomy, golf and wellness – targeted at corporate and shortbreak markets. Other development scenarios include the all-inclusive indoors visitor attractions, science or aquaparks. Although these establishments are independent of weather variations and seasons, there are a number of challenges. There are considerable capacity investments and maintenance expenses related to a spa or a theme park and operation costs may be increased dramatically if energy prices rise.

Tourism planners may decide upon focusing on low-tech, ideologically reconfigured retro-products or high-tech indoor experience arenas, according to how local stakeholders relate to future progress and adaptation. The choice of either of the two solutions implies a radical revision of how destination product portfolios are presented in marketing communication. Whichever development strategy actors subscribe to, they also vote for ceasing to rely on classical image and stereotyped products. The quest is not so much about revitalising declining sun-and-sea or ski resorts, but about designing hybrid concepts of destination products that mix both familiar and innovative elements into new thematic packages. Development strategies in temperate areas are mostly focused on special interests, including: gastronomy, golf, wellness, art, shopping and active holidays. Marketing efforts to attract MICE (meetings, incentives, conventions and exhibitions) and cruise visitors have been intensified in the past years. The strategic benefit of targeting these affluent segments is that they are relatively climate-insensitive and the majority of the product offerings are interchangeable and not season-dependent. The challenge is now to expand/renew these concepts so that they can attract an even more diversified and mobile market, including opportunities for local recreational and corporate visitors, in case other segments fail in the short term.

Another task related to thematic product developments is to change established employment structures and competence profiles. Coastal summer destinations and northern ski resorts are both highly dependent on an unskilled seasonal labour force. There might be problems in accessing the usual pool of seasonal employees in the off-season, because they typically have permanent occupations in urban areas (for example, students). Similarly, many small hotel and restaurant owners move their operations elsewhere in the low season. Many attractions are periodically closed and public transport runs on a less frequent basis, which means that tourist choice and access is limited in these periods. The destination must, therefore, be geared to welcome visitors on a different temporal basis (Baum and Hagen 1999). In order to meet the expectations of new market segments, destinations must also adjust their customer care skills. A growing number of international visitors necessitate the development of language skills of front-line employees. Furthermore, there is an urgent need to be conscious about different cultural understandings of service expectations and customer care. What domestic Scandinavian guests see as personal, informal service might be regarded impolite by overseas visitors. Some customer segments will need a higher level of coaching and information than others. Traditional segments, such as outdoor recreational visitors from Germany or Scandinavia are typically self-serviced, bringing their own provisions, maps and guidebooks. New market segments may be far less autonomous and their entire stay in the region must be facilitated by service providers. There is an urgent need for a destination-embracing hospitality training directed at accommodation providers as well as guides, instructors and outfitters. A formal training in customer care, complaint management as well as emissions management may not only standardise the level of welcome for a destination, but also create opportunities for customer relationship management, including loyalty schemes and regional bonus programmes.

Conclusions

'Global warming is an issue the tourism industry must – and I think will – pay more and more attention to in the near future,' said P. Wall after the release of the IPCC report (in Tidwell 2001). The early signs of irreversible environmental change are already confirmed from various locations, and it has become clear that on a global level, the tourism industry is standing on a 'burning platform', that is, a situation that is untenable even in the short term. This chapter has pointed out the potential future use of marketing agendas as a framework to influence market behaviour towards more energy-conscious forms of consumption. Furthermore, marketing tools may assist nature-based destinations ravaged by environmental change in defining alternative product development scenarios, for instance, by addressing contemporary malaises of post-modern society.

References

Aall, C. and Høyer, K.G. (2005) 'Tourism and climate change adaptation: the Norwegian case', in C.M. Hall and J. Higham (eds) *Tourism, Recreation and Climate Change*. Clevedon: Channel View Publications.

ACIA (2004) *Impacts of a Warming Arctic*. Susan Joy Hassol (ed.). Arctic Climate Impact Assessment. Cambridge: Cambridge University Press.

Allon, F. (2000) 'Nostalgia unbound: illegibility and the synthetic excess of place', *Continuum: Journal of Media & Cultural Studies*, 14(3) 275–87.

Baum, T. and Hagen, L. (1999) 'Responses to seasonality: the experiences of peripheral destinations', *International Journal of Tourism Research*, 1: 299–312.

Blomgren, K. B. and Sørensen, A. (1998) 'Peripherality - factor or feature? Reflections on peripherality in tourism research', *Progress in Tourism and Hospitality Research* 4(4): 319–36.

Bürki R., Elsässer, H. and Abegg, B. (2003) 'Climate change – impacts on the tourism industry in mountain areas', Proceedings, 1st International Conference on Climate Change and Tourism, Djerba, 9–11 April.

Cohen, E. (1972) 'Towards a sociology of international tourism', *Social Research*, 39: 164–82.

Cohen, E. (1979) 'A phenomenology of tourist experiences', *Sociology*, 13: 179–201.

Dewar, K. (2005) 'Everyone talks about the weather...', in C.M. Hall and J. Higham (eds) *Tourism, Recreation and Climate Change*. Clevedon: Channel View Publications.

Faulkner, B. (2001) 'Towards a framework for tourism disaster management', *Tourism Management*, 22(2): 135–47.

Giles, A.R. and Perry, A.H (1998) 'The use of a temporal analogue to investigate the possible impact of projected global warming in the UK tourism industry', *Tourism Management*, 19(1): 75–80.

Goulding, C. (2001) 'Romancing the past: heritage visiting and the nostalgic consumer', *Psychology and Marketing*, 18(6): 565–92.

Hamilton, J.M. (2004) 'Climate and the Destination Choice of German Tourists'. FEEM Working Paper No. 21. (February).

Higham, J. and Hall, C.M. (2005) 'Making tourism sustainable: the real challenge of climate change', in C.M. Hall and J. Higham (eds) *Tourism, Recreation and Climate Change*. Clevedon: Channel View Publications.

Hughes, G. (2004) 'Tourism, sustainability and social theory', in A.A. Lew, C.M. Hall and A.M. Williams (eds) *A Companion to Tourism*. Malden, MA: Blackwell Publishing.

IPCC (2001) *Climate Change 2001: Impacts, Adaptation, and Vulnerability*. Intergovernmental Panel on Climate Change, 3rd Assessment Report. Cambridge: Cambridge University Press.

Johnston, M.E. and Viken, A. (1997) 'Tourism development in Greenland', *Annals of Tourism Research*, 24(4): 978–82.

MacCannell, D. (1976) *The tourist: A new theory of the leisure class*. New York: Schocken.

Maddison, D. (2001) 'In search of warmer climates? The impact of climate change on flows of British tourists', *Climatic Change*, 49:193–208.

Parry, M.L. (ed.) (2000) 'Assessment of the potential effects and adaptations for climate change in Europe', The Europe ACACIA Project, Jackson Environment Institute, University of East Anglia.

Plog, S. (1987) 'Understanding psychographics in tourism research', in: J. Ritchie and C. Goelder (eds) *Travel Tourism and Hospitality Research*. New York: Wiley, pp.203–14.

Poon, A. (1993) *Tourism, Technology and Competitive Strategies*. Oxford: CAB International.

Prentice, R (2004) 'Tourist familiarity and imagery', *Annals of Tourism Research*, 31(4) 923–45.

Saarinen, J. (2004) 'Tourism and touristic representations of nature', in A.A. Lew, C.M. Hall and A.M. Williams (eds) *A Companion to Tourism*. Malden, MA: Blackwell Publishing.

Saïd, E.W. (1979) *Orientalism*. New York: Vintage Books.

Schellhorn, M. and Perkins, H.C. (2004) 'The stuff of which dreams are made: representations of the south sea in German-language tourist brochures', *Current Issues in Tourism*, 7(2): 95–133.

Shaw, J. (2001) 'Winning territory: Changing place and change place', in: J. May and N. Thrift (eds) *Timespace: Geographies of Temporality*. London: Routledge, pp.120–32.

Stewart, R.E., Walters, L.C. Balint, P.J. and Desai, A. (2004) *Managing Wicked Environmental Problems: A Report to Jack Blackwell, Regional Forester*, USDA Forest Service, Pacific Southwest Region.

Tidwell, M. (2001) 'Glaciers are melting, islands are drowning, wildlife is vanishing', *The Washington Post*, 9 September.

TRI (Travel Research International) (2003) 'The interrelations between tourism and climate change. Report prepared for WTO in association with the CRU at UEA', *Proceedings of the First International Conference on Climate Change and Tourism*, Djerba, Tunisia, 9–11 April.

Viner, D. and Amelung, B. (2003) Climate change, the environment and tourism: the interactions. *Proceedings of the ESF-LESC Workshop in Milan: 'Climate Change, the Environment and Tourism: the Interactions'*. Norwich: ÉCLAT, Climatic Research Unit.

15 US ski industry adaptation to climate change

Hard, soft and policy strategies

Daniel Scott

Introduction

The reduction of anthropogenic greenhouse gases (GHG) has been the main climate change policy focus of the international community. However, it is increasingly recognised that, even if current agreements to limit GHG emissions are successfully implemented, atmospheric GHG concentrations will not stabilise and some climate change is inevitable. With this realisation, climate change adaptation has been gaining importance in the international literature and policy making community as a crucial response to reduce the risks and take advantage of any opportunities associated with climate change (Smit *et al.* 2000).

The tourism and outdoor recreation sector is inherently sensitive to climate conditions, yet our understanding of how climate variability affects the sector and how the sector has adapted to climate remains very limited (Wall 1992; Perry 1997; Scott 2005). The US National Academy of Sciences has indicated that the tourism and recreation sector is more vulnerable to climate change than most other sectors of the economy because of its close interlinkages with climate and the natural environment. Nonetheless, relative to other economic sectors (for example, agriculture and forestry), tourism has been largely neglected by the climate change impact research community. Wall (1998) noted that tourism was not mentioned in the IPCC first assessment report, which was completed in 1990, reflecting the paucity of research available. More recently, Agnew and Viner (2001) and Scott *et al.* (2002) have commented that research on climate change and tourism is still in its formative stages. Until very recently climate change also garnered minimal attention from the tourism industry and tourism research community. Surveying 66 national tourism and meteorological organisations around the world in 1989, Wall and Badke (1994) found most respondents (81 per cent) felt climate was important to their country's tourism industry, but few were aware of any specific research or existing publications related to climate change and tourism. A decade later, Butler and Jones (2001: 300) reached a similar conclusion, stating:

> [Climate change] could have greater effect on tomorrow's world and tourism and hospitality in particular than anything else we've discussed ... The most worrying aspect is that ... to all intents and purposes the tourism and hospitality

industries ... seem intent on ignoring what could be *the* major problem of the century. (original emphasis).

With research on climate change and tourism still in its formative stages, it is not surprising that adaptation is not a well-developed theme within this field. Discussion of climate change adaptation in the tourism sector is most advanced for the ski industry (see Elsasser and Bürki 2002; Scott *et al.* 2003) and will be the focus of this chapter as well. Although the chapter will mainly concentrate on potential climate change adaptation by the US ski industry, discussion will occasionally be extended to the broader North American ski industry.

Adaptation to climate change in the ski industry is a diverse topic. In order to advance our understanding of climate change adaptation in the ski industry it is necessary to assess what types of adaptations are possible, who are the agents involved and what is their decision-making process, which are the most feasible options that are likely to occur, and what barriers (if any) need to be overcome to facilitate implementation of these adaptation options. This chapter will focus on the first of these questions, with the objective of developing a comprehensive inventory of climate change adaptation options available to the ski industry, including climate adaptations currently in use and other initiatives used in other climate sensitive economic sectors that could be used by the ski industry in the future. This chapter does not attempt to evaluate the relative merit – institutional feasibility, economic efficiency, technical feasibility or environmental sustainability – of the adaptation options in the inventory or assess the likelihood of implementation. Addressing these questions is beyond the scope of this chapter, but they remain important areas for future inquiry.

This chapter will begin with a brief overview of the US ski industry and how it is currently affected by climate variability. The next section will identify the broad range of climate change adaptations available to ski industry and has been organised into hard, soft and policy options.

Climate sensitivity of the US ski industry

In 2001, the North American ski market represented just over 20 per cent of the global ski market with approximately 72 million skier visits (57 million in the USA and 15 million in Canada) (Lazard 2002). Although there is a perception that the North American ski industry has been in a period of stagnation since the late 1980s, the industry has reinvigorated itself and, despite having 100 fewer ski areas in operation in the period of 2000–04 than 1986–90, has averaged approximately four million more ski visits per year (56,527,000 in 2000–04 and 52,753,000 in 1986–90) (National Ski Areas Association 2004).

According to Packer (1998), the US snow sports industry generates approximately US$12 billion in the economy annually. Because the economic impact of the industry is concentrated in the regions where ski areas operate, the industry is a very important part of the economies of some communities. For example, the small state of Vermont accounted for 8 per cent of skier visits in the USA in 2000–01 and the ski industry is estimated to contribute US$1 billion to the state economy each year (Smith 2001).

Ski areas are located in climatically diverse regions of the USA, ranging from Maine in the east and California in the west and from New Mexico in the south to Alaska in the north. Inter-annual climate variability affects the ski industry in each of these regions differently, but the aspects of climate variability that affect the ski industry are mainly fluctuations in snow cover and temperature. Availability of snow is a prerequisite for the ski industry and the unreliability of adequate natural snowfall gave rise to the invention and widespread implementation of snowmaking systems across North America. The adaptation of snowmaking largely overcame the ski industry's sensitivity to a lack of precipitation. However, snowmaking does not overcome two other impacts related to variable snowfall. A lack of snow in urban centres is thought to have a detrimental effect on skier demand. In discussions with ski management professionals, this 'urban snow effect' has been observed in ski regions across Canada and the USA. Conversely, too much snow is a problem for the ski industry because it can disrupt transportation systems, preventing skiers from reaching ski areas, and it can increase the avalanche hazard, putting skiers and ski area infrastructure at risk. The other main climate attribute affecting the ski industry is temperature. Very cold temperatures reduce skier demand, while temperatures above freezing adversely affect snow conditions and prevent efficient snowmaking. In some areas, wind can also affect ski area operations by rendering ski lifts unsafe for operation.

As Figure 3.1 demonstrated (Chapter 3), inter-annual climate variability has a major impact on the length of operating seasons in the US ski industry. Even with widespread use of snowmaking, the difference in the length of the operating season in the winters of 1995–96 and 1996–97 (40 per cent) illustrates how sensitive the ski industry in the mid-west and south-east ski regions remain. Putting this inter-annual variability into a business context, consider if the number of days a retail store was open for business varied by 40 per cent from year to year, with no way to predict the length of the business year. The already difficult challenge of marketing, cash flow and personnel planning would be compounded tremendously. It is within this climate-sensitive business environment that the US ski industry has evolved and developed a range of adaptation strategies to reduce its exposure to the impacts of climate variability.

Climate change adaptation strategies

There are many types of potential climate change adaptations that could be undertaken by the ski industry, including technical, financial, operational management, behavioural and policy options. There are also a wide range of stakeholders that can influence the adaptation options available to the ski industry including: individual ski areas, ski industry associations, governments – at local, state/provincial, and national levels – the financial sector – banks, investors and insurance companies – skiers and tourists more broadly, nearby landowners and communities, and environmentalists. Each of these stakeholders has different, yet often interrelated, roles that will contribute to the collective response of the ski industry to climate change.

Table 15.1 Types of climate change adaptation options available to the ski industry

Hard technological developments

- Snowmaking systems and additives

- Slope development

- Energy and water systems

- Cloud seeding

- Improved climate prediction

- Artificial ski slopes

Soft business practices

- Ski area operations

- Market and revenue diversification

- Marketing and public education

- Industry consolidation and regional diversification

- Private insurance and weather derivatives

- Industry-wide income sharing programme

Government and industry policy

- Government environmental regulatory frameworks

- Government energy policies

- Government subsidies

- Ski industry climate change policy

The inventory of potential adaptation strategies in this chapter has been organised according to the three main categories of adaptations set out in this volume: hard technological developments, soft business practices, and government and industry policy (Table 15.1). These main adaptation categories are not mutually exclusive. For example, a water reservoir constructed to provide water supply for snowmaking is part of the snowmaking system (technological developments), but may also function as water supply for a golf course or boating (diversification of business), and would be governed by changes in environment regulatory frameworks (government policy). Specific examples of adaptation options from each of the three main categories are discussed in the remainder of this section.

Hard technological developments

Snowmaking

Snowmaking is the most widespread climate adaptation currently used by the ski industry in North America. Snowmaking technology was first implemented at the

Grossinger Resort (Fahnestock, New York) in 1952 and over the past 30 years has become an integral component of the ski industry in North America. Figure 15.1 shows the historical diffusion of snowmaking technology from 1974–75 to 2001– 02 in five ski regions of the USA. In the mid-1970s there was a very distinctive east–west regional pattern in the use of snowmaking technology. More than 80 per cent of ski areas in the mid-west region had snowmaking systems. Similarly, the majority of ski areas in the north-east and south-east regions had snowmaking in place. The use of snowmaking was much lower in the higher alpine ski regions of western USA, with less than 20 per cent of ski areas in the Rocky Mountain region and less than 10 per cent in the Pacific West region using snowmaking.

A series of winters with poor snow conditions in eastern USA during the late 1970s and early 1980s motivated ski areas in these regions to invest millions of dollars to improve their snowmaking capacity. For example, the president of a major ski area in Vermont stated: 'We learned one thing this winter [1979–80]: how to operate entirely on machine-made snow' (Robbins 1980: 81). A historian of the ski industry, commenting on the early development of snowmaking in the north-east USA, stated: '[the lack of natural snow] got so bad it couldn't be ignored. By 1981–82, it became obvious that if you wanted to stay in business, you had to have top-to-bottom snowmaking'(Hamilton *et al.* 2003: 65). Consequently,

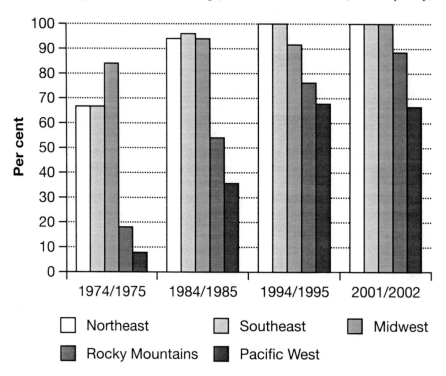

Figure 15.1   US ski areas with snowmaking systems

Sources: National Ski Area Association annual state of the ski industry reports

by the mid-1980s almost every ski area (greater than 93 per cent) in the south-east and north-east regions had snowmaking systems in place.

Over the past 20 years, the difference in the level of snowmaking between ski areas in eastern and western USA has been diminishing. Today, all ski areas in the north-east, south-east and mid-west ski regions use snowmaking, although the amount of skiable terrain covered by snowmaking in 2001–02 varied from 62 per cent in the north-east, to 95 per cent in the south-east, and 98 per cent in the mid-west. While the majority of ski areas in the Rocky Mountains (89 per cent) and Pacific West (66 per cent) regions now employ snowmaking, the proportion of skiable terrain covered by snowmaking remains much lower (13 per cent and 8 per cent respectively).

A similar east–west pattern for snowmaking implementation exists in Canada. In eastern Canada, 100 per cent of the skiable terrain in the Province of Ontario is covered with snowmaking, while in Quebec snowmaking coverage varies in the 50–90 per cent range of skiable terrain. The use of snowmaking is very different in the higher elevation alpine regions of western Canada, where only 50–75 per cent of ski areas have snowmaking systems and the proportion of skiable terrain is typically 25–50 per cent.

The importance of snowmaking as an adaptation to climate variability cannot be understated. In their analysis of six ski areas in eastern North America, Scott *et al.* (2005) found that snowmaking extended the average ski season from 55 to 106 days during the baseline period (1961–90) (Table 15.2). In addition to its current importance to the ski industry, Scott *et al.* (2005) documented the increased importance of snowmaking under projected climate change. Table 15.2 displays the average ski season in the 2020s under two climate change scenarios (low impact = NCARPCM-B2, high impact = CCSRNIES-A1). In almost every case the difference between the natural ski season and the ski season extended by snowmaking is greater than in the baseline period. In both the baseline period and the 2020s a financially viable ski season would not be possible at four of the six locations without snowmaking.

The development of sophisticated snowmaking systems that delivered reliable snow conditions had important synergies with other adaptation strategies. By providing a more predictable ski season length, snowmaking allowed ski areas to begin to diversify their business operations. With a reliable winter tourism product, investment in larger resorts became more feasible, as did the sale of recreational homes. Even as early as the late 1970s the relationship between reliable snow cover and resort development was noted by the president of the Sugarbush ski area (Vermont), who stated: '... the addition of a sports centre at the base of Sugarbush Valley and the construction of over 300 mountain side condominiums last summer were made possible because of the snowmaking' (Robbins 1980: 82). At a number of major ski resorts, the technological adaptation of snowmaking allowed for additional adaptations in business operations to take place, which further reduced the ski area's vulnerability to climate variability.

Scott *et al.* (2005) also found that the amount of snowmaking required to maintain ski seasons in their study areas could increase substantially under warmer

Table 15.2 Natural and snowmaking-enhanced ski seasons in eastern North America

Study area	Baseline (1961–90) days			Scenario	Climate change (2020s) days		
	Natural snow	With snowmaking	Difference		Natural snow	With snowmaking	Difference
Brighton, Michigan	6	115	+99	Low impact	1	109	+108
				High impact	0	83	+83
Orillia, Ontario	60	150	+90	Low impact	24	146	+122
				High impact	9	122	+113
Quebec City, Quebec	99	160	+61	Low impact	87	158	+71
				High impact	59	139	+80
Rutland, Vermont	13	119	+106	Low impact	4	113	+109
				High impact	1	89	+88
Ste. Agathe-des-Monts, Quebec	108	163	+55	Low impact	89	163	+74
				High impact	53	142	+89
Thunder Bay, Ontario	44	164	+120	Low impact	33	161	+128
				High impact	7	136	+129

Source: based on modelling outputs of Scott et al. 2005

climate change scenarios. For example, under the highest impact 2050s scenario (CCSRNIES-A1), snowmaking requirements were projected to more than double at four of the six ski areas.

The potential for large increases in snowmaking requirements raises interesting questions about the sustainability of this adaptation strategy in some locations. There have been concerns in several communities about the environmental impacts of water withdrawals for snowmaking. In most situations, the concern is that water withdrawals from natural water bodies (streams and lakes) can lower water levels at critical times during the winter, impacting fish and other marine species. Responding to this issue, the State of Vermont implemented some of the strictest water withdrawal rules for ski areas in the USA in 1996. The State of Vermont established a 'February Mean Flow' standard, which means that water withdrawals for snowmaking are not permitted when water supply in natural water courses are at or below the average mean flow. Ski areas with water withdrawal permits must comply with FMF standard and show that they will continue to meet this standard if applying to expand snowmaking capabilities. Studies of snowmaking impacts on fish populations by the Vermont Agency of Natural Resources found that if the FMF standard was complied with, no loss of fish populations above or below withdrawal sites occurred (Ski Vermont 2004). If snowmaking requirements by ski areas in Vermont double, as projected under a high impact 2050s scenario (Scott *et al.* 2005), the ability of ski areas to obtain adequate water supply for snowmaking and maintain the FMF standard may be severely tested. Ski areas may increasingly need to construct water reservoirs to address this challenge. For example, the Okemo ski area in Vermont built the largest snowmaking pond in New England (70 million gallons capacity), which enables them to 'stockpile' water from the Black River during the spring and summer, thus having the necessary water available during the winter.

Without adequate access to water resources for snowmaking, the adaptive capacity of ski areas to climate change will be dramatically reduced, forcing some ski areas out of business. Water supply is already a limiting factor for some ski areas. For example, the owner of the Ski Sunrise resort in Wrightwood, California (since 1942) recently tried to sell the resort and has been unable to because the ski area does not have snowmaking capabilities. The cost of installing a snowmaking system is not the principal barrier – it is the inability to obtain the necessary water-use permits that prevents the ski area from increasing its adaptive capacity. Without the adaptive capacity provided by snowmaking, the resort has had marginal business during most of the warmer than normal winters from 1998 to 2002 and the owner has reduced the sale price of the resort by 50 per cent since it first went on the market in 1996 (Ramage 2003).

A second aspect of sustainability raised by potentially large increases in snowmaking requirements is energy use and associated costs. Energy requirements for snowmaking are high and are inversely related to temperature – the warmer the temperature the more energy is required to make snow. Consequently, the impact of climate change on energy use (and costs) for snowmaking is two-fold: a greater volume of snow will need to be made and on average snow will need to be made at warmer temperatures.

The Sunday River ski area in the State of Maine will be used to explore the potential impacts of climate change on snowmaking-related GHG emissions. Sunday River ski area (957 metres above sea level (masl)) has 663 acres of skiable terrain, 92 per cent of which is covered by snowmaking. The ski area uses approximately 26 million kilowatt-hours of electricity in an average year (Hoffman 1998). The US Department of Energy (2002) estimated that for each kilowatt-hour of electricity used from the State of Maine grid, 0.85lbs of CO_2 emissions are produced. Current average levels of snowmaking at Sunday River generate an estimated 10,022 tonnes of CO_2 emissions annually (or about 0.019 tonnes per skier visit). This is the equivalent to the emissions from the electricity use of 3,600 average households in the State of Maine. A doubling in required snowmaking would more than double the related emissions unless the Maine electricity grid developed renewable energy sources or the ski area substantially improved the energy efficiency of its snowmaking system. Ironically, these increased CO_2 emissions would contribute to the process of global climate change for which this adaptation is increasingly required.

Snowmaking technology has been constantly improving since it was first developed and will continue to do so in the decades before the projected climate change scenarios outlined above may be realised. Continued improvements in snowmaking towers, water and air distribution systems, and computer control systems will improve the water, energy and cost efficiency of snowmaking. The development of new snowmaking additives – natural substances that act as nucleators to increase the nucleation temperature at which water droplets begin to form ice particles – will also allow ski areas to make snow more efficiently at warmer temperatures in the future. Although it is not currently possible to incorporate these technological advancements into an assessment of the sustainability of snowmaking as a climate change adaptation, some efficiency improvement is a certainty and the aforementioned changes in water and energy use are likely a worst-case situation.

Slope development

Slope development can be used to adapt to climate variability and change in a number of ways. Landscape planning can be used to preserve snow cover and lengthen the ski season by developing at least some north facing ski slopes and preserving forest cover to partially shade ski slopes. Land contouring or smoothening ski slopes can be used to reduce the amount of snow a ski area requires to operate, thereby extending the ski season and, where applicable, reducing the amount of snowmaking needed. Land contouring could also be used in conjunction with water reservoir development as part of a comprehensive water resource management plan to capture on-site precipitation and snow melt for snowmaking purposes.

Another form of slope development used to adapt to climate is the expansion of ski areas into higher elevations where snow cover is generally more reliable and a longer ski season is available. This appears to be the principal climate change adaptation strategy being considered in Switzerland (König and Abegg 1997; Elsasser and Bürki 2002) and Austria (Breiling and Charamza 1999). Elsasser and Bürki

(2002) noted that while Swiss ski area operators have tended to discount the threat posed by climate change, they have already used climate change as a justification for plans to expand ski areas to higher elevations. For example, 36 ski areas in Austria have applied for permits to expand into higher elevations in 2002–03 (Tommasini 2003 – personal communication). Citizens in both nations do not generally favour development in sensitive high-mountain environments and public opposition may pose a significant political barrier to this adaptation strategy.

There is limited potential to expand existing ski areas in eastern North America into higher elevations, but it is possible that lower elevation ski areas could be abandoned and more new high elevation ski areas constructed. Upslope expansion has more potential as a climate change adaptation in the western ski regions of North America but, as in Europe, may face significant opposition from environmental groups. A project to develop a world class four-season ski resort on Jumbo Glacier in southern British Columbia has been held up since 1991 by opposition from environmental groups and local residents (Greenwood 2004). In 1998, the radical environmental group Earth Liberation Front caused over US$12 million in fire damage at Vail Resort (Colorado) in response to the start of forest clearing for a 1,000 acre expansion of the ski area (Faust 1998).

Cloud seeding

Cloud seeding is a weather modification technology that has been used in an attempt to produce additional precipitation for agricultural regions or to reduce the potential of large hailstone development that could damage crops and property in urban areas. Scientific evidence that this weather modification technology makes a significant difference in precipitation is controversial and although there are strong proponents of the technology it is not widely used. Nonetheless, ski areas in North America and Australia have used this technology in an attempt to increase snowfall. Vail Ski Resorts (Colorado) paid US$170,000 during the 2003–04 ski season to have clouds seeded with silver iodide in order to increase snowfall. In 2004, the New South Wales State government approved a US$15 million, five-year cloud seeding project that will use ground-based generators – instead of the usual airplane based distribution – to scatter silver iodide into passing clouds in the hopes of increasing snowfall by 10 per cent in Australia's Snowy Mountains. Snowfall in the region had been declining by approximately 1 per cent per year over the past 50 years and a New South Wales State government report warned that projected climate change could have a devastating impact on the ski industry in the region (Hennessy *et al.* 2003). The objective of this cloud seeding project is to keep ski areas viable under a changing climate and also to capture the runoff from additional snowfall for hydroelectric generation and crop irrigation in the summer.

Weather and climate prediction

Another type of technological advance that would have important implications for adaptation in the ski industry is improved weather and climate forecasting abilities. Ski areas use daily and weekly weather predictions for operating decisions such as when to produce snow and levels of staffing. Improved seasonal forecasts could become useful for key business decisions, including when to plan to open the ski hill – based on snowfall predictions at the ski area and centres of skier demand – and whether to purchase weather insurance. Greatly improved models for long-term climate change projections could also reach a level of practical utility by ski area operators and ski corporations. Information on probable long-term changes in temperature and precipitation could be used for risk assessments and strategic business decisions, including which ski area properties to acquire or sell, the level of investment in a ski resort or snowmaking system, and the amount of water rights to acquire.

Artificial ski slopes

Artificial, non-snow ski surfaces were first developed in the 1970s. This technology has improved over the past three decades and today offers surfaces with good gliding and edging properties that will not damage ski equipment. This technology still has very limited application in ski areas of North America and is generally only found in sport or theme parks – both indoor and outdoor.

In the future, if the cost of snowmaking becomes prohibitive or the ability to make snow is no longer feasible early in the ski season, some ski areas may adopt this technology for use in high traffic areas (for example, lift lines and exit areas) or for teaching areas, snowboarding parks and snowtube-toboggan runs. This technology may prove to be particularly attractive to smaller ski areas near large urban markets that would otherwise have difficulties maintaining sufficient snow for an economically viable ski season. It remains to be seen if this technology would be economical in large-scale applications, such as an entire ski run, and whether skiers would respond positively to such applications. Market research would be needed to evaluate skier perceptions of the potential applications of this technology. Would skiers go to a nearby ski area where the snowboarding park was open year round? If given the choice would skiers travel four hours to reach a ski area with snow or one hour to a ski area with an artificial, non-snow surface?

Another technology that could provide much the same adaptive capacity as non-snow ski surfaces for smaller applications would be the underground refrigeration of small slope areas (for example, snowboarding parks and snowtube-toboggan runs) to preserve the snowpack. The technical and economical viability of such an adaptation has yet to be examined, but the capital and operating costs would likely limit it to the niche applications identified above.

Soft business practices

Operational changes

Climate change may alter the ways that ski area managers have traditionally operated ski areas. Changes in the timing of business operations are one possible climate adaptation. For example, as the cost of snowmaking increases and the opportunities to make snow in the early part of the season diminish in the future, ski areas that today endeavour to open in October or November, may have to be content with opening in early December. Other ski areas, which are traditionally open for the Thanksgiving holiday in the USA (the last week of November), may have to adjust their business operations to open for the Christmas–New Year holiday instead. Ski areas may also have to adapt their operations to function with less snow. For example, very rocky slopes that require greater amounts of snow to operate safely may need to be abandoned or smoothed. Ski areas that traditionally waited until they had a 60cm base before opening may have to adjust their operations to open with a 30cm base or even less.

Ski areas may also attempt to increase the intensity of their operations by either raising their lift capacity or closing some of their skiable terrain – either permanently or seasonally when sufficient natural snow is not available – to reduce operating costs. The number of skier visits each season is critical to the profitability of ski area operations. If the utilisation level of ski areas can be increased so that the same number of skier visits can be accomplished in a shorter period of time, then any adverse impact of a shortened ski season may be negated.

Marketing and public education

The importance of snow conditions for skier satisfaction was emphasised by the research of Carmichael (1996), who found that snow condition was by far the key attribute in tourist image and destination choice for winter sports holidays. Reporting current snow conditions is a crucial marketing strategy for ski areas and has been a source of tension between ski areas and the media. Although the ski industry generally has a better relationship with the media today than 10 or 20 years ago, almost all of the ski industry representatives the author has communicated with have concerns about how the media represents (or misrepresents) snow conditions. This was particularly troublesome when snowmaking technology was being implemented into the ski industry. The media would report on the lack of natural snow or focus on ski areas that did not have snowmaking and were not open, to the detriment of the ski areas that did have snowmaking and were open for business. During the 1980s and 1990s many partnerships were formed between the ski industry and the media to address this situation. In the Province of Ontario, for example, the government ministry responsible for tourism began a programme where ski areas would report their snow conditions and operational status to a central tourism office, which would then provide this 'official ski report' to media outlets across the province and other regional market areas. The development of the internet provided ski areas with a marketing approach to overcome any perceived or real misrepresentation by the media. By placing pictures from live 'snow cams' on a website, ski areas had direct marketing access to skiers and could provide a non-biased report of current snow conditions.

Another marketing strategy used by ski areas to attract skiers when there has been low snowfall has been to issue snow guarantees. In the winter of 1999–2000, the American Ski Company promised visitors to its six New England resorts a 25 per cent reduction on their next vacation if the ski area failed to open 70 per cent of its ski runs during the Christmas–New Year holiday period. Warm temperatures made snowmaking difficult and the company had to pay out the rebates to customers at three of the six resorts (Keates 2000).

One of the few climate change specific adaptations initiated by the US ski industry thus far is the 'Keep Winter Cool' public education campaign. In partnership with the Natural Resources Defence Council (a leading environmental organization) the National Ski Area Association (NSAA) in the USA has run the campaign in February 2003 and 2004. The objective of the campaign is to combat climate change through public education and GHG emission reductions within the ski industry. The public outreach component of the campaign discussed the potential impacts of global warming on winter recreation and encouraged guests to do their part in reducing GHG emissions. The 'Keep Winter Cool' campaign also showcased the various initiatives ski areas have taken to reduce GHG emissions in their operations.

Business diversification

Business diversification is an important climate adaptation strategy for ski areas, which has been driven by the high seasonality of the tourism sector, inter-annual climate variability and other non-climatic business stimuli. Over the past 20 years, most ski areas in North America have diversified their winter operations and diversified their recreation market beyond traditional skiers. Others have also diversified their operations into other non-skiing related business lines that enable the resort to operate year-round.

The Economist (1998) referred to the transition of major ski resorts in North America from ski areas to winter theme parks as the 'Disneyfication' of the winter sports industry. By broadening their operations into accommodations, skiing and snowboarding lessons and retail sales, ski resorts were able to capture more of the profits generated by skiing activities. Table 15.3 illustrates how typical ski area revenue sources have changed over the past 25 years. In 1974–75, lift tickets represented almost 80 per cent of revenues for the average ski area in the USA. The dependence on lift ticket sales has been reduced almost by half, while other forms of revenue – food and beverages, lessons, accommodations and retail – have each increased substantially. If only larger, more diversified ski areas were considered in this comparison, this tremendous change in revenue sources would be even more pronounced.

Though not included in Table 15.3, real estate development and management has also become an increasingly important source of non-skiing revenue for larger ski resorts. Many larger ski areas have profited from the sale of condominiums and other real estate as well as the management of these properties on behalf of their owners – helping owners rent out the properties when they are not using them. During the sales phase of new real estate developments, real estate has often replaced resort operations as the prime source of revenue at some ski areas.

The past 15 years have also seen a diversification trend in visitors to ski areas and the activities they engage in. One of the most significant new markets to develop for ski areas has been the tremendous growth in the popularity of snowboarding. In the late 1980s and early 1990s, a number of ski areas banned snowboards. According to

Table 15.3 Ski area revenue sources

Revenue sources	1974–75 %	2001–2 %
Lift tickets	79.0	47.4
Food and beverages	2.8	14.1
Lessons	2.8	9.8
Accommodations/lodging	1.8	9.4
Other	2.1	6.1
Retail	0	5.5
Rentals	4.5	5.3
Property operations	5.1	1.2

Source: National Ski Area Association annual state of the ski industry reports

the NSAA, only three ski areas in the USA continue this policy. Most ski areas embraced this growing market and the proportion of total skier visits in the USA by snowboarders has increased from 5 per cent in 1990–91 to 29 per cent in 2001–02 (National Ski Area Association 2004).

Ski areas also did not ignore the potential market that non-skiers represented. Williams and Dossa (1990) estimated that 20–30 per cent of visitors to ski resorts in Canada did not ski during their visit. A similar pattern was found for visitors to winter resorts in France (Cockerell 1994) and Switzerland (Wickers 1994). To capture the potential of the non-skier market, many ski resorts made substantive investments to provide alternate activities for non-skiing visitors – including snowmobiling, skating rinks, dog sled-ride, indoor pools, health and wellness spas, gym and fitness centres, squash and tennis courts, games rooms, restaurants and retail operations.

Another very important and common climate adaptation strategy aimed principally at overcoming the business challenges imposed by tourism seasonality has been the transformation of ski resorts into four-season resort operations. This transformation included several ski resorts removing 'ski' from their name and referring to themselves as four-season resorts. Diversification to four-season operation required ski resorts to provide warm-weather season recreational activities like golf, boating and white-water rafting, mountain biking, paragliding, horse riding and other business lines.[1]

Ski conglomerates and regional diversification

Another major trend in the North American ski industry over the past decade has been the consolidation of a substantial proportion of the industry into ski

1 As Scott *et al.* (2001) noted, because many ski areas in North America now operate as four-season resorts, any climate change assessment should also examine the implications of climate change on warm-weather operations.

conglomerates like American Skiing Company, Intrawest and Vail Resorts. Each of these conglomerates owns several ski areas (Table 15.4) and has become geographically diversified in its operations. Although not a planned climate adaptation, this business strategy has reduced the vulnerability of the larger corporation, and therefore each individual ski area, to the regional effects of climate variability. The probability of a winter with poor snow conditions and negative economic impacts in one ski region of North America is much higher than for the multiple ski regions in which these conglomerates now operate. In other words, the economic consequences of a poor ski season at a resort in one ski region (for example, the mid-west) could be buffered by the average or better than average economic performances of ski areas in other regions (for example, Rocky Mountains, Quebec or British Columbia). This regional diversification, together with greater access to capital for technological and business diversification adaptations, has made these multi-resort conglomerates less vulnerable to the impacts of climate change than individual ski areas or ski companies that own multiple ski resorts in only one region (for example, Powdr Corp, Aspen Skiing Company, Alpine Valley Holding Company or Peak Resorts). A future adaptation strategy by individual ski areas or companies with multiple ski resorts in one region would be to join one of the above regionally diversified conglomerates or examine establishing a co-operative business programme that would provide the benefits of regional diversification (for example, a ski industry income stabilisation programme).

Table 15.4 North American ski conglomerates

American Skiing Company		Intrawest	
Mount Snow	Vermont	Whistler Blackcomb	British Columbia
Killington/Pico	Vermont	Panorama	British Columbia
Sunday River	Maine	Tremblant	Quebec
Sugarloaf	Maine	Blue Mountain	Ontario
Attitash Bear Peak	New Hampshire	Mammoth Mountain	California
The Canyons	Utah	Snowshoe/Silver Creek	West Virginia
Steamboat	Colorado	Stratton	Vermont
		Copper Mountain	Colorado
		Vernon Mountain Creek	New Jersey
Vail Resorts		Booth Creek Inc.	
Beaver Creek	Colorado	Loon Mountain	New Hampshire
Breckenridge	Colorado	Cranmore Mountain	New Hampshire
Keystone	Colorado	Waterville Valley	New Hampshire
Vail	Colorado	Northstar-at-Tahoe	California
Heavenly	California/Nevada	Sierra-at-Tahoe	California
		The Summit	Washington

Ski industry income stabilisation programme

The ski industry is a very competitive business and, although precedents for co-operative initiatives within the industry exist (for example, government lobbying, marketing, environmental standards), there is no tradition of economic co-operation for the common good as there is in some other economic sectors. If future research shows climate change to be a significant risk to the majority of the US ski industry, the potential for economic co-operation within the industry might change. Broad industry participation in an income stabilisation programme would have the potential to spread the climate risk exposure of individual ski areas and reduce their vulnerability to climate change. The ski industry income stabilisation programme would be a voluntary, perhaps government subsidised, savings programme, from which ski areas could draw when their income falls below a threshold value because of adverse climatic conditions. With broad regional participation, an income stabilisation programme would provide individual ski areas with similar benefits to joining a regionally diversified ski conglomerate.

Weather derivatives and insurance

Despite its growth to economic prominence in the past 50 years, the tourism sector does not have the lobby power and tradition of government support of other economic sectors (for example, agriculture, forestry and mining). Consequently, tourism operators do not generally benefit from government subsidised weather insurance – although disaster relief funds are sometimes available. Instead any climate-related insurance for the ski industry would be a private sector initiative. Even private weather insurance for the tourism and recreation sector is a fairly recent development.

Snow insurance and weather derivative products have been made available to the ski industry by companies such as Société Générale SA and Goldman Sachs, and through the Chicago Weather Derivatives Exchange. During the 1999–2000 ski season, Vail Resorts in the State of Colorado bought snow insurance that paid the resort US\$13.9 million when low snowfall affected skier visits (Bloomberg News 2004). Insurance premiums have increased substantially in the past five years and major ski corporations like Intrawest and Vail Resorts have decided not to buy insurance because of the high cost. Interestingly, Vail Resort has instead opted to pay US\$170,000 in the 2003–04 ski season for unproven cloud seeding technology.

If larger ski companies like Intrawest and Vail Resorts find the current cost of weather insurance prohibitive, the cost is sure to exclude the small to medium size ski enterprises that are at greatest risk from climate change. A co-ordinated initiative by the ski industry to increase participation in weather insurance and thereby reduce premiums for the entire industry is an adaptation strategy that should be further examined. In the future, perhaps the NSAA (or even a larger international ski organisation) may be able to negotiate with a preferred insurance provider and governments where winter tourism is important to the economy to offer affordable weather insurance to ski operators.

Contraction

The alternative to a co-ordinated, co-operative ski industry approach to promote the financial sustainability of its members in an era of climate change would be the continuation of the current competitive business environment that would ensure what some might see as a 'healthy' contraction of the industry. A contraction of the US ski industry, in terms of operational ski areas, has been underway for the past two decades. The total number of ski areas in the USA has declined from 735 in 1983 to 494 in 2003 (Figure 15.2). Regionally, the New England Lost Ski Areas Project (2004) lists over 550 ski areas that have been 'lost' over the last four decades in the New England region and an additional 50 in other regions of the USA. In their analysis of the contraction of the ski industry in the State of New Hampshire (New England Region), Hamilton *et al.* (2003: 68) indicated that the 'extinction of the small [ski areas], and concentration of the industry into a few high-investment, high-elevation northern areas, was driven partly by a changing climate.'

Analysis of the impact of climate change on the New England ski area revealed that some of the most vulnerable ski areas were lower lying, more southerly ski areas near the major urban markets of Boston and New York (Scott 2004). While these ski areas are not major contributors to the ski industry in terms of overall skier visits, they play an important market development role as ski areas where people in major urban markets learn to ski, refine their skills and build the desire to visit the larger, more challenging ski areas in Vermont, New Hampshire, Quebec and Western North America. The ski industry must evaluate more closely the role these 'feeder hills' play in the development of skier demand and carefully assess the potential implications of their potential loss as a result of climate change. Can the ski industry get by without these feeder hills or can it accomplish market development in other ways? If the answer is no, then an unplanned, market-driven contraction may not be the best strategy for the ski industry and the ski industry may need to collectively devise a strategy to ensure that key feeder hills remain operational. The aforementioned income stabilisation programme or other direct economic support of these feeder hills by the ski industry (for example, snowmaking system improvements and training, emergency financial support) are some potential means of accomplishing this co-operative industry response to climate change.

Government and Industry Policy

Government Policy

Unlike other economic sectors (agriculture, forestry) where government policy has a direct role in climate change adaptation (for example, through subsidies, public insurance, marketing boards or research), government policy has a more limited role in the ski industry. The policy responses of government to climate change will not be inconsequential, however, because policy changes will facilitate or place constraints on the adaptation options available to the ski industry.

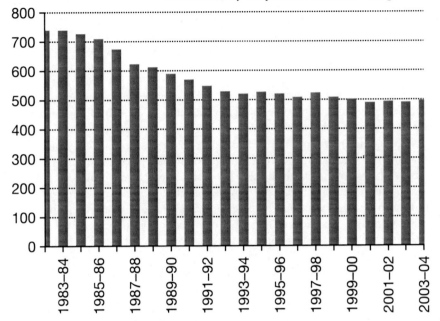

Figure 15.2 Number of ski areas operating in the USA (1983–2003)

Sources: National Ski Area Association annual state of the ski industry reports

Government resource management policies will affect land use, water use and energy prices, with potentially important implications for the ski industry. Land-use policy and environmental assessment processes will determine whether ski areas can expand into higher elevations. Water resource policy and programmes will determine whether ski areas can obtain the water withdrawal permits necessary for snowmaking. Energy policy, including climate change mitigation policies, will influence energy prices strongly, with implications for the operating costs of ski areas and, perhaps more fundamentally, tourism demand. The later point is critical for the tourism industry and has been discussed in more detail elsewhere in this volume. Its implications for the ski industry provide an interesting example of maladaptation to climate change.

Air travel, which is necessary for long-haul tourism, is the most GHG-intensive mode of transportation. Currently, the United Nations Framework Convention on Climate Change does not cover GHG emissions from international aviation. This is unlikely to remain the case given the substantive post-Kyoto Protocol GHG emission reductions required to achieve atmospheric CO_2 stabilisation. Some nations are already beginning to investigate options to reduce aviation-related GHG emissions. The UK Department of Transport (2003) has examined the use of economic instruments to reduce the growth in UK air passengers by 10 per cent before 2030.

The application of post-Kyoto GHG emission reduction strategies on air travel would have a greater impact on international tourists. Some ski areas in North America have concentrated on increasing their international markets – mainly from Western Europe and Japan – in recent years; the tourism market that is potentially the most vulnerable to post-Kyoto Protocol energy policy changes. Adaptation by ski areas to this energy policy uncertainty would focus market development at the local, regional or national level, where tourists will be less sensitive to increased travel costs.

Other than discounted electricity prices, long-term leases of public lands and infrastructure grants – subsidies that are available to other business sectors – governments in North America have not generally provided direct subsidies to the ski industry. In the future, in order to help sustain this economically important winter tourism sector under climate change conditions, governments may need to consider the use of direct subsidies. In the past, some state and provincial government-owned electricity providers provided electricity to ski areas for off-peak snowmaking at highly discounted prices. Although the level of electricity discount to the ski industry has declined in most North American jurisdictions over the past five to ten years, this policy could be used again by governments to help the ski industry adapt to climate change.

Including the tourism sector in disaster relief programmes is another specific example of a government policy change that would enhance the adaptive capacity of the tourism sector and the ski industry by decreasing the risk of climate-related income loss. When drought occurs in farm regions, there are often relief programmes available to assist the affected communities and businesses. Tourism is becoming increasingly important to the economies of many communities and a snow drought or warm winter temperatures that affect winter tourism can have an equally adverse impact on communities with tourism-based economies. Under projected climate change scenarios there may be some years when winter tourism becomes impossible and it is during such extreme situations that income loss compensation could be made available to tourism areas as it is for other climate sensitive economic sectors and communities.

Ski industry policy and political lobby

Writing in the industry journal *Ski Area Management*, Best (2003) observed that the ski industry has begun to acknowledge its vulnerability to a climate that is shifting in the wrong direction for ski operations and the need to confront climate change. 'This is a remarkable turnaround for an industry that just five or six years ago had largely shrugged off global warming' (Best 2003: 57).

A large proportion of ski areas in the USA adopted an Environmental Charter ('Sustainable Slopes') in 2000 to address environmental issues relevant to the ski industry. The charter identified global climate change as a key environmental issue and the climate change policy developed by the NSAA is outlined in Table 15.5. In support of its climate change policy, 65 ski areas lobbied government to increase political support for the proposed Climate Stewardship Act in the USA.

The only individual ski company and one of the very few tourism operators in the world to adopt a climate change policy is the Aspen Skiing Company. In 2001, the company indicated that it shared the concern of the scientific community about the

Table 15.5 National Ski Areas Association policy on climate change

To collectively address the long-term challenges presented by climate change and continue our commitment to stewardship under the Sustainable Slopes program, we hereby adopt this climate change policy. Through this policy, we aim to raise awareness of the potential impacts of climate change on our weather-dependant business and the winter recreation experience; reduce our own greenhouse gas emissions; and encourage others to take action as well. We are committed to working toward solutions that will keep both the environment and economy healthy and preserve quality of life. To this end, we will take the following actions:

- Educate the public and resort guests about the dependence of winter sports on natural ecosystems and the potential impacts of climate change on the winter recreation experience; educate guests on how they can help reduce GHG emissions
- Raise policy maker awareness of the dependence of winter sports on natural ecosystems and the potential impacts of climate change on the winter recreation experience
- Advocate the national reduction of GHG emissions through legislative, regulatory or voluntary measures
- Support sound, science-based solutions to climate change, including the use of renewable energy technologies
- Partner with appropriate organizations and agencies to assess opportunities to reduce resort emissions and increase energy efficiency; invest in new, more efficient products, practices and technologies; and measure our emission reductions.

linkages between increasing atmospheric CO_2 levels, increasing global temperatures and fossil fuel use. The company adopted two policy statements: (1) Aspen Skiing Company acknowledges that climate change is of serious concern to the ski industry and to the environment; and (2) Aspen Skiing Company believes that a proactive approach is the most sensible method of addressing climate change. More importantly, unlike the NSAA climate change policy, the Aspen Skiing Company established a climate change action plan, committing to the following:

- use of green development principles in new Aspen Skiing Company developments
- energy efficiency in old buildings through economically viable retrofits
- continued support of mass transportation and local employee housing
- annual accounting of GHG emissions
- a 10 per cent reduction in GHG emissions by 2010 based on a 1999 baseline.

The inclusion of annual monitoring of GHGs and an emission reduction target proportionally larger than that required of industrial nations (5 per cent) by the Kyoto Protocol, are particularly noteworthy.

Conclusion

The inventory of climate change adaptation options provided in this chapter is based mainly on the experience of the ski industry in North America. It is therefore considered a starting point only; to which it is hoped future contributions from

Europe, Japan, Australia and other nations will be added to provide a comprehensive overview of climate change options available to the international ski industry. The preceding inventory also provides an overview of historical trends in climate adaptation and offers some insight into how adaptation to climate change might proceed over the next 25 years.

The place- and context-specificity of each ski area means that the most beneficial combination of adaptation options varies for individual ski resorts. There is no single adaptation strategy that is optimal for all ski areas, in all ski regions and in all countries. There are clear geographic patterns in adaptation as the differential use of snowmaking across the major US ski regions illustrated (Figure 15.2). While virtually all skiable terrain in the ski regions in eastern USA and Canada is covered with snowmaking, the proportion of skiable terrain covered by snowmaking remains much lower in western North America (approximately 10–15 per cent) and even lower in Europe (estimated 3.6 per cent in Switzerland – Theus 1995). Research to better understand these regional patterns and their implications for future vulnerability to climate change is needed. The hundreds of millions of dollars invested by the North American ski industry in snowmaking over the past 25 years has paid big economic dividends for the industry and a comparison of recent studies in Europe (König and Abegg 1997; Breiling and Charamza 1999; Elsasser and Bürki 2002) and North America (Scott *et al.* 2003, 2005) appears to suggest that this investment in adaptation has made the ski industry of North America less vulnerable to climate change. Developing methods to assess this question of relative risk will be a significant step forward for researchers, but of course is also of great practical interest to tourism investors.

Taking into consideration the substantial uncertainty of climate change projections, the absence (until very recently) of credible climate change impacts assessments to inform the US ski industry of the potential magnitude of risk posed by climate change, and equally great uncertainty in the range of other non-climatic factors that will affect the business context in which the ski industry will operate in the 2020s or 2050s, it is not surprising that the US ski industry has not engaged in discussions about a co-ordinated climate change adaptation strategy for the industry. Instead, the ski industry has focused its climate change response on the development of public education programmes and GHG mitigation. This is not to exclude the possibility that forward-thinking stakeholders in the ski industry are not discussing climate change adaptation strategies. Currently, the potential for the development of an anticipatory, industry co-ordinated adaptation strategy for climate change appears very limited. Climate change adaptation by the US ski industry in the near-term is likely to be individualistic, reactive and incremental, focusing on 'no regrets' adaptation options – initiatives that enhance economic or environmental sustainability regardless of whether climate change occurs – that are modifications of current climate adaptations.

Without a substantive change in the current climate adaptation strategies employed by the US ski industry, the most likely scenario over the next 25 years, if climate change projections are accurate, is a continuation of the historical contraction in the number of operational ski areas. Twentieth-century climate change contributed to the loss of over 600 ski areas in the USA. The projected warming trend over the first five decades of the twenty-first century is also likely to have a pronounced impact, with the adaptive capacity of remaining ski areas determining the survivors. Ski areas with low adaptive capacity – low elevation ski areas in climatically marginal regions, that have less diversified revenue sources, and have less access to capital for business diversification and snowmaking technology, and fewer financial reserves to get them through poor business conditions – are likely to be put out of business by a combination of reduced average ski seasons and higher costs associated with snowmaking. Ski areas that have greater adaptive capacity – larger, well diversified resort operations with leading edge snowmaking systems, that are part of a larger company that can provide financial support if needed during poor business conditions – will be able to take advantage of this changing business environment. Under such a scenario it is conceivable that 30 years from now, a smaller US ski industry could be largely dominated by 5–10 large regionally diverse ski conglomerates which have a greater capacity to withstand a series of poor ski seasons in any ski region.

Innovative research is required to advance our understanding of the vulnerability of the global ski industry to climate change. Future inquiry into the adaptation decision-making process of key stakeholders, their perceptions of the relative risk posed by climate change, and the identification of barriers to the implementation of adaptation options would be important contributions to the tourism literature. This future research agenda can only be accomplished by greater collaboration between the climate change research community and ski industry stakeholders. This co-operation is only likely to occur once the ski industry believes that the risk of climate change is real enough to warrant consideration of a long-term adaptation strategy. Alternatively, the time line for this research agenda may be forced upon the ski industry by investors wary of the risk climate change poses. Considering recent trends in the financial sector, where large investors are requesting companies to examine their risk exposure to climate change and banks are incorporating climate change into credit risk assessments (Innovest Strategic Value Advisors 2003), pressure by investors for ski areas to assess the implications of climate change on their business may occur in the next 10 years.

Acknowledgements

The author is grateful to the Government of Canada's Climate Change Action Fund for partial financial support of this research and to all of the individual ski area managers in Canada and the USA who shared their time and insights on a wide range of issues relevant to their industry.

References

Agnew, M. and Viner, D. (2001) 'Potential impact of climate change on international tourism', *Tourism and Hospitality Research*, 3: 37–60.
Best, A. (2003) 'Is it getting hot in here?', *Ski Area Management*, May: 57–76.

Bloomberg News (2004) 'Operator betting on nature instead of snow insurance', *Financial Post*, 12 January: FP5.

Breiling, M. and Charamza, P. (1999) 'The impact of global warming on winter tourism and skiing: a regionalized model for Austrian snow conditions', *Regional Environmental Change*, 1(1): 4–14.

Butler, R. and Jones, P. (2001) 'Conclusions – problems, challenges and solutions', in A. Lockwood and S. Medlik (eds.) *Tourism and Hospitality in the 21st Century*, Oxford: Butterworth-Heinemann, pp.296–309.

Carmichael, B. (1996) 'Conjoint analysis of downhill skiers used to improve data collection for market segmentation', *Journal of Travel and Tourism Marketing*, 5(3): 187–206.

Cockerell, N. (1994) 'Market segments: the international ski market in Europe', *EIU Travel and Tourism Analyst*, 3: 34–55.

Elsasser, H. and Bürki, R. (2002) 'Climate change as a threat to tourism in the Alps', *Climate Research*, 20: 253–7.

Faust, J. (1998) 'Cloaks of secrecy surround activist group Earth Liberation', online. Available at: http://more.abcnews.go.com/sections/us/DailyNews/elf981022.html (accessed 22 September 2004).

Greenwood, J. (2004) 'Investors wait since 1991 on ski resort plan: regulatory faceoff', *Financial Post*, 12 January: FP1 and 5

Hamilton, L., Rohall, D., Brown, B., Hayward, G. and Keim, B. (2003) 'Warming winters and New Hampshire's lost ski areas: an integrated case study', *International Journal of Sociology and Social Policy*, 23(10): 52–73.

Hennessy, K., Whetton, P., Smith, I., Batholds, J., Hutchinson, M. and Sharples, J. (2003) *The Impact of Climate Change on Snow Conditions in Mainland Australia*. Aspendale, Australia: CSIRO Atmospheric Research.

Hoffman, C. (1998) 'Let it Snow', *Smithsonian*, 29(9): 50–8.

Innovest Strategic Value Advisors (2003) *Carbon Finance and the Global Equity Markets*. London: Carbon Disclosure Project.

Keates, L. (2000) 'Flake out: ski season off to slow start', *The Wall Street Journal*, 7 January: W7.

König, U. and Abegg, B. (1997) 'Impacts of climate change on tourism in the Swiss Alps', *Journal of Sustainable Tourism*, 5(1): 46–58.

Lazard, A. (2002) 'Ski winter: world flat', *Ski Area Management*, September: 24–7.

National Ski Areas Association (2004) Available at: www.nsaa.org (accessed 1 September 2004).

New England Lost Ski Areas Project (2004) Available at: www.nelsap.org (accessed 11 September 2004).

Packer, J. (1998) 'Everything you wanted to know about ski and snowboard tourists but were afraid to ask', *Journal of Vacation Marketing*, 4(2): 186–92.

Perry, A. (1997) 'Recreation and tourism', in R. Thompson and A. Perry (eds) *Applied Climatology*. London: Routledge, pp.240–8.

Ramage, J. (2003) 'Ski resort for sale: lack of snow is problematic', *Tribune Business News*, 13 February.

Robbins, P. (1980) 'Weather, gas, inflation – so what else is new in skiing?', *Ski Area Management*, January: 79–96.

Scott, D. (2004) 'Tourism plenary', Paper presented at the New England Governor's and Eastern Premiers Conference on Climate Change Impacts and Adaptation, Boston, March. CD-ROM available from New England Governors' Conference Inc. Available at: www.negc.org.

Scott, D. (2005 in press) 'Climate change and sustainable tourism in the 21st century', in J. Cukier (ed.). *Tourism Research: Policy, Planning, and Prospects*. Waterloo, Canada: Department of Geography Publication Series, University of Waterloo.

Scott, D., Jones, B., Lemieux, C., McBoyle, G., Mills, B., Svenson, S. and Wall, G. (2002) *The Vulnerability of Winter Recreation to Climate Change in Ontario's Lakelands Tourism Region*. Waterloo, Canada: University of Waterloo, Department of Geography Publication Series, Occasional Paper 18.

Scott, D., McBoyle, G. and Mills, B. (2003) 'Climate change and the skiing industry in Southern Ontario (Canada): Exploring the importance of snowmaking as a technical adaptation', *Climate Research*, 23: 171–81.

Scott, D., McBoyle G., Mills B. and Minogue, A. (2005 in press) 'Climate change and the sustainability of ski-based tourism in eastern North America: a reassessment', *Journal of Sustainable Tourism*.

Ski Vermont (2004) Available at: www.skivermont.com (accessed 15 September 2004).

Smit, B., Burton, I., Klein, R. and Wandel, J. (2000) 'An anatomy of adaptation to climate change and variability', *Climatic Change*, 45(1): 233–51.

Smith, M. (2001) 'Vermont sees year-over-year skier increase', *DCSki*, June 10.

The Economist (1998) 'Winter wonderlands', 31 January.

Theus, R. (1995) 'Klima und Seilbahnen im Wandel: Struktur und Strategien der Seilbahnunternehmungen', unpublished manuscript of a presentation at the Status Seminar organized by the Swiss National Research Programme, Zurich, November.

Tommasini, D. (2003) Personal communication. NORS- North Atlantic Regional Studies, University of Roskilde, Roskilde, Denmark.

United Kingdom Department of Transport (2003) *Aviation and the Environment: Using Economic Instruments*. London: Department of Transport.

US Department of Energy (2002) *State-level GHG Emissions Factors for Electricity Generation*. Available at: ftp.eia.doe.gov/pub/oiaf/1605/cdrom/pdf/e-supdoc.pdf (accessed 15 Sept 2004).

Wall, G. (1998) 'Climate change, tourism and the IPCC', *Tourism Recreation Review*, 23(2): 65–8.

Wall, G. (1992) 'Tourism alternatives in an era of global climate change', in V. Smith and W. Eadington (eds) *Tourism Alternatives*. Philadelphia, PA: University of Pennsylvania, pp.194–236

Wall, G. and Badke, C. (1994) 'Tourism and climate change: an international perspective', *Journal of Sustainable Tourism*, 2(4): 193–203.

Wickers, D. (1994) 'Snow alternative', *Sunday Times*, 27 November: 9.

Williams, P. and Dossa, K. (1990) *British Columbia Downhill Skier Survey 1989–90*. British Columbia Ministry of Tourism. Vancouver, Canada: Simon Fraser University, The Centre for Tourism Policy and Research.

16 The example of the avalanche winter 1999 and the storm Lothar in the Swiss Alps

Christian J. Nöthiger, Rolf Bürki and Hans Elsasser

Tourism can ... be significantly exposed to natural disasters, because of its attachment to high-risk areas with exotic scenery. The lure of snow-capped peaks brings the hazard of avalanches. Tropical beaches attract tourists to the potential paths of hurricanes ...

(Murphy and Bayley 1989: 36)

This quotation clearly illustrates the fact that natural hazards are no deterrent to tourism. On the contrary, the hazards of natural disasters are often particularly high in regions attracting large numbers of tourists. It is therefore surprising that, as Faulkner points out:

Relatively little systematic research has been carried out on disaster phenomena in tourism, the impacts of such events on the tourism industry and the responses of industry and relevant government agencies to cope with these impacts.

(Faulkner 2001: 136).

This pertinent statement refers not to natural disasters alone, but also includes man-made catastrophes like plane crashes and political crises.

Studies of the impacts of natural disasters on tourism do exist, dating from the 1970s and early 1980s, for example on avalanches and floods in Val d'Isère, France (Hanns 1975), on mudflows and avalanches in the federal state of Salzburg, Austria (Pipan 1977), or on the eruption of Mt Usu in northern Japan (Hirose 1982). However, such studies did not go beyond a description of specific occurrences. Murphy and Bayley's paper (1989) is also based on specific phenomena, namely the eruption of Mt St Helens, USA, in May 1980 and the forest fires in East Kootenay, British Columbia, Canada, in July 1985. In these cases general conclusions and recommendations were made for the tourism industry as to how to cope with natural disasters. In the 1990s, the few papers on this topic referred back to Murphy and Bayley, for example, Milo and Yoder (1991), who considered the role of travel journalists in repositioning tourist destinations after natural disasters, or Burby and Wagner (1996), examining how to protect tourists from tropical hurricanes on the coast of the Gulf of Mexico. The most in-depth investigation in the

1990s was conducted by Drabek (1994) on the correlations between tourism and natural disasters. His research focused mainly on risk awareness and precautionary planning by tourism suppliers (Drabek 1994), as well as on evacuated tourists' behaviour and their assessment of crisis management measures (Drabek 1996).

The earthquake in Assisi in September 1997 served as the basis of the first extensive study expounding how to calculate the financial impacts of natural disasters on the tourist industry (Mazzocchi and Montoni 2001). The principle employed is simple: the decrease of tourist frequency caused by the natural disaster is multiplied by a tourist's average daily expenditure. This yields the deficiency in receipts which can be put down to the disaster.

In Switzerland there is no scientific literature from before 1999 which deals with the topic of 'hazards of nature and tourism'. The avalanche winter 1999 and the storm Lothar altered this situation. The disaster analyses of the avalanche winter (SLF 2000) and the storm Lothar (WSL and BUWAL 2001) discussed various aspects of the impacts of the occurrences. Extended studies dealt with the impact of the avalanche winter on the tourist region of Elm, the consequences of the avalanches for the Swiss mountain and cable railway companies, and the effects of that winter on the tourist industry in Davos. The results of these studies, along with numerous literary references, can be found in Nöthiger (2003).

Basically, every natural disaster which could possibly occur in the Alps can also affect tourism. Preponderant among the natural hazards in alpine mountain regions are gravity-induced mass movements: floods, mudflows and, particularly, avalanches. There are also disasters which are not specifically alpine in nature, but which affect the countryside surrounding the alpine area as strongly as the actual mountain region, for example storms. Phenomena with a very low probability of occurrence, for example, devastating earthquakes, are looked upon as hardly relevant for the alpine tourism industry, because they are not considered in any long-term economic planning either.

Direct and indirect damage

Reports on the impacts of natural disasters often focus on direct damage – damage directly induced by the effect of the damaging process. Losses resulting from conditions altered by a damaging occurrence are described as indirect or consequential damage. The underlying assumption is that direct damage can be assigned a monetary value; direct damage requires an object potentially at risk, which is estimated at a certain value. Thus, direct damage can also be called direct costs. The formulation of direct damage is closely connected with the actual time when the occurrence took place. As regards indirect effects, this is merely the starting point and the period of effectiveness is theoretically indefinite. Furthermore, direct damage is limited to the process area of the natural disaster, whereas indirect effects can, in extreme cases, become globally manifest.

Indirect damage subsumes all resultant costs from a natural disaster which exceed the costs of clearing and repair work of direct damage. The main emphasis is on economic costs resulting from the occurrence. These are the deficiency in receipts caused by the natural disaster through loss or difficulties of production, loss of market shares and suchlike. Indirect costs can arise not only in the area directly affected by the natural disaster; companies further away can also be dependent on supplies from the destroyed businesses. On the other hand, it is also possible that certain firms receive new orders which can no longer be completed by the destroyed businesses, thus these companies achieve additional revenue because of the disaster.

Natural disasters can also be put to conspicuous touristic use after the event. The classic examples are the Roman cities Pompeii and Herculaneum, which were destroyed by the eruption of Mt Vesuvius. A more current example is the Mt St Helens National Volcanic Monument in the USA, which was erected two years after the devastating eruption on 18 May 1980 (Murphy and Bayley 1989).

The avalanche winter 1999 and the storm Lothar

In Switzerland, the direct costs of the avalanche winter totalled US$313 million. Approximately US$21 million or 7 per cent of that sum can be attached to damage of the tourist infrastructure, cable cars and ski lifts being the most affected. Considering the great importance of the tourist industry for the alpine economy, these costs nevertheless have to be characterised as minor. Agriculture, for example, was affected by direct costs to a far greater extent by comparison.

In February 1999, more than 40 tourist resorts in Switzerland were cut off from the outside world for up to 14 days in a row, resulting in indirect costs caused by the complete absence of day-trippers. Where roads had to remain closed for a particularly long time, overnight guests had to be evacuated by air, leading to even bigger losses. Forty-five per cent of Swiss cable cars and ski lifts could not run for an average of seven days. This meant that some day-trippers stayed away even

Table 16.1 Direct costs of the avalanche winter of 1999 in Switzerland

Category of damage	Costs (US$ million)
Buildings	139
Personal property	41
Streets	45
Railway lines	8
Cable cars/ski lifts	12
Protecting structures	6
Power supply system	19
Forests	33
Farm land	10
Total	*313*

Source: SLF 2000

Table 16.2 Loss of earnings for the tourist industry in the Swiss Alps caused by the avalanche winter of 1999

Category of expenses	Losses (US$ million)
Lodging	30
Catering	78
Retail	32
Cable cars/ski lifts	59
Other	16
Total	*215*

Source: Nöthinger 2003

when access roads were actually open. Not just cable car companies but also the rest of the tourist industry had to bear the financial consequences of the situation.

The direct damage from the storm Lothar on 26 December 1999 amounted to a total of US$1,216 million, US$6.4 million of which related to cable cars and ski lifts (Nöthiger 2003). It is not known to what extent buildings of the hotel and restaurant industry were affected. The enormous total damage caused by the storm Lothar is due to the fact that the principal damage occurred in the densely populated midland and not in the alpine region. In the case of the cable cars and ski lifts, the storm Lothar caused more damage events than the avalanche winter; the average damage costs per facility and the total damage sum, however, were lower. The indirect costs caused by the storm Lothar – the deficiency in receipts for the tourist industry in the Swiss mountain region – were estimated at US$90 million (Nöthiger 2003). Again the indirect costs were significantly higher than the direct costs for the tourist industry.

Measures

All measures aimed at preventing direct damage by natural disasters are at the same time preventive measures against indirect damage. Therefore, the question of whether an additional measure can be offset by the total of the prevented damage costs should necessarily take the indirect damage costs into account. Assuming low effectiveness may cause a measure to be rejected, even though consideration of the indirect costs would make it profitable. The difficulties which present themselves in calculating indirect damage are often the reason for not taking it into account.

The tourist industry is only of small significance regarding the preparation of preventive measures against direct damage by natural disasters (Wilhelm 1996). Cable car and ski lift companies are an exception, because they need specialist staff to assess the danger of avalanches, to trigger controlled avalanches and to close off ski runs when necessary. The World Tourism Organization (WTO 1996) recommends that tourism businesses inspect the emergency plans of the civil defence as a preventive measure. That would guarantee that the presence of a large number of tourists would be taken into consideration in the plans, if need be.

Besides prevention, the correct measures also play an important role during and after an occurrence in keeping indirect damage costs low. After the avalanche winter 1999 a survey was conducted among the Swiss cable car and ski lift companies. They were asked which special measures should be taken in future, based on their experiences, to lower indirect costs from natural disasters. More active media work (external communication) and better-informed customers (internal communication) were foremost among the responses (Nöthiger 2003). This shows that external and internal communication among businesses and organisations in tourism is of the utmost importance. The following recommendations from the WTO regarding external communication are, however, not only valid for the tourist industry: be quick, be honest and objective, be responsive to the media, be prepared (WTO 1996). Nevertheless, it is internal communication – on-the-spot communication – which should be prioritised, because it is directed at people who are potentially at risk from natural disasters.

The most important measure following the end of the acute phase of a natural disaster is the restoration and recommissioning of the tourist infrastructure. Tourism businesses have the possibility of taking out insurance against the financial consequences of natural disasters. So far as the direct costs are concerned, that means mainly insurance of buildings. Insurance against service interruptions and repercussive damage are recommended for indirect costs. However, insurance companies demand relatively high premiums on natural disasters if the probability of occurrence is not known. That is the reason why insurance against repercussive damage is not at all common in the tourist industry. This, in turn, explains why the indirect costs from natural disasters are predominantly borne by the businesses concerned.

The restoration of the infrastructure after a natural disaster does not completely rectify the situation for the tourist industry, because 'the recovery of destinations usually takes longer than the period required for the restoration of services to normalcy' (Faulkner 2001: 142). It is of the utmost importance that the first visitors return straight after the situation has been dealt with. As Murphy and Bayley (1989: 39) note:

> Such visitors will bear witness to the re-established safety and attractiveness of the area and will bring new revenue to assist in reconstruction. Successful visits can reinforce the fact that tourism has survived and that recovery is under way.

Again, external and internal communication are of great significance in assisting this process. One of the main difficulties of public relations work after a natural disaster is that a successfully mastered situation is of only small interest to the media. But for tourism the message that everything is back to normal and that visitors can return is particularly important. The faster the situation is normalised, the lower the indirect damage costs are, but the return to normalcy must be successfully communicated.

Climate change and avalanches

On the basis of its predictions for climate change, the IPCC (2001) anticipates a general increase of extreme occurrences. Although the economic consequences for tourism are usually locally or regionally confined, the overall threat to the tourist industry of an increase of natural disasters must not be underestimated. So far as the avalanche situation is concerned, for example, assuming warming of median temperatures, an increase of precipitation in winter and a rise in the frequency of extreme weather situations in winter because of global warming, the snow line in the Alps will climb by several hundred metres (OcCC 2003). Taking just the influence of warming into consideration, that would lead to less snow cover and shorter duration of snow cover. However, the result of an increase of precipitation in winter will be greater snow cover at high altitudes. If the precipitiation is not evenly distributed throughout the winter, but falls mainly during an extraordinary weather situation, the potential for extreme avalanche situations is heightened and a rise in the frequency of extreme weather situations also means an increasing probability of an extreme avalanche situation in winter. One of the consequences of climate change is that the winter tourism industry will increasingly open up high-altitude regions (Elsasser and Bürki 2002), and it is precisely those areas which will be even more imperilled by avalanches.

References

Burby, R.J. and Wagner, F. (1996) 'Protecting tourists from death and injury in costal storms', *Disasters. The International Journal of Disaster Studies and Practice*, 20(1): 49–60.

Drabek, T.E. (1994) 'Risk perception of tourist business managers', *The Official Journal of the National Association of Environmental Professionals (NAEP)*, 16(4): 327–41.

Drabek, T.E. (1996) 'Disaster evacuation behaviour. Tourists and other transients', Program on environment and behaviour, Monograph No. 58, Institute of behavioural science, natural hazards research and applications information centre, University of Colorado, Boulder.

Elsasser, H. and Bürki, R. (2002) 'Climate change as a threat to tourism in the Alps', *Climate Research*, 20: 253.

Faulkner, B. (2001) 'Towards a framework for tourism disaster management', *Tourism Management*, 22: 135–47.

Hanns, C. (1975) 'Die Lawinen- und Überschwemmungsschäden in der Wintersportstation Val d'Isère (Hochtarantaise). Folgen einer physiogeographische Gegebenheiten nicht berücksichtigenden Siedlungserweiterung', *Geographische Rundschau*, 9(75): 390–400.

Hirose, H. (1982) 'Volcanic eruption in northern Japan', *Disasters. The International Journal of Disaster Studies and Practice*, 6(2): 89–91.

Mazzocchi, M. and Montoni, A. (2001) 'Earthquake effects on tourism in central Italy', *Annals of Tourism Research*, 28(4): 1031–46.

Milo, K.J. and Yoder, S.L. (1991) 'Recovery from natural disaster. Travel writers and tourist destinations', *Journal of Travel Research*, 30(1): 36–9.

Murphy, P.E. and Bayley, R. (1989) 'Tourism and disaster planning', *Geographical Review*, 79(1): 36–46.

Nöthiger, C.J. (2003) *Naturgefahren und Tourismus in den Alpen. Untersucht am Lawinenwinter 1999 in der Schweiz*, Davos: Eidg, Institut für Schnee- und Lawinenforschung.

OcCC (Organe consultatif sur les Changements Climatiques) (2003) *Extremereignisse und Klimaänderung*, Bern: OcCC.

Pippan, T. (1977) 'Der Einfluss von katastrophalen Wetterereignissen auf den Saison-Fremdenverkehr und die behördlichen Kontrollmassnahmen am Beispiel des Landes Salzburg', *Mannheimer Geographische Arbeiten*, 1, Mannheim.

SLF (Eidg. Institut für Schnee- und Lawinenforschung) (2000) *Der Lawinenwinter 1999. Ereignisanalyse*, Davos: SLF.

Wilhelm, C. (1996) *Wirtschaftlichkeit im Lawinenschutz. Methodik und Erhebungen zur Beurteilung von Schutzmassnahmen mittels quantitativer Risikoanalyse und ökonomischer Bewertung*, Davos: Eidg, Institut für Schnee- und Lawinenforschung.

WSL (Eidg. Forschungsanstalt für Wald, Schnee und Landschaft) und BUWAL (Bundesamt für Umwelt, Wald und Landschaft) (2001) *Lothar. Der Orkan 1999. Ereignisanalyse*, Birmensdorf und Bern: WSL und BUWAL

WTO (World Tourism Organization) (1996) *Tourist Safety and Security. Practical Measures for Destinations*. Madrid: WTO

17 Tourists and global environmental change

A possible scenario in relation to nature and authenticity

Erika Andersson Cederholm and Johan Hultman

Introduction

The aim of this chapter is to discuss global environmental change (GEC) through the social construction of nature and notions of authenticity. We will discuss the social construction of nature in a specific context, namely that of different kinds of nature-based tourism. The argument will revolve around what we perceive as a shift in focus from the image of nature as *thing* to nature as *experience*, also expressed as a shift from nature as *place* to nature as *globalised locality*. We will sketch a development from 'traditional' nature tourism practices and rhetoric to the formation of another cultural economy of nature in order to discuss possible effects of GEC on tourist phenomenologies. By structuring our analysis around three aspects of authenticity, we will argue that different nature tourism operators stress the non-essentialist, experiential image of nature, where the destination *per se* seems to have a secondary role. We further suggest that the non-essentialist and therefore fluid concept of authenticity in tourist experiences might be adaptable to perceptions of change on a global scale. We aim to interpret GEC, although encompassing potentially devastating processes in a great number of ways, as also being able to be incorporated in common discourses of tourist experiences.

Global environmental change and the rearrangement of nature

One important aspect of tourism involves the social construction of nature. Websites, brochures, television programmes, travel magazines and other kinds of promotional material represent local natures and cultures and shape later tourist experiences – these kinds of textual and pictorial materials act as manuals to the interpretation of nature and culture. GEC is also a kind of production of nature in the sense that it is a process producing qualitative, quantitative and geographical rearrangements of nature. This is seldom seen as positive, it is commonly defined as a negative consequence of anthropogenic resource use practices. In order to make these two kinds of nature production practices inhabit the same argument, we will in the following give environment and nature the same meaning although there are good reasons not to do so. One of these is that environment in this context is a scientific concept, at the same time geographically abstract and measurable down to the molecular level, while nature is a romantic concept harbouring a host of images, feelings and experiences. It is nature that tourists seek, not environment, but GEC affects touristic nature, thus making it necessary to interpret the two concepts equally. In this text, however, GEC could as well have been termed global nature change.

It is interesting to note that while changes or trends in nature driven from outside the tourism industry (for example, GEC) are slow, changes or trends within the industry are comparably fast, unstable and subject to sudden shifts in symbolic meaning (Buckley 2000). At the same time, nature both in tourism and GEC discourses respectively is commonly represented as unstable and non-robust, sensitive and vulnerable (see Bandy 1996). Tourism on the other hand is a stable process, unstoppable, ever growing and robust – a sure and significant cultural force, an important form of consumption. Change in (at least western) culture is not so problematic, it is often expected and even cherished, but change in nature is commonly regarded as wrong, immoral and problematic. This is because nature is natural, and therefore untouchable. Nature shows us our origin, and furthermore a number of ecological imperatives which must be followed if we want to avoid imminent ecological collapse. But since tourism is the construction of nature, nature's 'naturalness' must here be treated as a cultural category and a subject for discourse (Demeritt 1994; Braun 2002; but see Cafaro 2001).

Nature-based tourism, specifically ecotourism, involves environmental management, for example, education and contribution to conservation. At the same time, many business actors and stakeholders strive for growth, and this is also true for government and municipal agencies. For example, tourist agencies see it as one of their primary roles to facilitate and help local business by attracting tourists (Andersson Cederholm *et al.* 2004). So tourism is basically a competition between stakeholders in the commoditisation and commercialisation of local nature and culture (see Luke 1998; McAfee 1999). These categories are then translated and represented in promotional material, which in turn act as carriers of meaning and expectations. In the same way, 'texts' about environmental change act as promotional material, promoting a trend, creating expectations and experiences (Burgess 1990; Catasus 2000; Bickerstaff and Walker 2003; Pedynowski 2003). Such 'texts' become manuals to the creation of meaning as well. There are large discrepancies between these two kinds of meaning constructions, so how might this issue act out from tourists' perspectives and tourists' perceptions of nature?

In a way, all nature-based tourism depends on scarcity – without special nature in short supply, why bother to label anything unique or eco-? The question here is what happens with touristically desirable natures when these are subjected to GEC? Is truly unique nature destroyed and made scarcer – a dystopic scenario whose ultimate consequences are discussed by Tonn (2002)? Or is nature – from the viewpoint of tourists – rather rearranged, redefined, displaced and thus available elsewhere? The tourism industry is after all driven by constant change and is in this sense discursively – not absolutely – structured. Trends and changes in travel styles, tourist practices and destination popularity follow no natural law.

Our purpose in this chapter is to analyse GEC's impact on tourism by discussing nature with this question in mind. Our wider framework is the concept of cultural economy in relation to nature, that is the contention that: 'the cultural construction of nature is one medium of translation between the biophysical world and economic systems of value and exchange' (Mansfield 2003: 329). Nature is in many ways incorporated into the economy either as a consumer object in itself – common, of course, in the tourism industry – or as an attribute of other consumer objects such as automobile motors, snacks or clothes. Thus, the cultural economy of nature articulates the inseparability of culture and economy if this relationship is analysed in terms of practice (Simonsen 2001). Nature is culturally constructed – that is given meaning – by the actual doings of, for example, nature-based tourism operators, and this meaning is transformed into cash flows through tourist practices. It is this construction of nature by touristic practices we will focus upon here.

This chapter is structured as an examination of a change having occurred in nature-based tourism representations and practices during the past few years. We will tentatively define this change as a shift in focus from nature as thing to nature as experience, and consequently from nature as place to nature as globalised locality. We will deal with this change by an analysis of three aspects of authenticity: the notion of the essentialist Origin, that is the image of nature as a unchangeable and untouchable thing; the notion of the Unique, indicating exclusivity in a commercial sense; and finally the notion of Existential authenticity, indicating the subjective and often spiritual experience of being part of a greater whole. The material we use comes from eco- and nature tour operators. We view such organisations as mediators or constructors of nature in the way they represent nature. We will sketch a development from 'traditional' nature-based touristic practices and rhetoric to what we perceive as a new cultural economy of nature in order to illustrate the change we have identified, eventually arriving at some conclusions with potential relevance for GEC's impact on tourists' perception of nature. The material is internet-based, since web-mediated interaction between operators and tourists is becoming an increasingly important means of pre-travel information gathering (Andersson Cederholm *et al.* 2004). This would then imply that websites are important elements in the formation of discourses about nature, and thus also important for the tourist's interpretation of nature and expectations of authentic experiences.

Tourism and dissociation from nature: nature as thing

Nature-based tourism and ecotourism rhetoric often revolves around the moral obligation of the individual tourist not to degrade nature and local culture. In the standard definition of ecotourism, proper ecotourist behaviour is translated as a sense of responsibility while being in nature: 'responsible travel to natural areas that conserves the environment and improves the well-being of local people' (ecotourism.org). There is an inscription written into this simple definition, hidden within the phrase 'conserve the environment', which highlights the fundamental paradox contained within ecotourism. On the one hand, ecotourism and much nature-based tourism are all about getting to know nature. This is evident in the rich textual and pictorial material that is used, for example, to construct ecotourism websites, or in quotes such as: 'Ecotourism travel gives visitors the *possibility to acquire knowledge* about issues related to nature, culture, environment and development' (ekoturism.org, our translation, emphasis in original). But at the same time the ecotourist must not bodily engage with nature in any invasive way; nature shall remain untouched, undisturbed. The trace of the tourist must be non-existent.

In order to resolve this paradox in practice, two things must happen. The first is that tourism operators must represent nature as distanced from the tourist. Nature is coded as a thing, an object to be worshipped; the goal is to: 'actively *exhibit* charismatic and rare or sensitive species in a non-invasive way' (ekoturism.org, our translation and emphasis). The other is that the tourist must distance herself from nature. Nature must be viewed, not messed with. Just as in many other kinds of tourisms, the camera lens is a primary filter through which the ecotourist views nature and culture. So the actual role of the ecotourist is often, or has at least been, that of the gazer (Urry 1990), and nature and local people have consequently been constructed as attractions in a way that mirrors mass tourist attractions, for example, buildings, monuments, London punk rockers, etc. Many nature-based and ecotourism websites bear witness to this in the way these virtual encounters with nature are structured around a style of photography that is geared towards capturing the sublime and awe-inspiring aspects of nature. The result is that nature, and local culture, is in several ways represented as fragile, pre-modern and mysterious, that is conceptually and geographically distant from modern society.

Interpreted in this way, tourism in nature is one mode of reproducing the modernist dichotomy between nature and society. Nature is a sphere outside or apart from society. Mass tourism is in effect a toxic leak from society to nature, but ecotourism is a gentle exploratory expedition into the secrets of nature. So while traditional ecotouristic rhetoric stresses that ecotourism is all about coming really close to nature and local culture in a respectful way, on nature's own terms, ecotourism practices have actually served to reproduce the distance between nature and western culture. And there is a strong normative agenda behind separating touristic nature and culture from western society. The explicit reason for ecotourism organisations to exist is that: 'all tourism related to nature or the combination of nature and culture is guided towards fulfilling ecotourist requirements' (ekoturism.org, our translation). This would, in effect, mean that all tourist practices would have to withdraw from close engagement with the natural or cultural Other – in a way this would mean a formalisation of the tourist gaze.

The ecotourist dichotomy of nature and (western) culture is paralleled in much of the sustainability discourse in general (Hultman 2003). For all the anxiety expressed about nature, the different kinds of we-are-a-part-of-nature rhetoric that suffuse this discourse, the result is still that nature has been firmly positioned as a system totally apart from the societal system (as evident and/or discussed in different ways in e.g. Muir 1994; Hornborg 1998; Gullone 2000; Sneddon 2000;

de Paiva Duarte 2001; Barr 2004; see also Hammond 2004; but see Luke (1998) on the fusion logic inherent in the commercialisation of nature). This is an ontological logic carried on by many GEC scenarios: society is active, dominant and ignorant/ malevolent, nature is passive, holistic, amoral but still normative and pure in essence (e.g. UNEP/WTO 2001; WTO 2003). In line with this, it has been argued that ecotourism has been governed by an approach informed by science and above all planning – as opposed to the unmanaged character of other kinds of tourism when it comes to environmental issues (Hughes 1995). Nature, and the tourist, must be managed in an organised way in order to remain attractive. Ecotourist travel: 'is conducted with the outmost care, and the least possible wear on the desti-nation's natural and cultural values, *with the purpose of conserving the biodiversity and cultural values* that the visitor has come to experience' (ekoturism.org, our translation and emphasis). Encounters between the tourist on the one hand and nature and local culture on the other must be controlled and directed. On several levels, the scientisation of nature (Urry 1999) in tourist contexts has thus acted to dissociate tourists from nature and local culture, quite contrary to the general rhet-oric. Nature has been objectified, a viewable thing.

What it also has done is in a way to situate nature and indigenous culture in the same position (Bandy 1996). Ecotourism allows the traveller to: 'explore rain-forests, mountains, deserts, tropical beaches, coral reefs ... guided by those who know them best – the people who live there' (tourismconcern.org.uk). Local people are part of nature; they have intimate knowledge of nature and all its secrets. In some ways nature and local culture have been fused together, they have been hybridised, and through this ontological arrangement runs a discourse of conservation because ecotourism is a touristic mode that: 'actively contributes to the protection of nature and safeguards cultural values' (ekoturism.org, our trans-lation). To expand on this in order to understand how GEC will be a factor in nature tourist industries, we turn to the concept of authenticity.

A sense of origin: nature as place and as globalised locality

Authenticity is much debated in studies of the tourist experience. Quite often the notion of authenticity is related to the experience of non-authenticity, and MacCannell's (1973, 1976, 1992) concept of 'staged authenticity' indicates the disillusionment when the tourist scene seems too adapted to the expectation of the tourist gaze (Pearce and Moscardo 1986). Studies on the search for authenticity quite often focused on the image of the cultural Other (Silver 1993; Albers and James 1988). However, the cultural Other is often portrayed, in tourist narratives as well as in marketing, as part of nature or at least close to nature – in a geograph-ical as well as a cultural sense. The fusion of culture and nature thus reflects a primitivistic image of the natural Other, common in tourist mythology (Andersson Cederholm 1999; Elsrud 2004). One aspect of authenticity is thus the notion of the Origin; an essentialist image of cultures and natures preserved – even conserved – and of time standing still. Quite often, the idea of the Origin acts to highlight a perception of time prevalent in late modern societies: the notion of acceleration of time. 'You have to go there before it is too late', is a quite common notion among tourists seeking the last reservoirs of authenticity in the world (Andersson Cederholm 1999). This argument is also evident in tourist marketing, where the explicit threat of modernisation is used to legitimise travel to authentic milieus (Kilroy Travels, autumn 2004 campaign: 'Go before it's too late'; authenticity here ranges from the African savannah to British industrial heritage). Thus, you have to run faster to be able to stand still, even though it might be just for a few days visiting the remote Other, in a geographical and/or cultural sense.

Peoples of Origin are also expected to have a traditional knowledge of nature and how to cultivate it in a small-scale and sustainable way. The knowledge they have acquired is regarded as inherited and traditional. A local guide is thus expected to convey knowledge about the local culture that even a well-trained non-local guide will never manage. Stressing the presence of a local guide is common in ecotourism marketing. In Nature's Best – a certifying organisation for Swedish ecotourism products – requirements directed to operators applying to use the brand, the connection to the local community is emphasised. Using a local guide not only benefits the local economy, but also gives the arrangement an aura of authenticity. Knowledge of nature and traditional culture are not only regarded as inherited, but mysterious, tacit and essentialist. It is natural and thus should be respected. One of Nature's Best's six main requirements is formulated as: 'Respect the limits of the destination – the least possible impact on nature and culture' and has as a subrequirement: '*Always respect* local rules and recommendations for protected areas' (naturensbasta.com, Document of Requirements 2002–2005, our translation and emphasis).

However, it is not always authenticity in an essentialist, primitivistic meaning that tourists search for. Another aspect of authenticity, prevalent in ecotourism discourse as well as in other forms of tourism, is the notion of *the unique*. It implies the search for the unique experience not as much in relation to a western material-istic society – as in the notion of Originality – but rather in relation to the tourist industry and the existence of other tourists. That is, a destination is regarded as authentic if the local way of life is not 'bought' or dominated by the tourist industry, even though it is obviously commoditised to serve the needs of the customer. However, it is important for the tourist not to be treated as *merely* a customer (Andersson Cederholm 1999), but as a *person*. This notion of *the unique* is echoed in the 'new tourism' (Poon 1993) directed towards the demand for tailor made, non-standardized personal solutions. Exclusivity is a key word in marketing and the concept of 'cutting edge product' is often used by Nature's Best: '[In this way] have several of Laponia's absolute top-notch attractions been transformed into bookable cutting edge products' (ekoturism.org, our translation). The notion of uniqueness implies uniqueness in the business arrangement as a whole, rather than the destination *per se*. Here, perhaps, can we glimpse a first clue as to how GEC could affect tourists' perception of nature: place might not be the most important aspect of nature experiences. Instead, it is the way in which experiences are discursively constructed and packaged that matters most for tourists' appreciation of nature.

Even though the notion of the Origin as well as uniqueness are cultural ideas existing side by side in tourist industry narratives, we argue that there has been a shift in focus from emphasis on the essentialist notion of authenticity, towards the experiential. Or, as Kurt Kutay of ecotourist company Wildland Adventures puts it:

> Some of our more novice clients still think authenticity is synonymous with travel to pristine natural areas and untrodden villages where native peoples retain traditional values. (…) However, what I find equally gratifying and meaningful is simply the truth. (…) Authentic experiences are just as available in popular tourism destinations like Costa Rica and Thailand, as they are in remote Mongolia or the Bolivian highlands. *It all depends on how we conduct our business* and integrate our tour operations from trained guides to informed guests.
>
> (wildland.com, our emphasis)

This director positions himself as being in the frontline of ecotourism business: 'some of our more novice clients *still* think'. Further, authenticity is not about the destinations, it is about conducting business and doing it well. This shift in focus is reflected by a Swedish certified ecotourism operator:

> Nature and the culture of people of nature have always fascinated the traveller: we seek backwards to the Origin and to the vital beauty of nature. This quest often brings us far away, to distant corners of the world, in spite of the fact that the same possibilities for experiences are present close by.
>
> (lapplandssafari.se, our translation)

A sense of uniqueness could be found nearby as well as in remote destinations. The product – described as 'cutting edge product' – is marketed and sold as an *experience*. It is the unique experience of nature that is in focus, not a specific nature – nature as place – or a specific culture. Only by this way of constructing nature can a horse riding holiday through 'the mythology of the deep, Swedish forest' – taking place in the southern part of Sweden and brought to the consumer wrapped in literary references to John Bauer and Astrid Lindgren – include as a specified attraction one night in a Sami tent complete with reindeer skins to sleep on (wildhorseriding.com). This is a non-essentialist type of authenticity at one level, but the notion of Origin is still prevalent as the background to the experiential type of authenticity implied in the concept.

When experiential authenticity is the key concept, the actual nature that is the setting for the experience could be anywhere in the world, as long as it has a local flavour and is represented as being far from modern society – at least in an discursive sense, if not a geographical one. However, it is a globalised, generalised locality. Locally produced food and local attractions with a long history and a place inhabited by people of Origin are important, but the local nature and culture in question could be localised anywhere in the world. It is not the place that is important, but a sense of place. A sense of place that gives you a certain kind of experience: 'Ecotourism works as well in the Laponia mountains and our archipelagos as it does in Nepal or New Guinea' (naturensbasta.com, Document of Requirements, our translation). It is ecotourism defined as experience-packed practice that is important, and this can take place *anywhere* local.

So even though the notion of Originality is implicit in the documents from Nature's Best and in marketing material from different ecotourism and nature tourism operators, it is a sense of place, a sense of tradition and, quite often, the personal encounter with the locals that are supposed to give you a unique experience, juxtaposed to a standardised or mainstream type of travelling. This would then further imply that tourists' experiences of nature are not so dependent on actual map coordinates since it is *the everywhere unique* tourists seek. Instead, it is how tourists position themselves in relation to each other that gives a sense of authenticity, something GEC would not necessarily have a great impact upon. It would highlight the relevance of how tourist natures are discursively constructed and likewise valued in relation to each other, something that would be affected by GEC. For example, the geographical distribution of desirable natures could change dramatically, even though the relative scale of value distinguishing between different natures would remain. The reason is that tourism is a form of consumption and consumption is a communicative practice and a primary source in the creation of cultural and social meaning.

Connecting to the greater whole

We will mention a third aspect of authenticity, even more experiential and less dependent on the actual destination of the traveller. It is the notion of *existential authenticity*, indicating a sense of belonging, an intimate relationship between the tourist and the world surrounding her, a non-reflexive attitude and a sense of flow (Csikzsentmihalyi 1991). This emotional state of authenticity could occur at any time and in any place in life, in those spheres of life which are socially institutionalised and framed as daily life as well as more extraordinary events like tourist trips, even though tourists describe these emotional states as being more frequent when they travel than in their daily life (Andersson Cederholm 1999).

The tourist trip is socially defined as extraordinary, but within this institutionalised liminal sphere (Turner 1969) individual islands of liminal experiences of intensity and flow become markers of the trip as a whole. They define the journey as an extraordinary event and make it even more unique in relation to the experiences of other tourists. In ecotourism marketing, nature is used as the medium for reaching the holistic experience of being part of something eternal – the encounter with nature works 'by bringing you in collective harmony with the web of life that surrounds us', something that initiates a process of reclamation: 'Rediscover a sense of belonging to something larger than yourself' (wildland.com). This is often juxtaposed to a society dominated by a stressful busy lifestyle, governed by the clock. As one of the Swedish ecotourist operators describes it: 'Discover the calmness, silence and freedom far beyond mobile phones and "technostress"' (lapplandsafari.se).

It is a way of feeling and being that is stressed, no matter where you are in the world. It is not the tourist gaze that is encouraged, but a sensual holistic experience. The background is often nature or peoples of nature, but it is not as much nature *per se* that is marketed, but nature as an environment that encourages a certain way of feeling and thinking. Even if a Sami village or bear safari is the exotic background that legitimises an expensive trip that is socially framed as extraordinary, the main attraction is a specific kind of authentic experience. The last two aspects of authenticity mentioned – the quest for uniqueness and existential authenticity – are experiential rather than essentialist. The experience of nature and culture is thus both individualised and commoditised. Nature's Best and its certified operators emphasise the experience, not nature *per se*: 'Geunja, Sami mountain lodge in roadless land. A creative and inspiring meeting place for development, group cooperation, leadership and fresh thinking' (lapplandsafari.se).

Nature as the exclusive experiential product

We would argue that what can now be discerned is the formation of new touristic discourses and practices related to nature, where tourism operators are in the process of situating nature in a new global cultural economy. The scientific approach to managing nature has given way to another logic: conservation through exclusivity, which can also be expressed as conservation through commoditisation. The way nature and local culture is commoditised is how these categories are translated as *experiences* instead of (scientific) objects. Nature moves from object to experience. It is clearly exclusive – the targeted consumers are obviously upscale – and the notion of sustainability is strikingly absent in the message. In a 2003 conference presentation of the Nature's Best concept, one of the organisation founders explicitly stated that the goal was to dissociate nature from environmental sustainability. Tourists were not seen as willing to pay anything for sustainability, whereas they were more than willing to pay for the added-value *experience* of nature. So within the context of nature-based tourism, nature is being constructed in new ways. Rhetorically and discursively, nature is developing towards a medium for entertainment and self-fulfilment, and thus away from the scientific object it is within 'traditional' ecotourism.

This meaning given to nature, in combination with how the notion of nature as specific place is exchanged for nature as globalised locality, forms the basis of what we perceive as a new cultural economy of nature. Nature and local culture are perhaps not becoming more democratic – we are after all talking about cutting edge products – but these touristic categories are being incorporated into tourist phenomenologies in new ways. In a sense, the ecotourist is allowed and encouraged to experience nature reflexively instead of objectively. Because nature is represented as a medium for sensual experiences and personal development, rather than an object *per se*, the actual destination might have a secondary role. The search for uniqueness as well as existential authenticity – as two important aspects of authenticity – are connected to the quality of the experience and the exclusivity of the business product, rather than the actual physical place.

The concept of Originality, however, as the first aspect of authenticity mentioned in this chapter, is inherently connected to physical place. As such, it is also more vulnerable to environmental change and the physical as well as the perceptual carrying capacity of a specific destination. However, the disconnection from the physical space significant for experiential notions of authenticity opens up endless variations of the tourist product.

To conclude our argument, GEC will certainly affect the tourism industry as a whole. Tourists' perceptions of nature will also change – although we find it difficult to relate perceptions of global change to discussions about perceptions of local change (see Hillery *et al.* 2001). For example, the discursive construction of nature as fragile might be reinforced as a result of a stronger and more explicit symbiosis between touristic and scientific narratives – in a way a variant of the almost total economisation of nature discussed by Luke (1998). But our discussion here would suggest that GEC impact on tourism might primarily become an issue of geographical rearrangements of natures and thus a spatial redistribution of tourists. Tourist discourses of nature, authenticity and experiences are possibly well able to incorporate GEC without this necessarily having a noticeable effect on tourists' motivation and willingness to travel and experience nature. In this sense tourists' perception of nature might not change very much, although the imperative 'Go before it's too late!' will continue to haunt us and remind us of the late modern condition. As long as nature is constructed as an exclusive experiential product, the consumption of nature will have the same communicative function as before, even if anthropogenic climate changes rearrange nature on a global scale.

References

Albers, P.C. and James, W.R. (1988) 'Travel photography: A methodological approach', *Annals of Tourism Research*, 15: 134–58.

Andersson Cederholm, E. (1999) *Det extraordinäras lockelse: luffarturistens bilder och upplevelser*. Lund: Arkiv Förlag.

Andersson Cederholm, E., Gyimóthy, S. and Hultman, J. (2004): *Kompetenceudvikling i Norra Öresunds turismeerhverv*. Rapport. Department of Service Management, Lund University, Campus Helsingborg.

Bandy, J. (1996) 'Managing the Other of nature: sustainability, spectacle, and global regimes of capital in ecotourism', *Public Culture*, 8: 539–66.

Barr, S. (2004) 'Are we all environmentalists now? Rhetoric and reality in environmental action', *Geoforum*, 35: 231–49.

Bickerstaff, K. and Walker, G. (2003) 'The place(s) of matter: matter out of place – public understanding of air pollution', *Progress in Human Geography*, 27(1): 45–67.

Braun, B. (2002) *The Intemperate Rainforest. Nature, culture, and power on Canada's west coast*. Minneapolis, MN: University of Minnesota Press.

Buckley, R. (2000) 'Neat trends: Current issues in nature, eco- and adventure tourism', *International Journal of Tourism Research*, 2: 437–44.

Burgess, J. (1990) 'The production and consumption of environmental meanings in the mass media: a research agenda for the 1990s', *Transactions of the Association of British Geographers NS*, 15(2): 139–61.

Cafaro, P. (2001) 'For a grounded conception of wilderness and more wilderness on the ground', *Ethics & the Environment*, 6(1): 2–17.

Catasus, B. (2000) 'Silent nature becomes normal', *International Studies of Management & Organisation*, 30(3): 59–83.

Csikszentmihalyi, M. (1991) *Flow. The Psychology of Optimal Experience*. New York: Harper Perennial.

Demeritt, D. (1994) 'The nature of metaphors in cultural geography and environmental history', *Progress in Human Geography*, 18(2): 163–85.

Elsrud, T. (2004). *Taking Time and Making Journeys. Narratives on Self and the Other among Backpackers*. Lund Dissertations in Sociology 56. Lund: Dept. of Sociology, Lund University.

Gullone, E. (2000) 'The biophilia hypothesis and life in the 21st century: increasing mental health or increasing pathology?' *Journal of Happiness Studies*, 1: 293–321.

Hammond, K. (2004) 'Monsters of modernity: Frankenstein and modern environmentalism', *Cultural Geographies*, 11: 181–98.

Hillery, M. *et al.* (2001) 'Tourist perception of environmental impact', *Annals of Tourism Research*, 28(4): 853–67.

Hornborg, A. (1998) 'Towards an ecological theory of unequal exchange: articulating worlds system theory and ecological economics', *Ecological Economics*, 25(1): 127–36.

Hughes, G. (1995) 'The cultural construction of sustainable tourism', *Tourism Management*, 16(1): 49–59.

Hultman, J. (2003) 'Natur och miljö – diskurs och materialitet', in L.J. Lundgren (ed.) *Vägar till kunskap. Några aspekter på humanvetenskaplig och annan miljöforskning*. Stockholm: Symposion.

Luke, T.W. (1998) 'The (un)wise (ab)use of nature: environmentalism as globalized consumerism', *Alternatives*, 23(2): 175–213.

McAfee, K. (1999) 'Selling nature to save it? Biodiversity and green developmentalism', *Environment and Planning D: Society and Space*, 17: 133–54.

MacCannell, D. (1973) 'Staged authenticity: Arrangements of social space in tourist settings', *American Journal of Sociology*, 79(3): 589–603.

MacCannell, D. (1976) *The Tourist: A New Theory of the Leisure Class*. New York: Schocken Books.

MacCannell, D. (1992) *Empty Meeting Grounds. The Tourist Papers*. New York: Routledge.

Mansfield, B. (2003) 'From catfish to organic fish: making distinctions about nature as cultural economic practice', *Geoforum*, 34: 329–42.

Muir, S.A. (1994) 'The web and the spaceship: metaphors of the environment', *Et Cetera*, summer.

Paiva Duarte, F. de (2001) '"Save the earth" or "Manage the earth"? The politics of environmental globality in high modernity', *Current Sociology*, 49(1): 91–111.

Pearce, P.L. and Moscardo, G.M. (1986) 'The concept of authenticity in tourist experiences', *The Australian and New Zealand Journal of Sociology*, 22: p.x.

Pedynowski, D. (2003) 'Science(s) – which, when and whose? Probing the metanarrative of scientific knowledge in the social construction of nature', *Progress in Human Geography*, 27(6): 735–52.

Poon, A. (1993) *Tourism, Technology and Competitive Strategies*. New York: CABI Publishing.

Silver, I. (1993) 'Marketing authenticity in Third World countries', *Annals of Tourism Research*, 20: 302–18.

Simonsen, K. (2001) 'Space, culture and economy – a question of practice', *Geografiska Annaler*, 83B: 41–52.

Sneddon, C.S. (2000) '"Sustainability" in ecological economics, ecology and livelihoods: a review', *Progress in Human Geography*' 24(4): 521–49.

Tonn, B.E. (2002) 'Distant futures and the environment', *Futures*, 34(2): 117–32.

Turner, V. (1969) *The Ritual Process. Structure and Anti-Structure*. London: Routledge and Kegan Paul.

UNEP/WTO (2001) *Climate change 2001: Impacts, adaption and vulnerability*. Report. Intergovernmental Panel of Climate Change. UNEP/WTO (accessed through www.grida.no/).

Urry, J. (1990) *The Tourist Gaze. Leisure and Travel in Contemporary Societies*. London: Sage Publications.

Urry, J. (1999) 'Sensing leisure spaces', in D. Crouch (ed.) *Leisure/tourism geographies*. London: Routledge.

WTO (2003) *Djerba Declaration on Tourism and Climate Change*. Madrid: WTO.

Internet references

www.ecotourism.org
www.ekoturism.org
www.grain.org
www.lapplandsafari.se
www.naturensbasta.com
www.tourismconcern.org.uk
www.wildhorseriding.com
www.wildland.com

18 Conclusion

Wake up … this is serious

Stefan Gössling and C. Michael Hall

Wake Up … This is Serious (headline in *The Age*)

(Fyfe 2005)

If our economies are to flourish, if global poverty is to be banished, and if the wellbeing of the world's people [is to be] enhanced – not just in this generation but in succeeding generations – we must make sure we take care of the natural environment and resources on which our economic activity depends.
(Gordon Brown, UK Chancellor of the Exchequer, 15 March 2005)

As the various chapters in this book have indicated, tourism is implicated in a number of key areas of global environmental change (GEC). Tourism is both a factor in GEC and is, in turn, affected by it. However, arguably one of the most remarkable aspects of tourism's relationship to GEC is the relative lack of attention paid to these issues not only by much of the tourism industry but also by many people who research tourism. Indeed, tourism is often portrayed as a significant contribution to the conservation of natural resources, particularly through the growth of 'ecotourism' and 'sustainable' tourism which serve as clarion calls for academic research as much as they do for government and industry. Indeed, the word 'sustainable' is now seemingly a 'standard' term to throw into tourism planning documents which is then seen to magically transform them into something that will make a better contribution to stakeholders in the tourism development process, whether they are willing stakeholders or not (Mowforth and Munt 1998). This is not to say that the editors are claiming that tourism is necessarily an 'evil' industry. Instead, the evidence of this book suggests that tourism is like any other industry, it does make some positive contributions to society, communities and the environment but it can also be extremely negative. Critically, tourism cannot claim to have any moral high ground. Furthermore, for all the writing about sustainable tourism since the mid-1980s, the global evidence clearly suggests that in terms of damage to the environment things have got worse, not better.

Issues of scale and perspective

Arguably, one of the key problems in understanding tourism's contribution to GEC is the issue of scale as it applies to both conceptualising tourism and undertaking empirical research. For example, conceptually the notion of sustainable tourism is highly problematic in terms of GEC, as it is highly likely that one can have sustainable tourism without other elements in a specific environment also being sustainable. Instead, the focus needs to be on the potential benefits of tourism within a wider context. But then, what should that context be? Most tourism research occurs at the destination (Hall 2005a). At first glance this would seem obvious – but then first looks can be deceiving. The reality is that it has long been recognised that a tourist trip occurs in various stages, which are usually divided into decision to travel, travel to destination, return travel, and the recollection stage once the traveller has returned 'home'. Why, therefore, should one focus just on the destination stage given that the effects of tourism will be felt at all those stages? This is not just an 'academic' observation but one that goes to the heart of understanding tourism's global contribution to environmental change. In fact, the further one travels the greater will be the environmental impacts of tourism away from the destination. For example, in their study of the ecological footprint of international tourism to the Seychelles, Gössling *et al.* reported that more than 97 per cent of the energy footprint was the result of air travel:

> This implies that current efforts to make destinations more sustainable through the installation of energy-saving devices or the use of renewable energy sources can only contribute to marginal savings in view of the large amounts of energy used for air travel.

> (2002: 208)

From a global perspective, transport produces approximately 20 per cent of carbon emissions and smaller shares of the other five greenhouse gasses covered under the Kyoto Protocol. According to International Energy Agency statistics, the transport sector's share of world greenhouse gas (GHG) emissions increased from about 19 per cent in 1971 to 23 per cent in 1997 (IPCC 2001), with CO_2 from combustion of fossil fuels the predominant GHG produced by transport, accounting for over 95 per cent of the annual global warming potential produced by the sector (IPCC 2001). The transport dimension is clearly critical in understanding the impacts of tourism because, although transport is responsible for about one fifth of worldwide energy consumption, approximately two thirds of transport energy demand in OECD countries is from passenger travel. In 1997, 54 per cent of the oil purchased by OECD countries was for transport, with an increase of 62 per cent projected by 2020 (Doering *et al.* 2002). Although nearly all modes of transport are showing growth in energy use air transport is the second largest, and most rapidly growing mode, accounting for about 12 per cent of current transport energy use (IPCC 2001). Thus, a key 'finding' of this book is that it is vital to look at the impacts of tourism over the totality of travel and travel decision making and not just in isolation. This not to deny that ecotourism, for example, can make a positive contribution to biodiversity conservation and community well-being at a specific location, but to point out that it should also be acknowledged that there are negative environmental outcomes arising from the travel to and from destinations that must be accounted for.

The environmental impacts of tourism also need to be understood over the life-span of the individual. Indeed, commentators such as Lanzendorf (2000) and Hall (2005b) have argued that such an approach needs to be undertaken in order to see how people exchange or substitute certain blocks of leisure mobility activity in time and space depending on their lifestyle, accessibility and mobility arrangement (e.g. see Ceron and Dubois (2003) for the implications of this with respect to leisure mobility contribution to climate change). Indeed, within tourism development there has generally been little attention to energy and environmental implications of the lifespan of infrastructure either. For example, Hall (2000) in discussing issues of designing sustainability in tourism identified the role of preservation in retaining inter-generational equity, noting that stadia, festival marketplaces and convention centres are often constructed with the likelihood of relatively short-term periods of use, in the order of 15–30 years, before they are replaced. Indeed, in terms of creating options for future use, one of the tenets of sustainability, Lynch (1972) writes of 'future preservation':

> Our most important responsibility to the future is not to coerce it but to attend to it. Collectively, [such actions] might be called 'future preservation', just as an analogous activity carried out in the present is called historical preservation.
>
> (Lynch 1972: 115)

Similarly, Brand (1997: 90) observed that: 'Preservationists have a philosophy of time and responsibility that includes the future'. In this sense, the preservation movement is creating a form of inter-generational equity through the maintenance and adaptive re-use of buildings and structures from one generation to another, while also contributing to substantial economic and energy savings. For example: 'Even extensive rehabilitation (services, windows, roof) typically costs 3 to 16 per cent less than demolishing and replacing an old building' (Rypkema 1992: 27), while preservation can also help conserve the 'embodied energy' of buildings and reducing the solid-waste burden of demolition (Rathje and Murphy 1992). However, with respect to preservation and recycling of old buildings, cultural and aesthetic arguments only go so far, 'economic' issues tend to remain at the fore-front of site preservation (Hall 2000).

In examining impacts the temporal and spatial scale of analysis therefore becomes highly significant (Lew and Hall 1998). Figure 18.1 indicates the importance of temporal and spatial resolution in assessing mobility related phenomena and highlights that the primary scale of research in tourism studies generally fails to intersect with the scale of analysis usually required to understand processes of change. Moreover, the more typical scale of analysis of research of tourism also fails to examine potential relationships between daily leisure mobilities and social and environmental change, even though such mobilities are also related to tourism. Figure 18.2 emphasises, for example, where tourism-related research is concentrated in terms of scales of analysis with respect to socio-economic systems, biodiversity and climate. Such clear failings in the underlying ontology and epistemology of much of tourism studies, at least with respect to potential contributions to understanding processes of global change and mobility, have begun to be substantially critiqued (e.g., Frändberg 1998; Coles *et al.* 2004). Nevertheless, the existence of this disconnect between scale of problem and of analysis may go some way to explaining the failure of much of tourism studies and, possibly, even the tourism industry to come to terms with issues of GEC. Perhaps just as critically, the relatively narrow scale within which tourism studies operates and a poorly developed though still utilised 'ergodic hypothesis' of tourism development – an expedient research strategy by which different areas in space are taken to represent stages in time in the 'evolution' of tourism and a tourism destination – also means that it is extremely difficult to adequately integrate data sets to identify 'at-risk' tourism destinations. (Table 18.1 does provide a list of such destinations; however, extreme caution needs to be applied to its use because of the extreme rate of variability of data in time and space that apply to different risk factors.)

Table 18.1 Most at-risk destinations

Land biodiversity loss	Marine biodiversity loss	Urbanisation
Polynesia/Micronesia	Polynesia/Micronesia	Coastal Mediterranean
Sundaland	Caribbean	Coastal southern China
California	Maldives	Coastal Malaysia
Mediterranean Basin	South China Sea	Coastal California
South African Cape region	Mediterranean	Florida

Water security	Sea-level rise	Regime change/fuel
South Africa	Mediterranean	Australia
Mediterranean	Gold Coast	New Zealand
Australia	Florida	Polynesia/Micronesia
Central America	Coastal China	South Africa
South-west USA	Polynesia/Micronesia	East Africa

Warmer summers	Warmer winters	Disease
Mediterranean	European alps	South Africa
California/western USA	Pyrenees	Mediterranean
North Queensland	Rocky Mountains	Western Europe
South Africa	Australian alps	USA
Western continental Europe	Eastern European alpine areas	Northern Australia

Cumulative
South Africa
Mediterranean
Queensland
South-west USA
Polynesia/Micronesia

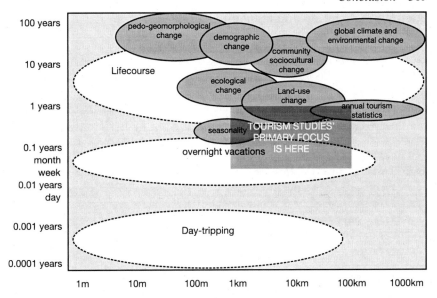

Figure 18.1 The influence of temporal and spatial resolution on assessing mobility-related phenomena

Socio-economic systems	Biodiversity	Climate
International	Global	macroclimate
Supranational	Continental	
National	Biome	
Regional	Bioregion	
	Landscape	
Local	Ecosystem	
Family	Stand/field/communities	mesoclimate
Individual	Individual species	microclimate

Figure 18.2 Scale in tourism analysis. The relative darkness of shading indicates the relative concentration of research at those scales

Adaption, mitigation and response

The tourism industry is increasingly having no choice but to engage with GEC issues. However, the nature of such engagement is highly variable. At a firm and resort level, interest is highest in those destinations that are already starting to be affected by GEC and climate change in particular (Scott *et al.* 2004; Bürki *et al.* 2005; see also Scott, Chapter 15, and Nöthiger *et al.*, Chapter 16, this volume). This means that many alpine winter resorts are starting to attempt to develop technological and product strategies that respond to changes in winter snow cover as well as greater variability in snow conditions (*The Observer* 2004). Similarly, some coastal destinations that are concerned about the impacts of high-intensity storm events are also starting to examine the impacts of GEC. Nevertheless, the reasons for such concerns do not necessarily arise out of a focus on the need to conserve the environment or even address possible changes in the behaviour of current or potential visitors. Instead, the tourism industry is often responding to external factors in the form of bank and lending institutions' interest in the viability of business operations and business risk, insurance company interest in assessing risk, and government interest in meeting international treaty obligations, development implications and the costs of responding to the impacts of GEC (Yohe *et al.* 1996). For example, it was not until 2003 that the World Tourism Organization organised its first conference on climate change and tourism.

At the level of individual firms, especially small business and entrepreneurs, it is also important to recognise that while GEC issues are significant they may not be ranked to be as significant as other, more immediate issues. For example, in a study of marine tourism operators in New Zealand, operators expressed substantial concern over commercial over-fishing, 'cowboy' operators and pollution of the marine ecosystem, and moderate concern as to the impacts of global warming. Interestingly, another area of significant concern was that the marine tourism industry would be overregulated (Orchistron 2004). Similarly, in a review of perceived threats to food and wine tourism operators in British Columbia, Western Australia, Victoria and New Zealand, Hall *et al.* (2003) reported that GEC issues such as biosecurity, climatic change and urbanisation were regarded as potential threats. Nevertheless, such threats were not regarded as being as significant as those that related to wine quality, service quality and increased competition. The relative ranking of threats to business will have substantial implications for how operators perceive the risks associated with GEC and are therefore willing to undertake mitigating responses that would otherwise be regarded as an unnecessary cost to business (see Burton 1997; Reilly and Schimmelpfennig 2000; Clark *et al.* 2001a, b; White *et al.* 2001; Yohe and Tol 2002; Berkhout *et al.* 2004).

Tourism and climate change are clearly interrelated: tourism contributes to climate change and it will be affected by climate change. Nevertheless, there appears to be a low level of support for measures that are aimed at reducing emissions of GHGs. This is, for the industry as a whole, highly irrational from an environmental point of view. In a recent study, Gössling *et al.* (2005a) analysed the eco-efficiency of tourism – emissions of CO_2-equivalents per unit of financial value generated – finding that eco-efficiencies can vary by several orders of magnitude. For example, in France, eco-efficiency ranges from 0.04 $kgCO_2$-e/ for Swiss tourists in the French countryside to 16.01 $kgCO_2$-e/ for Latin American visitors. Hence, the eco-efficiency of different tourist types in France varies by a factor of at least 400. The study also attempted to identify sectors of the tourism industry that are particularly harmful for the environment and economically less beneficial. The case studies included – from France, the Seychelles, Rocky Mountains National Park, Val di Merse/Italy and Amsterdam – suggest that the longer the travel distance, the less favourable become eco-efficiencies, confirming that air travel is the most important factor negatively influencing tourism's environmental sustainability. Long stays improve eco-efficiency, as do high expenditures per day, but these two factors need to be particularly high in order to counter the negative effect of long travel distances. In summary, large parts of tourism may be favourable from an environmental and economic point of view, but another share of tourism, generally based on long-distance air travel, is characterised by low value generation and high environmental damage. The study concludes that minor changes in the tourism production system could lead to great reductions in CO_2-e emissions at very low economic costs.

Currently, an eco-tax is under discussion to reduce the rapid growth in air travel and its environmental impact. However, it is questionable whether relatively small increases in prices for oil or eco-taxation would have a substantial impact on further growth of air travel. Rather, the growing perception in the developed world of air travel as a relatively cheap means of transport, fostered through campaigns by low-fare airlines to fly at zero cost or to give away seats, help create the perception that flying is cheap. This results in the paradox that oil prices hit a new high at US$55 a barrel for the first time in human history, while on the very same day a Swedish tabloid press headline can read 'fly wherever you want for 1 Swedish Crown' (1 Swedish Crown being equivalent to €0.11). As long as this perception of cheap flying dominates – in contrast to the perception that trains are expensive – there is little hope that only relatively small increase in prices, for example, in terms of 'temporary' aviation fuel price levies used by some major airlines in 2004 and 2005, will lead to a change in travel behaviour. On the contrary, currently ongoing changes in society, including air commuters – people living in one place and working in another – as well as children and a majority of young people becoming used to flying as a standard means of transport, will make it difficult to break with this trend and encourage substitution of transport modes.

There is little doubt that GEC will have far-reaching consequences for human activities (IPCC 2001). Options and opportunities for adaptation are thus of great importance. Global environmental change is both a long-term process, including, for instance, increasing temperatures, sea-level rises and biodiversity loss, as well as a process characterised by more sudden changes, as for example, land-use change, the rapid diffusion of disease, heat waves, storms and other weather extremes. While long-term processes such as increases in temperature are moderately slow and comparatively easy to consider in planning, extremes are not. Hence, it is far more difficult to deal with extremes, which are far more unpredictable but have a far greater impact (Changnon *et al.* 2001; Changnon 2003a, b; Hovarth 2003; Schär and Jendritzky 2004; Schär *et al.* 2004; Moberg *et al.* 2005). Nevertheless, any strategy dealing with adaptation needs to consider both processes. Stakeholders can basically act in two different ways: they can continue operations until they become unfeasible and react when extremes hit, or they can adopt a precautionary attitude. The first option will largely be based on technology, and is already widely used in specific environmental situations. For example, making snow with snow cannons has now become standard practice in most low-level alpine resorts (e.g. Scott *et al.* 2003, 2004). However, it is clear that technological fixes are usually cost- and energy-intensive and thus hardly sustainable in the medium or long term. As Hasselmann *et al.* (2003: 1923) reported:

> Even if forcefully implemented, currently available low-cost technologies have limited capacity for substantial global emission reduction and will not be able to counter the rising emissions projected for the long term. Future emissions will be driven mainly by the expanding populations of the developing world, which strive to achieve the same living standards as the industrial countries. An emissions reduction of 50% applied to a projected BAU increase in this century by a factor of four ... still leads to a doubling of emissions, far from the long-term target of near-zero emissions. Furthermore, the mitigation costs for today's technologies are estimated to rise rapidly if per capita emissions are reduced by more than half ... although the Kyoto protocol will boost technologies that are cost-effective in the short term, further emission reductions in the post-Kyoto period could be limited by prohibitive costs. Without affordable new technologies capable of higher global emission reductions, stricter emission reduction targets will be considered impossible to meet and will not be adopted.

In this context, it should also be noted that there might be great differences in the adaptation capacity of tourists, tour operators and tourist infrastructure stakeholders, such as hotels and resorts. Tourists might be the most flexible of these categories, being less spatially constrained in terms of mobility – they have the capacity to adapt both spatially and temporally to GEC (Madison 2001; Elsasser and Bürki 2002; Lise and Tol 2002; Hall and Higham 2005). For example, if desired conditions no longer exist during a certain month of the year, the tourists might simply choose to travel to that place in the main season. When destinations no longer provide desired conditions, tourists can substitute these places for other places. Tour operators might be able to direct tourist flows to new destinations, making contracts with new hotels and redirecting flights. Hotels and resorts might be the least flexible in this process, as they might not be able to move when, for example, snow fall is greatly reduced, and they represent substantial amounts of sunk costs. However, they might, particularly if privately owned, be able to create and find new visitor attractions and marketing strategies, as for example under prolonged summer seasons. Once again, there is great uncertainty, because the level of expected changes is not easily predicted. For example, snow fall reduction might go along with increasing rainfall, which might cause landslides. These could become by far more relevant to tourist industries. However, the most relevant parameter might be the specific characteristics of destinations.

Destinations might be unique in the sense that they offer sights or landscapes that cannot be found anywhere else in the world. For example, the Taj Mahal is an important World Heritage Site in India, and people might continue to visit it even if general conditions for vacation making in the area become less favourable, thereby leading to new seasonal demand patterns. Similar is true for a great range of unique sights worldwide. Such unique places, which will often be human made, can be assumed to generally be more resilient as long as they are not hit by extremes. Coral reefs, on the other hand, can be found all around the tropics, and as long as they are not very special in character – for example, with respect to specific species – diving and snorkelling tourists might seek to visit those places where reefs are still intact (Gössling *et al.* 2005b). The capacity to substitute destination attributes will therefore become an extremely important factor in the longer-term competitiveness of attractions and destinations under GEC (Hall and Higham 2005). Other variables will also continue to shape travel decisions, however. For example, many people have favoured destinations, where they have visited regularly over long periods. Such destinations relying on a large number of recurrent tourists might be able to maintain the tourist system even under less favourable conditions because of customer loyalty. Similarly, certain types of tourism related to business travel and visiting friends and relations is also likely to be reasonably resilient in the face of GEC, although new patterns of visitation over time are still likely to emerge.

In this context, the relative costs of vacation making will need to be considered. At the moment for example, within Europe, long-distance travel is still cheap in comparison to most inner-European destinations. This might change when prices for air travel increase along with the possibility of new taxation regimes, although such regimes are meeting substantial opposition from airlines (Seager 2005). For example, Giovanni Bisignani, Director General and CEO of IATA (International Air Transport Association) commented:

> Environmental responsibility is a pillar of the air transport industry ... All industries are being challenged to do better. But air transport is under particular attack. We must set the record straight on our industry's excellent environment record with facts and figures.
>
> (Bisignani 2005)

As yet, little is known of how such increasing prices will affect tourism, because there is some uncertainty about the price-elasticity of certain categories of tourists (Brons *et al.* 2002). Adaptation might then become relevant in terms of lifestyles, where recreation at home or nearby will become relevant. Furthermore, in the longer term, destination substitution might be difficult in cases where local and regional markets are replaced by long distance markets because mitigation policies are likely to focus on reductions of GHGs within specific jurisdictions. Because air travel causes substantial emissions, taxes and increasing energy prices might turn such substitution strategies into cases of maladaptation.

This *is* serious

In concluding a book such as this the editors would love nothing more than to be able to say that government and industry are willingly implementing the solutions that are in hand. Unfortunately, that is not the case. The media focus on the Kyoto Protocol coming into force in early 2005 seemed to suggest that something had been done – an example of good media coverage is that in *The Guardian* (2005). Indeed, it had, but it was only a very short-term holding mechanism which the largest energy consumer and GHG emitter in the world, the USA, deliberately chose to ignore, while it also did not apply to two of the world's fastest growing energy consumers, China and India. Moreover, from more of a direct tourism perspective, aviation fuel is exempt from the Kyoto Protocols. As Hasselmann *et al.* (2003: 1924) have stressed:

> Because the global political-economic system exhibits considerable inertia, a transition to a sustainable climate can be achieved without major socioeconomic dislocations only if the introduction of appropriate measures addressing the long-term mitigation goals is not delayed.

However, it is significant to note that since Hasselmann *et al.* (2003) gave their warning the situation regarding rapid climate change in the near future has begun to look more serious (Arctic Climate Impact Assessment 2004; Cox *et al.* 2004; Kerr 2004; Leemans and Eickhout 2004; Parry *et al.* 2004; Challenor *et al.* 2005; Leemans and van Vliet 2005; Pachauri 2005; Parry 2005; Rapley 2005; Stainforth *et al.* 2005). For example, Leemans and van Vliet (2005) reported that in the previous decade more ecological consequences have occurred than expected from the observed average 0.7°C warming trend and that current impact assessments of climate change are likely to underestimate ecological impacts and vulnerabilities.

Given the overview of GEC that this book presents it is readily apparent that the world, and tourism as a part of the global socio-economic system, is facing some grave challenges. Although there is an international system of governance in place for global environmental change concerns such as climate change, biodiversity, health and wetlands (Young 2002; Jagers and Stripple 2003), the place of tourism within this is poorly understood. Perhaps of even greater concern is that the tourism industry, for the greater part and for all the official statements regarding sustainable tourism and ecotourism, does not even seem unduly concerned by global environmental change issues. The fundamental goals of the World Tourism Organization (2001) and the World Travel and Tourism Council (2003) are to encourage and promote tourism mobility, perhaps with somewhat of a green tinge so as to assuage industry and individual guilt, because then you can travel to help people through pro-poor tourism or help the environment.

Please forgive the editors for what may seem academic cynicism. It is not. There is probably nothing more that the authors would want in relation to tourism than for it to contribute to alleviating poverty and encouraging a healthier environment. However, the reality is that concentrating on tourism alone, and by that we mean the tendency to focus just on what is happening at the destination, is one of the great problems with sustainable tourism. For tourism to really contribute towards security and sustainable development it needs to be placed within the bigger picture of human mobility, lifestyle, consumption and production. The consumption and production system that seeks to use 'pro-poor tourism' by those from the developed countries to help those in the developing world is the same consumption and production system that has often led to the situations that have contributed to inadequate development practices and poverty in the first place. The most sustainable forms of tourism in many cases may well be no tourism at all, rather focussing on other dimensions of development and a full consideration of alternatives.

The forecasts suggested for travel and tourism in the foreseeable future (Table 18.2) are just not sustainable if there is also the wish to mitigate the contribution of tourism to global climate change, biodiversity loss, health and disease impacts as well as other aspects of GEC. There is a desire in many people in developed countries, which is where the majority of tourists come from, to consume more sustainably. But to do so means to think about the totality of their consumption, where it occurs and its overall environmental and economic impact (Alfredsson 2002, 2004). This may well mean changing not only lifestyles but also seeking to change the consumption and production systems that, at times, actually give us very little real choice about what we consume and its impacts. Sustainable tourism does not necessarily mean no travel, but it does mean full costings of the impact of when you do travel and seeking to use low per capita energy pathways. In terms of the tourism industry it is also time that tourism was treated realistically like any other industry in terms of its costs and benefits and that its impacts were also fully charged for (Gössling 2005). This will not necessarily mean that there will be less tourism and leisure but it is likely that it will have different patterns to those that exist at present (Høyer 2000). Although it is time that the tourism industry became fully responsible for its actions this is not to single the tourism industry out for criticism. Indeed, issues surrounding religious and moral aspects of population growth, the role of major multinational corporations, particularly in the energy industry, government inaction, and the lack of sufficient media attention on GEC are all part of the reason why GEC is such a serious challenge to human welfare and security. And when, as at the time of writing, the trial of Michael Jackson receives more media coverage on most 'news' channels and evening reports than the empirical consequences of climate change, biodiversity loss and emergent diseases, such as avian flu in Asia, then you know that you have a problem.

Human evolution has been profoundly influenced by natural processes of environmental change. The human genus diverged from that of the other Great Apes in Africa 5–6 million years ago as the cooling of the Pliocene produced an ecological niche 'for an ape able to survive mostly out of the forest' (McMichael 2001: 39). It is now generally agreed that a succession of Homo species migrated out of Africa

Table 18.2   United States passenger and travel forecasts

Year	Total passenger traffic to/ from the US (US and foreign flag carriers)	US commer- cial air carriers actual and forecast system average passenger trip length	US commer- cial air carriers: total scheduled US passenger traffic system revenue passenger miles	US commercial air carriers: total scheduled US passenger traffic system revenue passenger enplane- ments	Total jet fuel and aviation fuel consump- tion US civil aviation aircraft*	Total jet fuel and aviation fuel consump- tion US civil aviation aircraft*	Total CO_2 equivalent emissions
	(millions)	*(miles)*	*(billions)*	*(millions)*	*(millions of gallons)*	*(millions of litres)*	*(million kg)*
1999	131.4	979.9	652.4	665.8	20,743	78,521	614,671
2000	140.6	995.7	694.6	697.6	21,350	80,818	632,652
2001	128.8	1,011.6	691.4	683.4	21,094	79,849	625,067
2002	120.8	1,008.8	631.3	625.8	18,761	71,018	555,937
2003	120.0	1,010.3	648.6	642.0	19,128	72,407	566,810
2004	134.0	1,042.1	717.4	688.5	19,372	73,331	574,043
2005	145.4	1,056.1	757.8	717.5	20,332	76,965	602,490
2006	155.0	1,067.1	805.5	754.9	21,536	81,523	638,171
2007	163.2	1,075.6	845.0	785.6	22,533	85,297	667,714
2008	170.8	1,083.3	878.6	811.0	23,405	88,598	693,555
2009	178.2	1,090.8	913.4	837.4	24,278	91,902	719,419
2010	185.6	1,098.5	950.1	864.9	25,199	95,389	746,716
2011	193.1	1,105.1	986.4	892.6	26,124	98,890	774,122
2012	200.7	1,111.8	1,023.9	921.0	27,069	102,467	802,123
2013	208.4	1,118.2	1,063.1	950.7	28,038	106,135	830,836
2014	216.4	1,125.2	1,104.9	981.9	29,063	110,015	861,210
2015	224.5	1,132.7	1,149.4	1,014.7	30,144	114,107	893,242
2016	232.9	1,139.4	1,194.8	1,048.6	31,237	118,245	925,635

* includes both passenger (mainline air carrier and regional/commuter) and cargo carriers.

Source for passenger and fuel consumption numbers
US Department of Transportation Federal Aviation Administration (FAA) (2005) *FAA Aerospace Forecasts: Fiscal Years 2005–16*, US Department of Transportation Federal Aviation Administration Office of Aviation Policy and Plans.

and in many cases co-existenced until around 27,000 years ago, or perhaps even more recently in some parts of the world. Here again climate and environmental change, along with competition, played a major role. Since that time *Homo sapiens* has been the sole survivor of the Homo genus and, as McMichael (2001) poignantly observes, if the current scale of anthropogenic change continues and species and ecosystems come under increasing pressure, it could become the sole representative of the whole great ape family in the very near future.

Increased mobility has been central to the capacity of humans to move to almost every environment on the planet. Mobility also lies at the heart of global anthropogenic environmental change, with tourism being a significant contributor to such change even though it often promotes itself as being environmentally friendly and a key factor in species conservation through 'ecotourism'. It would therefore be strangely ironic if the impact of tourism mobility also becomes the factor that leads to irreversible environmental change that will take not only many species and ecosystems with it, but possibly even humans themselves.

References

Alfredsson, E. (2002) *Green Consumption, Energy Use and Carbon Dioxide Emission.* GERUM kulturgeografi 2002:1, Umeå: Department of Social and Economic Geography, Umeå universitet.

Alfredsson, E.C. (2004) '"Green" consumption – no solution for climate change', *Energy*, 29(4): 513–24.

Arctic Climate Impact Assessment (2004) *Impacts of a Warming Arctic.* Cambridge: Cambridge University Press.

Berkhout, F., Hertin, J. and Gann, D.M. (2004) *Learning to Adapt: Organisational Adaptation to Climate Change Impacts.* Tyndall Centre Working Paper 47. Norwich: University of East Anglia.

Bisignani, G. (2005) 'Aviation and the environment', *Environment Summit*, 17–18 March, online. Available at: www.iata.org/pressroom/speeches/2005-03-17.htm (accessed 19 March 2005).

Brand, S. (1997) *How Buildings Learn: What Happens After They're Built.* London: Phoenix Illustrated.

Brons, M., Pels, E., Nijkamp, P. and Rietveld, P. 2002: 'Price elasticities of demand for passenger air travel: a meta-analysis', *Journal of Air Transport Management*, 8(3): 165–75.

Brown, G. (2005) 'Fulltext: Gordon Brown's speech on climate change,' *Guardian Unlimited* March 15. Available at: www.guardian.co.uk/climatechange/story/0,12374,1438296, 00.html (accessed 16 March 2005).

Bürki, R., Elsasser, H., Abegg, B. and Koenig, U. (2005) 'Climate change and tourism in the Swiss Alps', in C.M. Hall and J. Higham (eds) *Tourism, Recreation and Climate Change.* Clevedon, Channelview, pp.155–63.

Burton, I. (1997) 'Vulnerability and adaptive response in the context of climate and climate change', *Climate Change*, 36: 185–96.

Ceron, J.-P. and Dubois, G. (2003) 'Mobility patterns prospects and their impact on climate change'. Paper presented at Climate Change and Tourism: Assessment and Coping Strategies, Warsaw, 6–8 November, NATO Advanced Research Workshop.

Challenor, P., Hankin, R. and Marsh, B. (2005) 'The probability of rapid climate change'. Paper presented at International Symposium on the Stabilisation of Greenhouse Gases, Hadley Centre, Met Office, Exeter, 1–3 February.

Changnon, S.A. (2003a) 'Present and future economic impacts of climate extremes in the United States', *Environmental Hazards*, 5: 47–50.

Changnon, S.A. (2003b) 'Measures of economic losses from weather extreme', *Bulletin of the American Meteorological Society*, 84: 437–42.

Changnon, S.A., Changnon, J.M. and Hewings, G. (2001) 'Losses caused by weather and climate extremes: a national index for the US', *Physical Geography*, 23: 1–27.

Clark, W.C., Dickson, N., Jäger, J. and Eijndhoven, J.v. (eds) (2001a) *Learning to Manage Global Environmental Risks Volume 1: A Comparative History of Social Responses to Climate Change, Ozone Depletion, and Acid Rain*. Cambridge, MA: The MIT Press.

Clark, W.C., Dickson, N., Jäger, J. and Eijndhoven, J.v. (eds) (2001b) *Learning to Manage Global Environmental Risks Volume 2: A Comparative History of Social Responses to Climate Change, Ozone Depletion, and Acid Rain*. Cambridge, MA: The MIT Press.

Coles, T., Duval, D. and Hall, C.M. (2004) 'Tourism, mobility and global communities: New approaches to theorising tourism and tourist spaces', in W. Theobold (ed.) *Global Tourism: The Next Decade*, 3rd ed. Oxford: Butterworth Heinemann, pp.463–81.

Cox, P.M., Betts, R.A., Collins, M., Harris, P.P., Huntingford, C. and Jones, C.D. (2004) 'Amazonian forest dieback under climate-carbon cycle projections for the 21st century', *Theoretical and Applied Climatology*, 78: 137–56.

Doering, D.S., Cassara, A., Layke, C., Ranganathan, J., Revenga, C., Tunstall, D. and Vanasselt, W. (2002) *Tomorrow's Markets: Global Trends and Their Implications for Business*. A collaboration of World Resources Institute, United Nations Environment Programme, World Business Council for Sustainable Development.

Elsasser, H. and Bürki, R. (2002) 'Climate change as a threat to tourism in the Alps', *Climate Research*, 20: 253–7.

Frändberg, L. (1998) *Distance Matters: An inquiry into the relation between transport and environmental sustainability in tourism*. Humanekologiska skrifter no.15. Göteborg: Section of Human Ecology, Göteborg University.

Fyfe, M. (2005) 'Wake up – this is serious', *The Age* 12 February.

Gössling, S. (2005) 'A framework for the assessment of the global environmental costs of tourism', in B. Amelung, K. Blazejczyk, A. Matzarakis and D. Viner (eds) *Climate Change and Tourism: Assessment and Coping Strategies*. Dordrecht: Kluwer Academic Publishers (in press).

Gössling, S., Hansson, C.B., Hörstmeier, O. and Saggel, S. (2002) 'Ecological footprint analysis as a tool to assess tourism sustainability', *Ecological Economics*, 43: 199–211.

Gössling, S., Helmersson, J., Liljenberg, J. and Quarm, S. (2005a) 'Diving tourism and global environmental change. A perception case study in Mauritius', *Tourism Management*, submitted.

Gössling, S., Peeters, P., Ceron, J.-P., Dubois, G., Pattersson, T., and Richardson, R. (2005b) 'The Eco-efficiency of Tourism', *Ecological Economics*, (in press).

Hall, C.M. (2000) *Tourism Planning*. Harlow: Prentice-Hall.

Hall, C.M. (2005a) 'Time, space, tourism and social physics', *Tourism Recreation Research*, 30(1): 93–8.

Hall, C.M. (2005b) *Tourism: Rethinking the Social Science of Mobility*, London: Prentice-Hall.

Hall, C.M. and Higham, J. (eds) (2005) *Tourism, Recreation and Climate Change*, Clevedon: Channelview.

Hall, C.M., Sharples, E. and Smith, A. (2003) 'The experience of consumption or the consumption of experiences?: Challenges and issues in food tourism', in C.M. Hall, E. Sharples, R. Mitchell, B. Cambourne and N. Macionis (eds) *Food Tourism Around the World: Development, Management and Markets*. Oxford: Butterworth-Heinemann, pp.314–35.

Hasselmann, K., Latif, M., Hooss, G., Azar, C., Edenhofer, O., Jaeger, C.C., Johannessen, O.M., Kemfert, C., Welp, M. and Wokaun, A. (2003) 'The challenge of long-term climate change', *Science*, 302: 1923–5.

Hovarth, R.B. (2003) 'Catastrophic outcomes in the economics of climate change', *Climate Change*, 56: 257–63.

Høyer, K.G. (2000) 'Sustainable tourism or sustainable mobility? The Norwegian case', *Journal of Sustainable Tourism*, 8: 147–61.

IPCC (Intergovernmental Panel of Climate Change) (2001) *Climate change 2001: the Scientific Basis*, contribution of the working group I to the third assessment report of the Intergovernmental Panel of Climate Change. Cambridge: Cambridge University Press.

Jagers, S.C. and Stripple, J. (2003) 'Climate governance beyond the state', *Global Governance*, 9: 385–94.

Kerr, R.A. (2004) 'Three degrees of consensus', *Science*, 305: 932–4.

Lanzendorf, M. (2000) 'Social change and leisure mobility', *World Transport Policy & Practice*, 6(3): 21–5.

Leemans, R. and Eickhout, B. (2004) 'Another reason for concern: regional and global impacts on ecosystems for different levels of climate change', *Global Environmental Change*, 14: 219–28.

Leemans, R. and van Vliet, A. (2005) 'Responses of species to changes in climate determine climate protection targets'. Paper presented at International Symposium on the Stabilisation of Greenhouse Gases, Hadley Centre, Met Office, Exeter, 1–3 February.

Lew, A. and Hall, C.M. (1998) 'The geography of sustainable tourism: lessons and prospects', in C.M. Hall and A. Lew (eds) *Sustainable Tourism: A Geographical Perspective*. Harlow: Addison Wesley Longman, pp.199–203.

Lise, W. and Tol, R. (2002) 'Impact of climate on tourist demand', *Climatic Change*, 55(4): 429–49.

Lynch, K. (1972) *What Time Is This Place?* Cambridge, MA: MIT Press.

Madison, D. (2001) 'In search of warmer climates? The impact of climate change on flows of British tourists', *Climatic Change*, 49: 193–208.

McMichael, A.J. (2001) *Human Frontiers, Environments and Disease: Past Patterns, Uncertain Futures*. Cambridge: Cambridge University Press.

Moberg, A., Sonechkin, D.M., Holmgren, K. Datsenko, N.M. and Karlén, W. (2005) 'Highly variable Northern Hemisphere temperatures reconstructed from low- and high-resolution proxy data', *Nature*, 433: 613–7.

Mowforth, M. and Munt, I. (1998) *Tourism and Sustainability: New Tourism in the Third World*. London: Routledge.

Orchistron, C. (2004) 'Marine Tourism in New Zealand', unpublished Master's thesis, Dunedin: Department of Tourism, University of Otago.

Pachauri, R.K. (2005) 'Avoiding dangerous climate change: 2nd keynote address'. Paper presented at International Symposium on the Stabilisation of Greenhouse Gases, Hadley Centre, Met Office, Exeter, 1–3 February.

Parry, M. (2005) 'Avoiding dangerous climate change: Overview of impacts'. Paper presented at International Symposium on the Stabilisation of Greenhouse Gases, Hadley Centre, Met Office, Exeter, 1–3 February.

Parry, M.L., Rosenzweig, C., Iglesias, A., Livermore, M. and Fischer, G. (2004) 'Effects of climate change on global food production', *Environmental Change*, 14: 53–67.

Rapley, C. (2005) 'Antarctic Ice Sheet and sea level rise'. Paper presented at International Symposium on the Stabilisation of Greenhouse Gases, Hadley Centre, Met Office, Exeter, 1–3 February.

Rathje, W. and Murphy, C. (1992) *Rubbish!* New York: HarperCollins.

Reilly, J. and Schimmelpfennig, D. (2000) 'Irreversibility, uncertainty and learning: portraits of adaptation to long-term climate change', *Climatic Change*, 45: 253–78.

Rypkema, D. (1992) 'Making renovation feasible', *Architectural Record*, January: 27.

Schär, C. and Jendritzky, G. (2004) 'Hot news from summer 2003', *Nature*, 432: 559–60.

Schär, C., Vidale, P.L., Lüthi, D., Frei, C., Häberli, C., Liniger, M.A. and Appenzeller, C. (2004) 'The role of increasing temperature variability in European summer heatwaves', *Nature*, 427(2 January): 332–6.

Scott, D., McBoyle, G. and Mills, B. (2003) 'Climate change and the skiing industry in Southern Ontario (Canada): Exploring the importance of snowmaking as a technical adaptation', *Climate Research*, 23(2): 171–81.

Scott, D., McBoyle, G., Mills, B. and Minogue, A. (2004) 'Climate change and the ski industry in Eastern North America: A reassessment', in *Proceedings of the International Society of Biometeorology Commission on Climate, Tourism and Recreation*. 9–12 June, Kolimbari, Crete, Greece.

Seager, A. (2005) 'Airlines warn of fuel tax meltdown', *The Guardian*, February 7.

Stainforth, D.A., Aina, T., Christensen, C., Collins, M., Faull, N., Frame, D.J., Kettleborough, J.A., Knight, S., Martin, A., Murphy, J.M., Piani, C., Sexton, D., Smith, L.A., Spicer, R.A., Thorpe, A.J., and Allen, M.R. (2005) 'Uncertainty in predictions of the climate response to rising levels of greenhouse gases', *Nature*, 433: 403–6.

The Guardian (2005) 'What is this Kyoto thing all about anyway?', *The Guardian*, February 3.

The Observer (2004) 'Fight to the last resort as Alpine crisis looms', *The Observer*, September 19.

White, G.F., Kates, R.W. and Burton, I. (2001) 'Knowing better and losing even more: the use of knowledge in hazards management', *Global Environmental Change B: Environmental Hazards*, 3: 81–92.

WTO (World Tourism Organization) (2001) *Tourism 2020 Vision – Global Forecasts and Profiles of Market Segments*. Madrid: WTO.

World Travel and Tourism Council (2003) *Blueprint for New Tourism*. London: World Travel and Tourism Council.

Yohe, G., Neumann, J.E., Marshall, P.B., Ameden, H., (1996) 'The economic cost of greenhouse induced sea level rise for developed property in the United States', *Climatic Change*, 32: 387–410.

Yohe, G. and Tol, R.S.J. (2002) 'Indicators for social and economic coping capacity – moving toward a working definition of adaptive capacity', *Global Environmental Change*, 12: 25–40.

Young, O.R. (2002) 'Evaluating the success of international environmental regimes: where are we now?' *Global Environmental Change*, 12: 73–7.

Index